NONTARGET EFFECTS OF BIOLOGICAL CONTROL

COVER PHOTO CREDIT:

A native Hawaiian koa bug carries eggs laid by the tachinid fly, *Trichopoda pilipes,* which was introduced into Hawaii for biological control of the alien pest bug *Nezara viridula* (upper left). Another *Nezara* parasitoid, *Trissolcus basalis*, attacks eggs of the koa bug (lower right). Photos by Marshall W. Johnson.

NONTARGET EFFECTS OF BIOLOGICAL CONTROL

edited by

Peter A. Follett
USDA-ARS Pacific Basin Agricultural Research Center
Hawaii

Jian J. Duan
University of Hawaii

KLUWER ACADEMIC PUBLISHERS
Boston / Dordrecht / London

Distributors for North, Central and South America:
Kluwer Academic Publishers
101 Philip Drive
Assinippi Park
Norwell, Massachusetts 02061 USA
Telephone (781) 871-6600
Fax (781) 871-6528
E-Mail <kluwer@wkap.com>

Distributors for all other countries:
Kluwer Academic Publishers Group
Distribution Centre
Post Office Box 322
3300 AH Dordrecht, THE NETHERLANDS
Telephone 31 78 6392 392
Fax 31 78 6546 474
E-Mail <services@wkap.nl>

Electronic Services <http://www.wkap.nl>

Library of Congress Cataloging-in-Publication Data

Nontarget effects of biological control / edited by Peter A. Follett, Jian J. Duan
p. cm.
Includes bibliographical references.
ISBN 0-7923-7725-7 (alk. paper)
1. Biological pest control agents. 2. Biological pest control agents--Environmental aspects. 3. Agricultural pests--Biological control. I. Follett, Peter A., 1957- II. Duan, Jian J., 1963-

SB975 .N66 1999
632'.96--dc21

99-048110

Printed on acid-free paper.

Printed in the United States of America

CONTENTS

PREFACE

Throughout its history, the practice of biological control has been viewed as a progressive and environmentally friendly method of insect and weed pest management. Biological control has many benefits including essentially permanent management of the target species, no harmful residues, nonrecurrent costs, host specificity, and, for successful programs, a favorable cost-benefit ratio. In addition, it may be one of the few methods of reducing pest numbers over a broad geographical range. Now, biological control practitioners are on trial to justify the use of introduced organisms given the potential for unintended environmental effects. Important areas of concern include the irreversibility of alien introductions, the possibility of host range expansion to include innocuous native or beneficial species (nontargets), dispersal of the biological control agent to new habitats, and the lack of research on the efficacy and environmental impact of previous biological control attempts.

The debate over nontarget effects has been polarized strongly between biological control advocates and conservationists. The strict conservationist's point of view of no intentional introductions of alien species whatsoever has proved hard to defend because evidence for nontarget effects of arthropod biological control introductions is thin and often circumstantial. As a result, some biological control practitioners have been quick to dismiss the importance of adverse nontarget effects. However, the lack of available information appears to reflect the difficulties in evaluating the impact of biological control agents, which include the need to anticipate where nontarget effects may occur in order to gather pre-impact data, as well as our poor attempts at documenting nontarget effects after agent introductions.

In the United States, Executive Order 11987 requires the U.S. Department of Agriculture, in cooperation with the Department of the Interior, to restrict the introduction of exotic species unless it has been determined that the introduction will not have an adverse effect upon the natural ecosystem. In addition to this requirement, the

Animal and Plant Health Inspection Service (APHIS) must comply with a variety of complex environmental statutes, including the National Environmental Policy Act and the Endangered Species Act, which require regulators to take a "hard look" at potential environmental effects of permitting decisions before authorizing the release of biological control agents. Similar statutes are in place in other countries. These laws drive the need for specific data on target specificity and the potential host range of the candidate species for introduction.

Scientifically based information is needed to form a foundation on which risk assessment protocols can be based for future programs. Past experience is often one of the best bases for decisions on an unknown future. Careful, critical, constructive analysis of the information available on the process, procedures, and outcomes from past biological control programs in which nontarget effects appear to be a component will lead to a better basis for future biological control decisions.

Toward this end, we organized a symposium at the Annual Meeting of the Entomological Society of America in 1997 entitled "Biological Control for a Small Planet: Nontarget Effects of BC." This book is an outgrowth of the symposium. To the eight presentations at our symposium we have added nine others. As editors, we have attempted to achieve a balanced treatment of the topic through a collection of diverse viewpoints and approaches. We hope this volume will stimulate thinking and activity in this neglected area of applied ecology.

Peter A. Follett
Jian J. Duan

ACKNOWLEDGMENTS

We are deeply grateful to our colleagues and contributing authors who reviewed chapter manuscripts: Toni Riggin-Bucci, David Hawthorne, George Heimpel, George Roderick, Mandy Heddle, Svata Louda, David Horn, Les Ehler, Amy Sheck, Steve Welter, Kirsten Hural, Wayne Brooks, Jeffery Lockwood, Christine Griffin, Marjorie Hoy, Jay Rosenheim, Tom Unruh, Mickey Eubanks, Peter Stiling, M. Tracy Johnson, Roger Vargas, Steven Hight, Angelika Hilbeck, Richard Reardon, Jack Pinkerton, Tania Zaviezo, Claudio Gratton, John Losey, Christine Nalepa, James Ott, Pedro Barbosa, Keith Hopper, and Dale Evans.

CONTRIBUTORS

Mary Barbercheck
Department of Entomology
North Carolina State University
Raleigh,
North Carolina 27695-7634, USA

Barbara I. P. Barratt
New Zealand Pastoral Agricultural
Research Institute Ltd.
Invermay Agricultural Centre
Private Bag 50034, Mosgiel,
New Zealand

Linda Butler
Division of Plant and Soil Sciences
West Virginia University
Morgantown,
West Virginia 26505-6108, USA

Eric M. Coombs
Oregon Department of Agriculture 635
Capitol Street NE
Salem, Oregon 97301-2532, USA

Jian J. Duan
Department of Entomology
Kauai Agricultural Experiment Station
University of Hawaii
Honolulu, Hawaii 96822, USA

Les E. Ehler
Department of Entomology
University of California
Davis, California 95616, USA

Norm C. Elliott
USDA-ARS Plant Sciences Research
Laboratory
1310 North Western Rd.
Stillwater, Oklahoma 74078, USA

Colin M. Ferguson
New Zealand Pastoral Agricultural
Research Institiute Ltd.
Invermay Agricultural Centre
Private Bag 50034, Mosgiel,
New Zealand

Carlos Garcia-Salazar
Department of Entomology,
Pennsylvania State University
State College, Pennsylvania 16802, USA

Kristopher L. Giles
Department of Entomology and Plant
Pathology
Oklahoma State University
Stillwater, Oklahoma 74075, USA

Stephen L. Goldson
New Zealand Pastoral Agriculture Research Institute Ltd., Canterbury Agriculture and Science Centre
PO Box 60, Lincoln, New Zealand

Donald J. Hannah
Environmental Risk Management Authority
PO Box 131, Wellington, New Zealand

Ann E. Hajek
Department of Entomology
Cornell University
Ithaca, New York 14853-0901, USA

Peter A. Follett
USDA-ARS Pacific Basin Agricultural Research Center
PO Box 4459, Hilo, Hawaii 96720, USA

Douglas A. Landis
Department of Entomology
Michigan State University
East Lansing, Michigan 48824, USA

Jeffrey A. Lockwood
Department of Entomology
University of Wyoming
Laramie, Wyoming 82071-3354, USA

Svata M. Louda
School of Biological Sciences
University of Nebraska
Lincoln, Nebraska 68588-0118, USA

Peter B. McEvoy
Department of Entomology
Oregon State University
Corvallis, Oregon 97331-2907, USA

Rachel McFayden
Queensland Department of Natural Resources
Alan Fletcher Research Station
PO Box 36, Sherwood, Q 4075, Australia

Jennifer Marohasy
Canegrowers
GPO Box 1032, Brisbane, Q 4001, Australia

Jane Memmott
School of Biological Sciences
University of Bristol
Woodland Road
Bristol, BS8 1UG, United Kingdom

Russell H. Messing
Department of Entomology
Kauai Agricultural Experiment Station
University of Hawaii
Honolulu, Hawaii 96822, USA

Leah C. Millar
Department of Entomology
Box 7634
North Carolina State University
Raleigh, North Carolina 27695-7634, USA

Jeffrey C. Miller
Department of Entomology
Oregon State University
Corvallis, Oregon 97331-2907, USA

James R. Nechols
Department of Entomology
Kansas State University
Manhattan, Kansas 66506, USA

John J. Obrycki
Department of Entomology
Iowa State University
Ames, Iowa 50011, USA

David B. Orr
Department of Entomology
North Carolina State University
Raleigh, North Carolina 27695, USA

Craig B. Phillips
New Zealand Pastoral Agriculture
Research Institute Ltd., Canterbury
Agriculture and Science Centre
PO Box 60, Lincoln, New Zealand

Dan Simberloff
Department of Ecology and Evolution
University of Tennessee
Knoxville, Tennessee 37921, USA

Peter Stiling
Department of Biology
University of South Florida
Tampa, Florida 33620, USA

Toni Withers
New Zealand Forest Research Institute,
Ltd.
Sala Street, Private Bag 3020,
Rotorua, New Zealand

PART 1

PERSPECTIVES

1

CRITICAL ISSUES RELATED TO NONTARGET EFFECTS IN CLASSICAL BIOLOGICAL CONTROL OF INSECTS

L. E. Ehler
Department of Entomology, University of California, Davis, California

INTRODUCTION

Nontarget effects of introduced natural enemies are now part of the agenda of classical biological control. This development has been documented in a number of books and reviews, including Harris (1990), Howarth (1991), Ehler (1990a, 1991), Simberloff (1992), Hokkanen and Lynch (1995), and Simberloff and Stiling (1996a, 1996b). Much of the debate centers around host specificity of introduced agents and the possibility of host shifts to nontarget species (McEvoy 1996, Secord and Kareiva 1996). The response from the biological-control community to this latest challenge has varied from outright hostility to calls for more reasoned dialog between conservation biologists and practitioners of classical biological control. The ongoing debate also raises critical issues relative to the future of classical biological control, and this will be the major theme of this chapter. The emphasis will be on predators and parasites of insect pests. I will restrict my comments to three issues: choice of target pest, introduction strategy, and evaluation of impact on nontarget species.

CHOICE OF TARGET SPECIES

Historically, the justification to import natural enemies against insect pests was based on the fact an exotic species had established, was free of its coevolved natural enemies, and had attained pest status. The decision to commit to an importation program was relatively simple, and it did not necessarily involve all stakeholders. In recent years, however, such decisions have become more complex, primarily because of concerns over possible nontarget effects. Thus, in future projects, we can expect that the decision to sanction an importation program will be based on a broader range of issues, reflecting the concerns of all stakeholders. Some of these issues, reflecting ethical and (or) ecological considerations, are discussed below.

Ethical Considerations

Over thirty years ago, R. L. Doutt described the "ethical dimension" of biological control and noted that, in civilized societies, there was a need to protect and preserve natural and native values, including natural environments (Doutt 1967). However, Doutt (1967) was apparently concerned primarily with the ethical justification of employing biological control in order to prevent pesticidal contamination and/or disruption of natural areas. Ethical questions over nontarget effects of introduced natural enemies was not galvanized until the 1980s when conservation biologists began challenging the environmentally pure image of classical biological control. More recently, the journal Agriculture and Human Values devoted a special issue of volume 14 (1997) to "Ethical Issues in Biological Control." Much of the debate concerns the relative ethical justification of the various approaches to biological control.

Lockwood (1996) argued that not all types of biological control are ethically defensible. He further suggested that, in an ethical context, natural biological control was the most desirable, followed (in order) by augmentative, classical, and neoclassical biological control. Lockwood (1996) defined classical biological control as using an exotic agent to control an exotic, coevolved pest, compared to neoclassical in which an exotic agent is used to control a native pest. In the latter case, a novel host/enemy association (i.e., not coevolved) is employed. Neoclassical biological control is considered to be the least ethically defensible course of action because of (1) uncertainty over adverse effects of novel associations and (2) potential loss of ecological functions of native species.

The term "neoclassical" has not been popular among practitioners, who evidently consider this approach to be nothing more than a form of classical biological control. However, because the former approach relies on novel host/enemy associations, it is worth considering as a method quite distinct from employing coevolved host/enemy associations. (Whether or not it deserves the adjective "neoclassical" is another matter.) However, importation of an exotic agent against a native host is not the only way that a novel association can come about. Here, I have in mind a situation in which an exotic host is exploited by an exotic agent that is obtained from a related host species in a region where the exotic host does not occur. The empirical record in classical biological control reveals a number of such cases (Clausen 1978, Hokkanen and Pimentel 1984). Would these not also represent neoclassical biological control? Thus, controlling native pests and employing novel associations are separate, but related, matters; however, novel associations do not necessarily involve native pests and exotic enemies.

In assessing the ethics of the proposed use of a novel host-enemy association, we should distinguish among at least four scenarios. First, an exotic enemy is employed against an exotic pest and the enemy is restricted to the exotic host in the exotic habitat. Second, the exotic enemy in the previous case is not restricted to the exotic host in the exotic habitat. Third, the exotic enemy is employed against a native host and is restricted to that host. Finally, the exotic enemy in the previous case is not restricted to the native host. In these types of cases, our primary concerns should be host specificity of the introduced agent and the ethical (and ecological) implications of suppressing a member of the native flora or fauna. However, these should be addressed on a case-by-case basis, rather than relying on a broad generalization that ranks the ethical desirability of employing novel associations.

In the future, we must be prepared to recognize a variety of ethical postures among the stakeholders attendant to a proposed importation project. The philosophical basis for a given ethical position will vary according to one's particular vision of "moral duty and obligation." The resolution of these ethical conflicts is within the realm of possibility, and must be addressed. However, because of competing moral values, resolution will not be easy (see also Lockwood 1997, Duan and Messing 1997, Bentley and O'Neil 1997). In the end, no one's ethical analysis, regardless of how scholarly it may be, should be accorded logical primacy over that of another.

Ecological Considerations

The habitat of the target pest is one of the most important ecological factors to consider in determining the suitability of a pest for classical biological control. Historically, classical biological control has typically been employed against insect pests in perennial or stable environments, such as orchards, forests, and woody ornamentals; pests in annual or unstable habitats, such as annual crops, have received considerably less attention (Gilstrap 1997). Indeed, most of the complete successes in classical biological control of insects have been achieved in perennial crops and on ornamental trees (cf. Clausen 1978, van den Bosch et al. 1976, Gilstrap 1997). However, because of the movement toward integrated pest management (IPM) in crops such as cotton, wheat, corn, and soybean, there has been increased interest in biological control, including importation of natural enemies against both native and exotic pests. To be an effective biological-control agent in an annual crop, a natural enemy should have three attributes listed by Ehler (1990b):

1. colonizing ability, which permits the enemy to keep pace with spatial and temporal disruption of the habitat;
2. temporal persistence, so that following colonization, the enemy can maintain its population, even in the absence of the target pest species; and
3. opportunistic foraging, which enables the enemy to rapidly exploit a pest population.

Unfortunately, this normally means a natural enemy that is not host specific (see also Wiedenmann and Smith 1997). Among native North American species, some of the better examples are predators, including (1) omnivorous hemipterans, such as *Orius* spp. (Anthocoridae), *Geocoris* spp. (Lygaeidae), and *Nabis* spp. (Nabidae), and (2) larvae of green lacewings in the genera *Chrysopa* and *Chrysoperla* (Chrysopidae). Such opportunistic predators, which are so important in natural biological control of insect pests in annual crops, make poor candidates for classical biological control because of their potential nontarget impact. In such cases, we should expect a correlation between impact on the target pest and on nontarget species. The recent introductions of seven-spotted ladybird beetle (*Coccinella septempunctata* L.) and multicolored Asian ladybird beetle (*Harmonia axyridis* [Pallas]) into North America appear to be good examples of this; however, neither species is restricted to aphids in annual crops (see Obrycki et al., this volume).

One alternative to importation against pests in annual crops is augmentative biological control (Obrycki et al. 1997). There are now over 100 commercially available, biological-control agents in North America (Hunter 1997), and many of these are suited for use in augmentative release in annual crops. However, due to the relatively large acreage involved, such releases may not be economically feasible. This problem can be expected to intensify with the advent of new classes of insecticides that are affordable and better suited to IPM. Such materials are sometimes referred to as "Goldilocks" compounds because they are not too hard on the environment, not too soft on the target pest, but are "just right" (OTA 1995). This provides a new challenge for the discipline of biological control, especially in annual crops where knowledge of the relative importance of naturally occurring enemies (hence which ones to conserve) is often fragmented. At the same time, there is an opportunity for closer interaction with the agrochemical industry in the design, development, and use of more ecologically sound insecticides.

Another alternative to classical biological control in annual crops is to conserve and enhance incumbent natural enemies, primarily native species that are well adapted to the target habitat. However, this will require a good understanding of the relative importance of the predators, parasites, and pathogens in a given crop system so that species or complexes of species can be prioritized for conservation/enhancement efforts. Unfortunately, we often lack the kind of data needed to prioritize these natural enemies. Also, data that do exist are typically biased toward parasites and pathogens because it is easier to evaluate the number of hosts killed by these agents as opposed to hosts eaten by predators. Entomologists' general fascination with parasites (particularly Hymenoptera) has also contributed to this data gap. The time has come to make a comprehensive evaluation of natural enemies in major annual crops, including the relative impacts of all species at each developmental stage of the pest. In California cotton, for example, such studies strongly suggest that opportunistic, general predators are the key natural enemies of several insect pests, rather than parasites and pathogens (Ehler and Miller 1978, Rosenheim et al. 1993). There is growing interest in the use of "conservation biological control" (cf. Barbosa 1998, Pickett and Bugg 1998); however, a considerable gap between research and implementation remains. Unfortunately, this approach to applied biological control has yet to reach its proper place as an equal partner with classical and augmentative approaches (Ehler 1998). There are also parallels between the goals of conservation biological control and those of conservation biology (Letourneau 1998). Clearly, there is an opportunity here for interdisciplinary research, and one that should be exploited by both disciplines in the future.

Setting Priorities

Louda et al. (1998) suggested a hierarchical approach to pest management, in which the least risky options are implemented first. In their scheme, classical biological control, because of the attendant risks, would be reserved for the worst pest problems with the lowest probabilities of nontarget effects. Such an approach may very well come to dominate the agenda for importation programs, especially when natural ecosystems and/or rare and endangered species are involved. In such cases, prioritizing management tactics will involve ethical and ecological issues and economic ones as well. Given the current regulatory climate and the shortage of resources for classical biological control, the biological-control community should also consider the economic wisdom of committing to certain types of projects in the first place. Examples might include importation of specialist natural enemies against pests in annual crops and importations against pests for which there is a relatively low economic threshold. One might also question the economic wisdom of building new importation facilities in regions where the major pest problems are either native or occur in annual crops. In the end, however, the decision to sanction a contested importation program will most likely be a political one. Perhaps this is the way it should be.

INTRODUCTION STRATEGY

Historically, classical biological control of insect pests was carried out in a largely empirical or trial-and-error manner. Although this approach has yielded a large number of successes worldwide, it has been criticized because it is not predictive and it ignores the complexity of organisms, communities, and ecosystems (Ehler 1990b). In recent years, the empirical approach has been derided by some as a "willy nilly" or "Rambo" method, and there was newfound incentive to develop less empirical, more predictive approaches to introducing natural enemies. Predicting the species or combination of species to be released and the ecological impact of released species on the target pest are two of the major challenges. There has been good progress in the development of the attendant ecological theory (Murdoch and Briggs 1996), and practical guidelines are beginning to emerge (Waage 1990, Ehler 1995). The transition to a more predictive science of biological control also requires more investment of time and resources in pre-introductory investigations. Although some question the wisdom and utility of this (e.g., Gonzalez and Gilstrap 1992), the movement is gathering momentum. However, the growing concern over possible nontarget effects of classical biological control threatens to halt this momentum and shift the debate to predicting impact on nontarget rather than target species.

Louda et al. (1998) suggested that the best introduction strategy would be to release "the fewest and most effective agents with the lowest probability of nontarget effects." This of course assumes that effective agents that minimize risk are available. The ideal strategy would be to simultaneously minimize nontarget impact and maximize impact on the target pest. But would the introduction of highly host specific natural enemies (that minimize nontarget impact) necessarily provide the greatest impact on the target pest? For pests in annual crops, where successful natural enemies tend to be less specialized, the intuitive answer to this question would be "no." For pests of perennial crops or woody ornamentals, where successful natural enemies tend to be more specialized, the intuitive answer would be "yes." Although the relationship between impact on the target pest and risk to nontarget species has generally not been addressed in classical biological control of insects, there may be a few case histories that shed some light on the issue. The successful control of obscure scale (*Melanaspis obscura* [Comstock]) in northern California is a case in point.

Obscure scale is a diaspidid that is native to the eastern U.S. where it is associated primarily with oaks and pecan and is exploited by a complex of parasites (Ehler 1992, 1995). The parasite community includes native primary and secondary species, as well as exotic primary species. Four aphelinids are dominant members of the community: *Ablerus clisiocampae* (Ashmead), *Coccophagoides fuscipennis* (Girault), *Encarsia aurantii* (Howard) and *Physcus varicornis* (Howard). The former one is considered a hyperparasite whereas the latter three are primary parasites. Among the primary parasites, only *C. fuscipennis* is strictly host specific. In 1962, an infestation of obscure scale was discovered on oaks in Sacramento. Efforts to eradicate this population were not successful, and it became a target for classical biological control in 1981. Following pre-introductory investigations in the native home of the scale, a decision had to be made: seek to minimize risk by introducing *C. fuscipennis* or maximize impact by introducing *E. aurantii*. The latter species has a wide host range among the Diaspididae, and while there were no "protected" diaspidids in the target zone, there was concern that this parasite might restructure parasite guilds associated with other diaspidids. However, *E. aurantii* was also predicted to have the greatest ecological impact on the target pest. This prediction was based on its thelyotokous mode of sex determination and egg load at emergence. The latter was almost twice that for *C. fuscipennis*, which is an arrhenotokous species. (Because these parasites are short lived, the egg load at emergence is a good indicator of their lifetime fecundity). Based on this information, I introduced *E. aurantii* in 1988; it has now brought obscure scale under complete biological control (Ehler 1995, 1997), with no obvious nontarget effects. Importation of *C. fuscipennis* may well have produced the same result; unfortunately, this introduction strategy can no longer be tested. In summary, the results from this project suggest that an introduction strategy that maximizes impact on the target pest (e.g., *E. aurantii*) does not necessarily minimize impact on nontarget species. Similarly, a strategy that minimizes impact on nontarget species (e.g., *C. fuscipennis*) does not necessarily maximize impact on the target pest.

Prioritizing candidate species for introduction based on perceived risk to nontarget species raises both ecological and intellectual concerns. Because species that pose the least risk may not be the most effective biological control agents, multiple releases of such species could clutter the environment with agents that might provide (at best) an intermediate level of pest control. The presence of such a complex could then affect our ability to establish additional agents later. In addition, assessing candidate natural enemies according to potential nontarget effects should not come at the expense of assessing candidates according to impact on the target pest. Indeed, this may have occurred in biological control of weeds, where concern over nontarget effects is of prime importance. According to Harris (1991), there has been considerable progress made in strategy and techniques for host-specificity testing (so as to minimize nontarget effects), but much less progress has been made in the selection of effective agents. Because of this, most projects employ a combination of intuition and empiricism to select agents for introduction, and tend to experience a high failure rate (Harris 1991). Even today, some workers remain less than enthusiastic over the prospects of applying ecological theory to the selection of agents for biological weed control (e.g., MacFadyen 1998). The "state of the art" as described by Harris (1991) begs the question: has the emphasis on possible nontarget effects slowed the intellectual development of this discipline?

EVALUATION OF THE IMPACT ON NONTARGET SPECIES

The evaluation of ecological impact of an introduced natural enemy on nontarget species should not be restricted to species that we are trying to preserve, but should include nontarget pest species as well. These impacts can be either trophic or nontrophic in nature (Simberloff 1992, McEvoy 1996). Although this discussion is restricted to ecological impact on nontarget species in the conventional sense, we must also recognize the possibility that exotic natural enemies (including host-specific ones), because of their abundance or behavior, might be considered a nuisance by some.

Nontarget Pest Species

The empirical record in classical biological control reveals a number of cases in which exotic natural enemies that were not strictly host specific established and provided some level of control of both the target pest and a nontarget pest. Such an effect on a nontarget pest should be viewed as a form of "fortuitous biological control." This term originally referred to the impact of accidentally introduced natural enemies on native or exotic pests and the impact of native natural enemies on exotic pests (DeBach 1971). It should now be broadened to include all types of unplanned impacts on pests. These "positive nontarget impacts" typically result when a stenophagous natural enemy exploits both the target pest and a taxonomically related nontarget pest. For example, *Metaphycus helvolus* (Compere) (Encyrtidae) and *Coccophagus cowperi* Girault (Aphelinidae), two parasites originally introduced and established against black scale (*Saisettia oleae* [Olivier]) in southern California, also exploited European fruit lecanium (*Parthenolecanium corni* [Bouche]). Bartlett (1978) observed that these parasites may have contributed to the general decrease in abundance of the latter scale in southern California; however, the evidence is largely anecdotal. There are apparently a number of similar instances involving insect pests in the literature (cf. Clausen 1978), and these should be quantitatively evaluated. Unfortunately, classifying impacts as fortuitous may be problematic because it is not always clear from case histories just how "fortuitous" the events actually were. Perhaps the researchers were reluctant to admit in print that suppression of another pest species was not part of the original plan. In future projects, it would behoove biological-control workers to be alert to such positive nontarget impacts, and to document them whenever possible.

There may also be indirect or nontrophic effects of introduced natural enemies on nontarget pests. Of special concern is "pest replacement," in which a nontarget pest increases in importance following the successful biological control of the target pest. For example, following complete biological control of olive scale (*Parlatoria oleae* [Colvee]) by imported parasites in the Central Valley of California (see Huffaker et al. 1962, Huffaker and Kennett 1966), increases in black scale were observed (Sibbett et al. 1976, Shoemaker et al. 1979). In this case, the replacement was due primarily to cessation of chemical insecticides that were previously directed at olive scale, but which also suppressed black scale. The latter is also exotic, but those parasites that were established in the valley at that time were not sufficient to provide the level of biological control required. This led to a resumption of importations against black scale, resulting in establishment of additional parasites (Kennett 1986, Daane et al. 1991). This scale is now under substantial to complete biological control on olive (Daane and Caltagirone 1989). Pest replacements should be expected when the nontarget pest is also introduced free of its natural enemies and is being held in check by control measures directed at the target pest and/or interspecific competition with the target pest.

Another type of indirect effect would be when an introduced natural enemy causes a secondary outbreak of a nontarget pest by somehow disrupting the biological control of the latter. This is an intriguing idea, and although there are suggestions that it has occurred in classical biological control (Howarth 1991), there is little or no evidence to support this claim. The search for such enemy-induced, secondary pest outbreaks might be more profitably directed to augmentative biological control, especially cases involving inundative release of generalist agents.

Rare and Endangered Species

In assessing the impact of an introduced natural enemy on a rare or endangered species, it is important to distinguish between feeding (or parasitizing, infecting) and more serious consequences such as extirpation. We can recognize a series of trophic effects along a continuum as follows:

1. feeding (parasitizing, infecting), but no development of progeny,
2. as in 1, above, but incomplete development of progeny,
3. as in 1, above, but with development from egg to fecund adult,
4. as in 3, above, leading to reduction in population density,
5. as in 4, above, but leading to local extinction, and
6. as in 5, above, but leading to global extinction.

The former three outcomes can be assessed under laboratory conditions and observed in the field. However, these are not sufficient conditions for the latter three outcomes, which are more difficult to demonstrate and are of immediate concern.

In evaluating categories 4, 5, and 6 above, it is essential that investigators employ the same standards of scientific evidence that are expected for claims of pest population reduction in classical biological control and/or the local or regional eradication of an exotic pest. For types 4 and 5 above, long-term, before-and-after comparisons and manipulative experiments that exclude natural enemies can be very convincing. Unfortunately, relatively few biological-control "successes" have been evaluated in a manner that would provide such convincing evidence, so this is a challenge for both classical biological control and conservation biology. Obtaining convincing evidence of global extinction is more problematic. In addition, the whole story of an extinction episode should be told, for even though a biological-control agent may be implicated, other factors, such as loss of critical habitat, may be just as important.

Extinction can be a complicated process, as illustrated by the plight of the large blue butterfly (*Maculinea arion* [L.]) in Great Britain (Ratcliffe 1979, Moore 1987). The larvae of the large blue complete their development in underground nests of ants (*Myrmica* spp.) that nest only in well-grazed habitats. Grazing by the European rabbit (*Oryctolagus cuniculus* L.) was critical to maintaining this interaction and, when the *Myxoma* virus was introduced for rabbit control, it was so successful that ant numbers declined, leading to the extinction of the large blue. (The introduction is believed to have been unsanctioned). This case has been cited as an example of extinction brought about by classical biological control (Simberloff 1992, Simberloff and Stiling 1996a, Howarth 1991). However, it also turns out that the rabbit is exotic to Great Britain; it is native to southwestern Europe and was introduced into England by the Normans in the eleventh century (Moore 1987). In other words, an introduced species was critical to the survival of an endangered species. Habitat loss was also implicated in the decline of the butterfly (Moore 1987). Although conservation biologists should rightly be concerned with the accelerated extinction of species through human activity (Ratcliffe 1979), they might also consider the implications of decelerated extinction through highly unnatural means.

CONCLUDING REMARKS

In the future, we should expect continued pressure for tighter regulations on exotic predators and parasites of insect pests. The notion that candidate natural enemies should be considered "guilty until proven innocent" (cf. Simberloff and Stiling 1996a) may prove appealing to environmental extremists. This could lead to obstructionism in the regulatory process or outright calls for moratoriums on importations (see also Messing, this volume). Extremism of this type could also lead to unsanctioned releases by those who become totally frustrated with the regulatory process. This must be avoided, for such releases could well lead to serious impacts on nontarget species, defeating the goals of conservation biologists and damaging the image of biological control. Thus, there is good reason for both conservation biologists and biological-control workers to seek a common ground when it comes to regulation of the importation process. Finally, as Doutt (1967) put it over 30 years ago, there is "the need to strike a sensible balance." In future projects, a "sensible balance" may well include a given level of impact on nontarget species in the recipient community. If we cannot reach a sensible balance between economic reality and environmental ethics, then classical biological control may become an endangered scientific discipline.

REFERENCES

Barbosa, P. (ed.). 1998. Conservation biological control. Academic Press, San Diego, California.

Bartlett, B. R. 1978. Coccidae. *In* C. P. Clausen (ed.), Introduced parasites and predators of arthropod pests and weeds: a world review. U.S. Dept. Agric., Agric. Handbk. no. 480. Washington, D.C. p. 57–74.

Bentley, J. W., and R. J. O'Neil. 1997. On the ethics of biological control of insect pests. Agric. Human Values 14: 283–289.

Clausen, C. P. (ed.). 1978. Introduced parasites and predators of arthropod pests and weeds: a world review. U.S. Dept. Agric., Agric. Handbk. no. 480. Washington, D.C.

Daane, K. M., and L. E. Caltagirone. 1989. Biological control of black scale in olives. Calif. Agric. 43(1): 9–11.

Daane, K. M., M. S. Barzman, C. E. Kennett, and L. E. Caltagirone. 1991. Parasitoids of black scale in California: establishment of *Prococcophagus probus* Annecke and Mynhardt and *Coccophagus rusti* Compere (Hymenoptera: Aphelinidae) in olive orchards. Pan-Pac. Entomol. 67: 99–106.

DeBach, P. 1971. Fortuitous biological control from ecesis of natural enemies, p. 293–307. *In* Entomological essays to commemorate the retirement of Professor K. Yasumatsu. Hokuryukan Publ. Co., Ltd., Tokyo.

Doutt, R. L. 1967. Biological control, *In* W. W. Kilgore and R. L. Doutt (eds.), Pest control: biological, physical, and selected chemical methods. Academic Press, New York. p. 3–30.

Duan, J. J., and R. H. Messing. 1997. Biological control of fruit flies in Hawaii: factors affecting non-target risk analysis. Agric. Human Values 14: 227–236.

Ehler, L. E. 1990a. Environmental impact of introduced biological–control agents: implications for agricultural biotechnology. *In* J. J. Marois and G. Bruening (eds.), Risk assessment in agricultural biotechnology: proceedings of the international conference. Univ. Calif., Div. Agric. Nat. Res., Publ. no. 1928. p. 85–96.

Ehler, L. E. 1990b. Introduction strategies in biological control of insects. *In* M. Mackauer, L. E. Ehler, and J. Roland (eds.), Critical issues in biological control. Intercept, Andover, UK. p. 111–134.

Ehler, L. E. 1991. Planned introductions in biological control. *In* L. R. Ginzburg (ed.), Assessing ecological risks of biotechnology. Butterworth-Heinemann, Boston. p. 21–39.

Ehler, L. E. 1992. Guild analysis in biological control. Environ. Entomol. 21: 26–40.

Ehler, L. E. 1995. Biological control of obscure scale (Homoptera: Diaspididae) in California: an experimental approach. Environ. Entomol. 24: 779–795.

Ehler, L. E. 1997. Obscure scale declines after parasitic wasp introduced. Calif. Agric. 51(6): 29–32.

Ehler, L. E. 1998. Conservation biological control: past, present, and future. *In* P. Barbosa (ed.), Conservation biological control. Academic Press, San Diego. p. 1–8.

Ehler, L. E., and J. C. Miller. 1978. Biological control in temporary agroecosystems. Entomophaga 23: 207–212.

Gilstrap, F. E. 1997. Importation biological control in ephemeral crop habitats. Biol. Control 10: 23–29.

Gonzalez, D., and F. E. Gilstrap. 1992. Foreign exploration: assessing and prioritizing natural enemies and consequences of preintroduction studies. *In* W. C. Kauffman and J. R. Nechols (eds.), Selection criteria and ecological consequences of importing natural enemies. Entomol. Soc. Am., Lanham, Maryland. p. 53–70.

Harris, P. 1990. Environmental impact of introduced biological control agents. *In* M. Mackauer, L. E. Ehler, and J. Roland (eds.), Critical issues in biological control. Intercept, Andover, UK. p. 289–300.

Harris, P. 1991. Classical biocontrol of weeds: its definition, selection of effective agents, and administrative-political problems. Can. Entomol. 123: 827–849.

Hokkanen, H., and D. Pimentel. 1984. New approach for selecting biological control agents. Canadian Entomol. 116: 1109–1121.

Hokkanen, H. M. T., and J. M. Lynch (eds.). 1995. Biological control: benefits and risks. Cambridge Univ. Press, Cambridge, UK.

Howarth, F. G. 1991. Environmental impacts of classical biological control. Annu. Rev. Entomol. 36: 485–508.

Huffaker, C. B., and C. E. Kennett. 1966. Studies of two parasites of olive scale, *Parlatoria oleae* (Colvee). IV. Biological control of *Parlatoria oleae* (Colvee) through compensatory action of two introduced parasites. Hilgardia 37: 283–335.

Huffaker, C. B., C. E. Kennett, and G. L. Finney. 1962. Biological control of olive scale, *Parlatoria oleae* (Colvee), in California by imported *Aphytis maculicornis* (Masi) (Hymenoptera: Aphelinidae). Hilgardia

32: 541–636.

Hunter, C. D. 1997. Suppliers of beneficial organisms in North America. Calif. EPA, Dept. Pesticide Regulation, PM 97-01.

Kennett, C. E. 1986. A survey of the parasitoid complex attacking black scale, *Saissetia oleae* (Olivier), in central and northern California (Hymenoptera: Chalcidoidea; Homoptera: Coccidae). Pan-Pac. Entomol. 62: 363–369.

Letourneau, D. K. 1998. Conservation biology: lessons for conserving naturalenemies. *In* P. Barbosa (ed.), Conservation biological control. Academic Press, San Diego. p. 9–38.

Lockwood, J. A. 1996. The ethics of biological control: understanding the moral implications of our most powerful ecological technology. Agric. Human Values 13(1): 2–19.

Lockwood, J. A. 1997. Competing values and moral imperatives: an overview of ethical issues in biological control. Agric. Human Values 14: 205–210.

Louda, S. M., D. Simberloff, G. Boettner, J. Connor, D. Kendall, and A. Arnett. 1998. Insights from data on the nontarget effects of the flowerhead weevil. Biocontr. News Info. 19: 70N–72N.

McEvoy, P. B. 1996. Host specificity and biological pest control. BioScience 46: 401–405.

McFadyen, R. E. C. 1998. Biological control of weeds. Annu. Rev. Entomol. 43: 369–93.

Moore, N. W. 1987. The bird of time. Cambridge Univ. Press, Cambridge, UK.

Murdoch, W. W., and C. J. Briggs. 1996. Theory for biological control: recent developments. Ecology 77: 2001–2013.

Obrycki, J. J., L. C. Lewis, and D. B. Orr. 1997. Augmentative releases of entomophagous species in annual cropping systems. Biol. Control 10: 30–36.

Office of Technology Assessment (OTA). 1995. Biologically based technologies for pest control. OTA-ENV-636, U.S. Govt. Printing Office, Washington, D.C.

Pickett, C. H., and R. L. Bugg (eds.). 1998. Enhancing biological control. Univ. Calif. Press, Berkeley, California.

Ratcliffe, D. 1979. The end of the large blue butterfly. New Scientist 8: 457–458.

Rosenheim, J. A., L. R. Wilhoit, and C. A. Armer. 1993. Influence of intraguild predation among generalist insect predators on the suppression of an herbivore population. Oecologia 96: 439–449.

Secord, D., and P. Kareiva. 1996. Perils and pitfalls in the host specificity paradigm. BioScience 46: 448–453.

Shoemaker, C. A., C. B. Huffaker, and C. E. Kennett. 1979. A systems approach to the integrated management of a complex of olive pests. Environ. Entomol. 8: 182–189.

Sibbett, G. S., J. E. Dibble, and J. D. Babcock. 1976. Black scale now a major olive pest. Calif. Agric. (Nov): 12–13.

Simberloff, D. 1992. Conservation of pristine habitats and unintended effects of biological control. *In* W. C. Kauffman and J. R. Nechols (eds.), Selection criteria and ecological consequences of importing natural enemies. Entomol. Soc. Am., Lanham, Maryland. p. 102–117.

Simberloff, D., and P. Stiling. 1996a. How risky is biological control? Ecology 77: 1965–1974.

Simberloff, D., and P. Stiling 1996b. Risks of species introduced for biological control. Conserv. Biol. 78 (1–2): 185–192.

van den Bosch, R., O. Beingolea G., M. Hafez, and L. A. Falcon. 1976. Biological control of insect pests of row crops. *In* C. B. Huffaker and P. S. Messenger (eds.), Theory and practice of biological control. Academic Press, New York. p. 443–456.

Waage, J. 1990. Ecological theory and the selection of biological control agents. *In* M. Mackauer, L. E. Ehler and J. Roland (eds.), Critical issues in biological control. Intercept, Andover, UK. p. 135–157.

Wiedenmann, R. N., and J. W. Smith, Jr. 1997. Attributes of natural enemies in phemeral crop habitats. Biol. Control. 10: 16–22.

2

NONTARGET EFFECTS OF BIOLOGICAL CONTROL: WHAT ARE WE TRYING TO MISS?

Jeffrey A. Lockwood
Department of Entomology, University of Wyoming, Laramie, Wyoming

AN HISTORICAL PERSPECTIVE ON A MOVING (NON)TARGET

The history of pest management in general (Perkins 1982), and biological control in particular (Coppel and Mertins 1977, Caltagirone 1981, Goeden 1988, Van Driesche and Bellows 1996), has been thoroughly analyzed in terms of the dominant practices and how these actions relate to competing philosophies of ecological intervention (e.g., integrated *vs.* total pest management). While it is evident how environmentalism, as a social phenomenon, has altered the forms and concepts of pest management, it is not so clear how ecology, as a scientific discipline, might change the field of pest management. A great deal of work has been dedicated to the notion of "nontarget" effects, particularly as they relate to chemical control methods, but only recently have we begun to critically assess what exactly constitutes a "nontarget." From a simple perspective this would seem to be self-evident—all organisms other than the pest are nontargets. However, this tautological approach does little to advance the science of pest management, and it is clear that in practice, the nature of the targets and nontargets is far from simple. The (re)conceptualization of the target can lead to important changes in pest management (e.g., defining the target based on an economic threshold, USDA

1996). Extending these principles to our understanding of nontargets promises to open new avenues for monitoring and protecting those elements that we wish to protect from the hazards of pest management practices in general, and biological control in particular.

Before exploring the concept of targets and nontargets, I should note that the reader will find a strong bias to North American ecosystems and grasslands in particular. Most of my examples are drawn from the management of rangeland grasshoppers. The reason for this is that my scientific expertise and pest management experience are focused on these lands and insects. To the extent that I can draw upon broader representations of flora and fauna and wider perspectives of geography without overextending my arguments, I shall attempt such generalizations. However, the reader is encouraged to apply and adapt my reasoning to his or her context, with the understanding that sometimes the arguments will fit neatly into a new system, often they will need to be modified, and occasionally they may not work at all.

Humans: The Original Nontarget

Although adverse impacts of pesticides had been studied for years, the emergence of the "nontarget" concept in public and political discourse came with Rachel Carson's *Silent Spring*. The vast majority of the text was dedicated to vertebrates (birds being the defining basis of her book's title), and there can be little question that the focus of her concern was the wellbeing of humans. Frequent reference to humans being the toxicological endpoint of the misuse of pesticides leaves little doubt that the focus of concern was our own welfare. In the final chapter of the book, Carson advocates the introduction of exotic vertebrate predators for biological control and clearly places human health above concerns for native species and ecosystems.

While *Silent Spring* effectively focused our attention on the misuse of broad-spectrum synthetic pesticides, the accepted need to control organisms that compete for our food and endanger our health persisted. As such, it is not difficult to understand why biological control received an enthusiastic public greeting. The use of living organisms as our allies in controlling pests was hailed as an ecologically based, environmentally sound alternative to the abuses of pesticides. The immediate anthropocentric benefits of biological control, relative to the historical misuses of chemical pesticides, were incontrovertible. After all, biological control represented no toxicological threat to human health. Consequently, the biological control industry grew into the most powerful under-regulated (and in many instances, unregulated) environmental technology in the United States (Miller and Aplet 1993).

Having "Missed" Humans, the Nontarget Scope Widens

Using a biocentric, rather than anthropocentric, context, Howarth (1983, 1985, 1991, 1992) pioneered the movement to assess nontarget effects of biological control. By defining environmental impacts to include native species, rather than exclusively focusing on human health, he radically expanded the scope of concern to include nontarget arthropods and plants. This concern for organisms having no conditional value (i.e., being of no immediate or known utility to human interests) provoked a revolution in the field of biological control that has continued for 15 years and has yet to be resolved. While admitting the tremendous benefits to human health, at least in the short-term, Howarth recognized that the environmental risks of biological control are of a kind and degree that few agricultural technologies, including agrichemicals, have been able to match. Because biological control makes use of living organisms which are often intended to reproduce and spread from the point of release, this technology has the potential to permanently and irreversibly alter ecosystems on a continental, if not global, scale (Lockwood 1993a,b, 1996, Office of Technology Assessment 1993).

Chemical and other pest control technologies are spatiotemporally limited in their benefits and harms (Lockwood 1996), although highly residual compounds can be transported from the site of application. As such, environmental alterations due to the use of chemical, mechanical, and cultural tools were usually localized and were almost certain to disappear within ecological time. This assessment should not be construed to somehow diminish the serious problems which have arisen from the application of these technologies in aggregate, particularly pesticides in the context of human health. However, when biological control involves the establishment of an organism with the potential to track its host in time and space, the permanent suppression of pest across entire ranges is possible and represents the primary allure of this approach. However, when the agent attacks a nontarget species, there may also be the same severity and pervasiveness of impact. This degree of change is virtually impossible with other pest control technologies, other than successful eradication programs. Thus, along with the potential for indefinite benefit to human interests comes the potential for permanent, unintended, and largely unpredictable impacts on nontarget species—including extinction (Howarth 1991, 1992).

Most recently, Howarth's dire warnings have been substantiated in a surprising context. The biological control of weeds, which is far more carefully conducted and regulated (albeit voluntarily) than that of insects, has provided a striking cases of nontarget effects in the United States. *Rhinocyllus conicus* Froelich, a weevil introduced to control exotic thistles has begun to have an impact on native thistles, some of which are rare (Louda et al. 1997, Strong 1997). While biological control practitioners contend that this weevil would not have been released under today's standards, this claim does not diminish the compelling ecological evidence that biological control can cause serious harm to native species. Ironically (and perhaps cynically), one might suggest that the solution to this problem is at hand, as New Zealand scientists have found that the parasitic wasp, *Microctonus aethiopoides* Loan, originally released to control exotic weevils in alfalfa has switched to *R. conicus* in some areas, thereby biologically controlling the biological control agent (Ferguson et al. 1995, Barratt et al. 1997). Unfortunately, a number of native and endemic weevils are also serving as hosts for the wasp (Barratt et al. 1998; see also the Barratt et al. chapter in this book).

Process Ecology: The (Non)Target Moves Again

There can now be little doubt that native species have been and are being placed at risk through the introduction of biological control agents. The extent and severity of this risk has been dismissed by those who assert that extinction—the loss of a species as physical entity—is a relatively rare consequence of biological control. However, this debate is based on a common and long-standing (but mistaken, or at least limited) conceptual tradition in western science, that of materialism. We might argue about what species or individuals constitute nontarget organisms and whether these entities are likely to disappear, but there is general consensus that "things" are what we mean when we refer to nontargets. As such, the suppression of nontarget densities, local extirpations, and other quantitative impacts are largely dismissed as not constituting a basis of concern as long as the species in question persists.

This reductionistic focus probably derives, or at least is reinforced, by the most powerful piece of environmental legislation in the US—the Endangered Species Act. Even though this act alludes to ecosystems and their processes, the focus is clearly on species as material entities. Hence, it is not surprising that concerns regarding the nontarget effects of biological control have paralleled the dominant legal pathway through which this technology might be regulated (Miller and Aplet 1993). Moreover, the growing field of Biodiversity dedicates the overwhelming share of its resources and attention to species-as-entities, with vague allusions to the ecological services that they provide (Wilson 1988). The first and foremost goal appears to be taxonomic inventory (Wilson 1988), with an understanding of ecological function being relegated to a distant and indistinct future. Such a divide-and-conquer approach to assessing either biodiversity or nontarget impacts risks eroding the whole while assuring that all of the pieces are still in existence.

The materialist perspectives have dominated ecology and other scientific disciplines, and there is no doubt that material existence is an element of the nature. However, this perspective is analogous to monocular vision—the view is "valid" but it is ultimately incomplete and lacking in depth. It now appears that a re-perception of the natural world is underway with the emergence of the field of complexity (Waldrop 1992, Lewin 1993). Although a review of this revolution is beyond the scope of this paper, the central paradigm shift of relevance to the topic of biological control involves understanding ecosystems (and other complex systems) as networks of *processes*, rather than objects. For example, catastrophe theory treats populations as processes of growth and decline, with the physical beings serving as the manifestations of the underlying dynamics. The traditional ecologist may object to this approach as being a "black box" (Sussman and Zahler 1976, 1978; Berryman and Stanseth 1984), which is an understandable reaction from the materialist perspective that views the real world as comprised of objects upon which processes act or from which actions arise. However, there is no scientifically compelling reason to structure the world in such a fashion, and it is equally revealing to perceive an organism, species, population, or ecosystem as "being" what it *does* (Lockwood 1997a). In context of our example, a population is not the collection of individuals but rather the energy and nutrient flows that are manifest through the organisms.

This approach was developed by Johnson (1991) in his pursuit of understanding what constitutes a living entity. He noted that populations as collections of organisms are virtually impossible to define because their composition and boundaries are forever changing. However, as processes of reproduction, competition, etc. populations are unique and self-identifying (Johnson 1991). As such, perceiving a population as the dynamic processes of outbreaks and crashes or consumption and excretion is as ecologically compelling as defining a population as a collection of individuals

(Lockwood 1997b). In the latter, we pursue models of populations as physical states; in the former, we develop models of population dynamics as processes. Hence, if we perceive the world as processes manifest through physical changes, dynamic models become lucid. At least they are no more mysterious than object-oriented models, in which objects are taken as a given, and how processes arise from them is assumed (e.g., the traditional box-and-arrow approach).

If processes, rather than material entities, are taken to be ecological realities, then species (and their possible extinction) becomes something of a red herring. Nobody would doubt that extinctions are environmentally relevant and/or ethically objectionable,[1] but extinction is an extreme criterion upon which to base assertions of nontarget impacts of biological control. While the loss of a species is the basis for concern in a world perceived as being comprised of material entities [the biological inventory concept that dominates Biodiversity discourse (Plotkin 1988, Wilson 1992)], from the perspective of ecological processes, the numerical suppression of a species or its extirpation from a portion of its range may be a matter of grave concern. The ongoing reintroduction of the wolf into Yellowstone National Park, although initiated under the aegis of the Endangered Species Act, was not precipitated by a global concern for the species, as it is abundant in Canada and recovering in the Great Lakes region. Rather, the loss of species from the ecosystem, and more particularly the absence of its *ecological process* of predation, became the rationale for the reintroduction. In principle, serious changes in ecosystem processes could even result without any change in species' abundance, if the behavior of a guild or keystone species was sufficiently disrupted (e.g., an introduced pathogen could have sublethal effects reducing the aeration of soil by invertebrates, thereby allowing soil compaction and reducing water infiltration with the attendant consequences to plant vigor, hydrologic cycles, etc.). So, what does the notion of a "target/nontarget process" mean for our understanding of biological control?

THE FOUNDATION OF IPM: PESTS AS PROCESSES

The Goal of Pest Management: Reducing Losses, Not Densities

Before assessing the potential and realized effects of biological control on nontarget processes, let us consider how this approach has been applied in the more familiar context of the target system. In an unusual juxtaposition, it seems that empiricism (pest management) was well ahead of ecological theory in recognizing and adopting process-oriented perspective. The primacy of process has a long-standing tradition within applied ecology and particularly integrated pest management (IPM). Although there are many definitions of IPM, the central concept is most often expressed in terms of an economic threshold (ET), which is the point at which the cost of a management intervention is less than the damage that can be expected if the pest is left untreated. Although the ET is usually expressed in terms of a pest density, this measure is simply a surrogate for a particular point in an underlying dynamic that constitutes the process of damage. Hence, the goal of pest management is to avoid the losses that are associated with a particular intensity of a biological process—most often herbivory, although parasitism, hematophagy and other behaviors are certainly relevant.[2]

As such, under the idealized (but perhaps unusual [Stiling 1987, 1988]) scenario of biological control, the agent acts in a density-dependent manner to suppress the target process as it increases towards the economic injury level (EIL, the point at which damage exceeds the costs of control). By analogy, we can view the system as being a radio, and our goal is to modulate the sound, not destroy the speaker. This concept explicitly recognizes that the fundamental target is a process, rather than an object, as we judge a biological control program to be successful when damage by the pest is prevented, rather than as a function of the number of hosts parasitized or prey killed or target species eradicated.

Uncontrolled Biological Control: When Targets Become Nontargets

The ideal formulation of biological control as a density-dependent process that prevents a pest population from damaging a resource is a compelling concept. However, if goal of IPM is to avoid economic injury, then a pest ceases to exist—at least in terms of its being a process that must be controlled—when its feeding is less than that associated with the EIL. Thus, using population density as a surrogate measure of the process of damage, low-density populations represent nontarget processes. One might argue that when dealing with exotic (nonindigenous) species or some area-wide management systems, eradication is acceptable, perhaps even desirable. In this context, the species is always a legitimate target. This situation is most applicable to classical biological control (the use of an exotic agent to control an exotic target), which is usually premised on the establishment of a self-sustaining agent. As such, the application of either an EIL or ET is irrelevant as there is no increased cost of controlling harmless populations of the "target" organism.

When the target organism is a native species, however, the situation may become more complex. With conservation, augmentative, and inundative biological controls (and chemical control), the intervention can be spatiotemporally or ecologically limited, so suppression can be terminated when the pest process is eliminated. However, the use of self-sustaining biological control agents, particularly in context of neoclassical (i.e., the use of exotic organisms to control native species as described by Lockwood [1993a, 1996], which is a specific manifestation of the broad strategy termed "new associations" by Hokkanen and Pimentel [1984, 1989]) programs, is problematical if the agent continues to suppress a native species and its ecological processes even when the economic injury has ceased. This concern is not merely hypothetical given (1) that most insect parasitoids exhibit density-independence or inverse density-dependence (Stiling 1987, 1988), (2) the interest in using exotic biological control on native species (Hokkanen and Pimentel 1984, 1989), and (3) the potential for native species to play important ecological roles at sub-economic population densities (Howarth 1985, 1991, Lockwood 1993). Even if density-dependent biological control is manifest, the agent's response may not correspond with the density at which the pest process appears or beneficial processes cease.

Application to Rangeland Grasshoppers: The Changing Target

The notion that a process, rather than an object, is the target in biological control of rangeland grasshoppers is well established. Onsager (1988) persuasively argued that in assessing biological control agents of grasshoppers, specifically *Nosema locustae* Canning, the focus must be on the process that we are attempting to manage, rather than the "target" insects per se. According to this contention, the goal of grasshopper management is to save forage; killing grasshoppers is only one, crude means to this end. Therefore, traditional measures of management efficacy, such as short-term mortality, risk overlooking entirely successful control programs that protect the commodity via nonlethal changes to the target process. Indeed, *N. locustae* is effective in reducing feeding, delaying development, and increasing egg parasitism (Ewen and Mukerji 1980, Oma and Hewitt 1984, Johnson and Pavlikova 1986, Lockwood and DeBrey 1990). Hence, forage loss (a process) is reduced without a corresponding reduction in the density of the target (an object). Most recently, the National Grasshopper Management Board defined the goal of rangeland grasshopper pest management to be one of "keeping good stewards on the land." This assertion represents an even more radical departure from the notion that the target of pest management is a species or other material.

To further elucidate the target-nontarget relationship in the rangeland agroecosystem, one must recognize that grasshoppers both damage and sustain the economic value of the plant community. The possibility that an organism that is damaging at high densities (i.e., a pest) but beneficial at low densities is exemplified by several species of grasshoppers. In assessing the risks of biological control of grasshoppers through the introduction of exotic pathogens and parasitoids (Lockwood 1993a, b), I advocated adopting the null hypothesis that at an ecosystem scale, over long time periods, native species should be considered as being beneficial (nontargets) in sustaining ecological processes. However, even narrow economic definitions of beneficence (e.g., control of noxious weeds) suggest that some of the classic "pest" species (e.g., *Melanoplus sanguinipes* [F.], *M. occidentalis* [Thomas], and *M. femurrubrum* [De Geer]) that are associated with economic damage at high densities prefer to feed on weedy and low-value forbs while at low densities. Hence, the pest status of these species is a function of their ecological processes (feeding on grasses or forbs), for which population density is a plausible surrogate. The chronic, density-independent (or even worse, inverse density-dependent [Stiling 1987, 1988]) suppression of these species would likely result in greater harm than benefit to western rangelands. Thus, the suppression of rangeland grasshoppers at low-density populations would constitute a potentially serious impact on nontarget, even beneficial, processes.

THE NATURE ON NONTARGET PROCESSES

A Pragmatic Rationale for Consideration of "Function over Form"

The analysis of biological diversity includes aspects of both richness and evenness. While the former measure is the most common expression of diversity, the latter measure is a much more sensitive measure of community dynamics. For richness to change, a species must be added to or lost from a system, but evenness changes as the relative frequencies of extant species vary. In a similar sense, changes in ecosystem processes may precede noticeable changes in species richness. The accumulation of detritus will likely precede the measurable loss of detritivores; a decline in fruiting will indicate a decrease in pollinators, and an increase in herbivory will often be noticed before a reduction in parasitoids is obvious. Given that the primary economic value of biological diversity is a function of ecological processes, rather than existence value (an extremely difficult value to estimate), there is further basis directing our attention not only to detecting extinction but also towards changes in the flow of energy, the efficiency of pollination, the competition between organisms, and the distribution of nutrients. In most agroecosystems, our goal is to sustainably extract nutrients or fiber, so it is ultimately the disruption of these processes (and supporting actions and associations) that constitute the existence of a pest.

At least in extensive agroecosystems, where rich communities of species interact to sustain primary production, soil fertility, nutrient/water cycles, it may be easier to monitor these processes than the complex array of species responsible for them. Indeed, we often do not know which species and interactions are directly and indirectly responsible for these processes, so identifying and tracking nontarget species may be difficult or impossible both in principle and in practice. For example, the guidelines for assessing rangeland health (Busby 1994), make no reference to insects or other nontarget species as indicators. Rather, the condition of the system is assessed via the status of ecological processes as measured directly or via surrogate ecosystem states (e.g., soil condition).

An example of how a nontarget ecological process may be altered causing serious economic loss and ecological disruption before a nontarget species is markedly, or perhaps even measurably, impacted was raised in the context of concerns over the neoclassical biological control of rangeland grasshoppers (Lockwood 1993a, b). *Hesperotettix viridis* (Scudder) was found to be within the host range of the proposed biological control agent, and this grasshopper species is known to primarily feed on snakeweeds, *Gutierrezia* sp. , along with a number of other poisonous plants (Parker 1984). The weedy hosts of this grasshopper cause estimated annual losses of $118 million in the Great Plains biome (DeLoach 1976). Indeed, a <1% expansion in snakeweed would cost the livestock industry more in terms of poisoning and control costs than the average annual cost of rangeland grasshoppers in the USA. The possibility of such losses substantiated by reports that insecticide control programs locally suppressed *H. viridis*, leading to a serious outbreak of snakeweed (Thompson in Lockwood 1993a). In this case, only after the snakeweed populations increased (via suppression of grasshopper herbivory) did we begin to look for changes in the frequency of the nontarget species, *H. viridis*. Thus, the disruption of nontarget processes (herbivory) may be most effectively monitored via a material surrogate (weed density), rather than by monitoring a nontarget species (grasshopper). This suggests that the nontarget species concept may not be the only or most effective approach to assessing the ecological effects of biological control (Fig. 1).

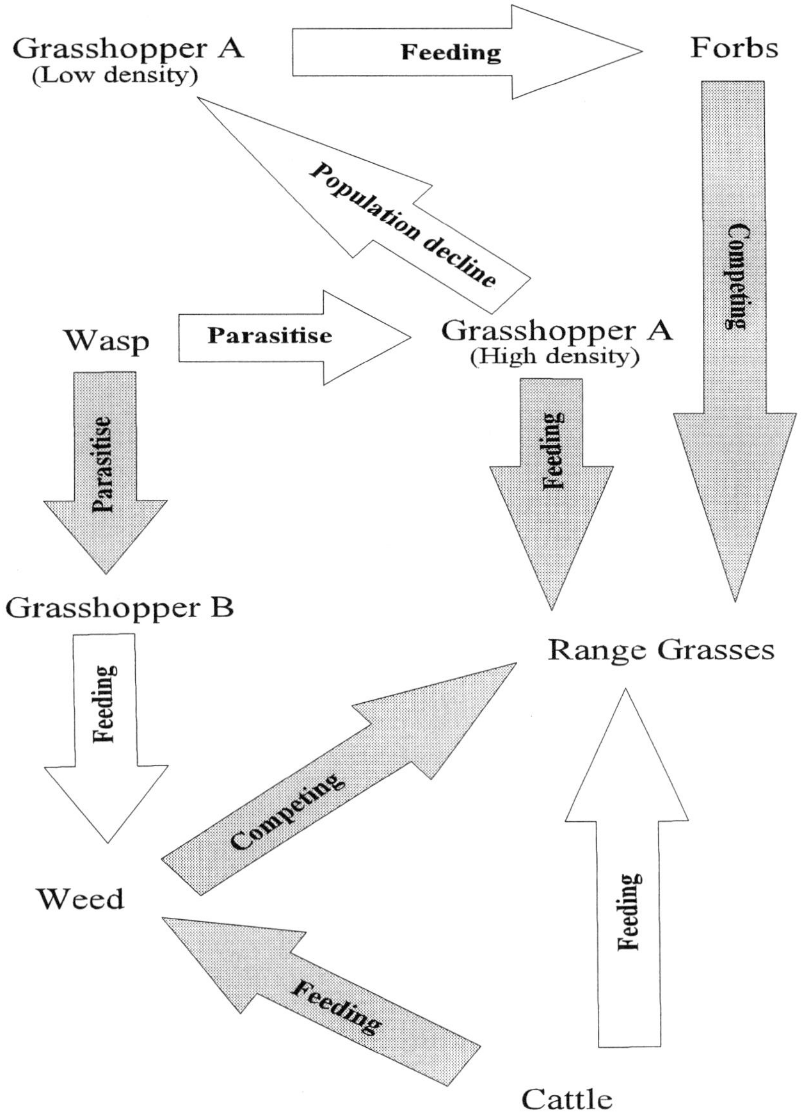

Figure 1. Simplified set of ecological entities and their relationships (arrows) in a rangeland habitat, illustrating the interacting nature of nontarget (open arrows) and target (shaded arrows) processes. Grasshopper species A (e.g., *M. sanguinipes*) at low densities feeds on forbs that compete with range grasses, but at high densities this species is a pest, feeding on valuable forage. Under the latter conditions, a parasitic wasp would be considered a beneficial biological control agent, but the wasp becomes a detriment when it attacks grasshopper species B (e.g., *M. occidentalis*), which feeds on a weed that competes with rangeland grasses

A Theoretic Rationale for Consideration of "Function over Form"

Perhaps the most serious limitation of using nontarget species as the measure of environmental impacts of a biological control program is the existence of nonlinearities between population densities and ecological processes. For both targets and nontargets, the relationship between the degree or extent of the process and the population density of the species involved in the process will be linear over only a limited range of values (Fig. 2). Thus, putative pest or nontarget densities may initially increase or decrease with little or no change in the associated ecological processes. Similar nonlinear effects may occur at when densities are at the opposite extremes, as well.

Nonlinearities can arise through a number of well-defined processes (Zeeman 1976, Saunders 1980). First, threshold dynamics or the existence of modality (with intermediate, inaccessible states) can give rise to sudden changes in a system. Such changes can be understood, and even predicted, but only through assessing the process—the material state provides no indication of the impending catastrophe. Next,

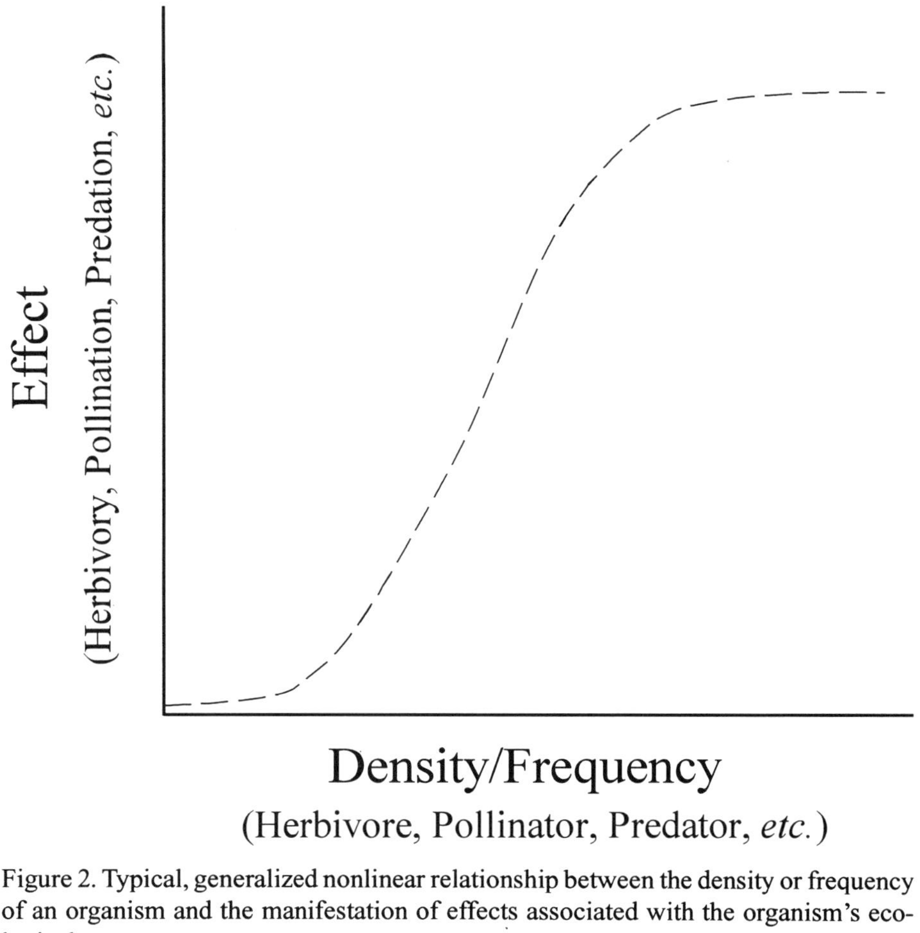

Figure 2. Typical, generalized nonlinear relationship between the density or frequency of an organism and the manifestation of effects associated with the organism's ecological processes.

systems may show divergence or sensitivity to initial conditions, so that relatively minor alterations at one time can be manifest as extreme differences at a later point. Hence, the state of such a system may not provide a good indication of impending changes in the underlying processes. Finally, processes may exhibit hysteresis (backtracking the sequence of events that lead from one state to another does not return the system to the original state), such that the dynamics are functionally irreversible. This critical property can not be predicted from the material state of the system. Thus, the density of a nontarget species may not be an accurate or informative basis for understanding the underlying dynamics of concern. Because nonlinearities or state-and-threshold effects are now widely recognized in ecological and other complex systems (Waldrop 1992, Lewin 1993), it is clear that an incremental or gradual change in an organism's density can not be taken to suggest a corresponding continuous change in ecological processes. Using our earlier analogy of a radio, the size of a speaker is, at best, a crude measure of sound volume at any point in time.

The study of grasshopper ecology has provided a particularly compelling basis for considering nonlinear associations between states and processes. Outbreak dynamics have been shown to exhibit the "symptoms" of discontinuous dynamics, and catastrophe theory has provided valuable insights to population dynamics (Lockwood and Lockwood 1991). More recently, grasshopper population dynamics have been found to be consistent with the existence of self-organized criticality[3] (Lockwood and Lockwood 1997), further substantiating the nonlinear nature of population dynamics and hence target and nontarget processes (e.g., forage loss and nutrient cycling). Perhaps most importantly, habitat types within the rangeland ecosystem also exhibit nonlinear dynamics (Lockwood and Lockwood 1993). Thus, it is essential that our concept of "nontargets" in extensive agroecosystems include not simply the existence or condition of selected species but the vital processes that ultimately determine these states.

Which Nontarget Process(es) Should be Monitored?

The choice of nontarget processes to monitor in the course of a biological control program is not as familiar as the selection of nontarget species. However, at least some of the "classic" nontarget organisms are based on their association with important ecological processes (e.g., bees are often chosen as nontarget species because of their importance in pollination). Other ecological processes (e.g., nutrient cycling or soil development) are rather more difficult to associate with particular species. In such cases, using surrogate or indicator species may be ineffective, and the process itself may need to be directly monitored. In intensive agroecosystems, the diversity of processes will be greatly reduced, so the challenge of monitoring essential dynamics should be relatively tractable.

In general, a hierarchical approach to identifying nontarget processes seems to be a rational strategy. For example, on rangeland agroecosystems the output process of interest is livestock herbivory (Fig. 3). This process can be directly affected by the processes of grass production (positively) and poisonous/deleterious plant production (negatively). Hence, these two subsidiary processes are reasonable nontargets and are amenable to monitoring. The complexity of monitoring increases as one moves deeper

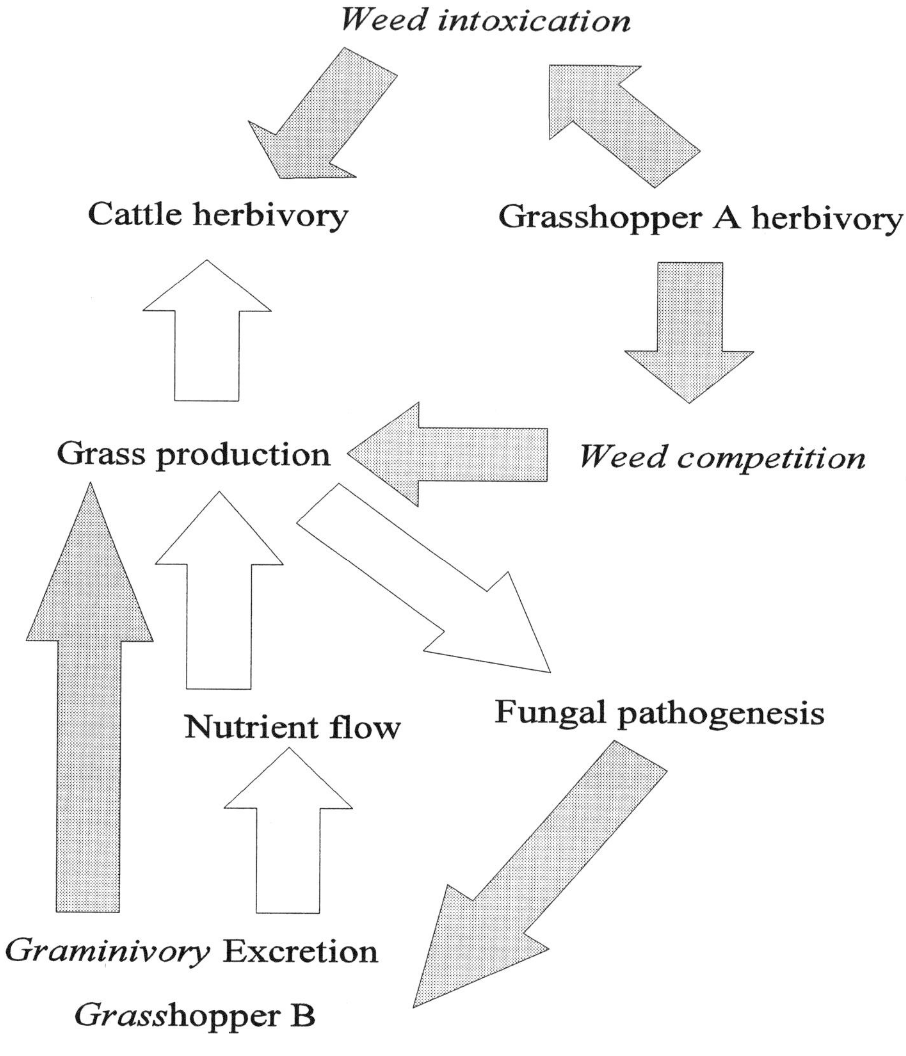

Figure 3. Process-based model showing the effects (shaded arrow = decreasing; open arrow = increasing) of selected target (*italics*) and nontarget (normal font) relationships in a rangeland agroecosystem. Grasshopper species A feeds on weeds that cause intoxication of cattle or compete with grasses needed by livestock. High grass production increases fungal pathogens that attack grasshopper species B—an organism that both competes with cattle for forage and enhances nutrient cycling, which thereby improves grass production.

into the hierarchy, but some contributing processes may be particularly relevant in terms of likely nontarget effects (e.g., grasshopper consumption of forbs and weeds). Thus, in some cases the nontarget processes can be monitored using the changing states of the system (e.g., cattle weight gain provides a measure of herbivory and plant nitrogen may provide a measure of grasshopper-mediated nutrient flow), while other processes may be most reasonably assessed using carefully chosen species (e.g., weed densities provide an indication of grasshopper herbivory and grasshopper densities may provide an indication of fungal epizootiology).

Depending on the land management goals, a wide range of other ecological processes may bear consideration as nontargets during a biological control program. Where wildlife production is of concern monitoring may need to reflect the role of grasshoppers in nutrient cycling (via litter creation, feces deposition, food for vertebrates, etc.), microbial transmission (distribution of mycorrhizal fungi, vectoring of avian tapeworms, etc.), structuring of plant communities, and competition with other herbivores. It would appear that few of these nontarget processes are readily amenable to the substitution of nontarget species, but they may be effectively monitored using surrogate ecological states (e.g., nutrient cycling may be altered with changes in the amount of grasshopper feces).

THE LIMITS AND OPPORTUNITIES OF RECONCEPTUALIZING NONTARGET EFFECTS

The purpose of this paper is not to advocate a wholesale conversion of assessment programs to "function over form", as ecosystems are comprised of objects as well as dynamics. Rather, the goal is to suggest that a singular reliance on nontarget species as the means of assessing adverse effects of biological control (or any other form of pest management) is unnecessarily myopic. Moreover, this traditional tactic risks missing potentially serious ecological changes. It seems reasonable that the more tools available to assess both target and nontarget effects of a biological control program, the more likely we are to draw sound and defensible conclusions about the benefits and costs of this environmental technology. As such, monitoring nontarget impacts of biological control agents through changes in ecological processes has some potentially powerful advantages over the use of nontarget species, but it is evident that this approach faces some significant methodological hurdles.

The Limits of Monitoring Nontarget Processes: Tracking Ghosts?

The limits of focusing our attention on processes derive primarily from a strong historical bias in western science to explore and value material existence. Hence, our analytical and theoretical capacities are heavily invested in a worldview in which physical existence is "real" and processes are the mysterious and invisible transitions between material forms. In this context, there are several limitations to monitoring processes, some of which may preclude application of process-based assessments in particular situations.

Simply put, it may be extremely difficult to assess ecological dynamics. Empirical ecological methods to directly monitor processes are often cumbersome and costly, and theoretical approaches are still relatively unsophisticated. As such, we most often resort to surrogate measures using material indicators. The limitation of this approach is that we can reasonably expect there to be lags between a change in a process and its physical manifestation, and the nature and extent of such lags are poorly understood. Despite these difficulties, there are good reasons to focus attention on the processes of concern, as the species or other material items that we use to reflect the processes will not necessarily be the classical or otherwise selected nontarget species. At the very least, acknowledging the importance of ecological processes will help narrow the choice of nontarget species, a selection that currently appears to be often a matter of tradition, expediency, and convenience.

The Opportunities of Monitoring Nontarget Processes: Discovering Our Roots

Adopting a process-oriented perspective in assessing nontarget impacts gives rise to important opportunities. The foundations of IPM lie in the recognition that it is the process of damage, not the existence of a species, that creates a pest. Aligning nontarget monitoring with this concept puts the assessment of nontarget effects on a common and demonstrably effective footing with analysis of target effects. It also draws biological control to the forefront of ecology and promises to move this technology from art to science. Thus, the use of nontarget processes is both conservative (i.e., consistent with established IPM concepts) and progressive (i.e., relevant to recent advances in ecological theory).

The focus on material existence may also account for the disproportionate attention that has been directed towards the development of classical and augmentative biological control. These tactics involve the release of species—easily seen and manipulated objects employed as tools to suppress the target "objects." While this approach has demonstrated its viability, it has overshadowed the more sophisticated and potentially more widely applicable approach of conservation biological control. This tactic does not necessitate that species become the objects of conservation. Rather, conservation biological control allows, even encourages, a more process-oriented approach. Indeed, its purpose is to maintain those processes that normally prevent the damage that defines the existence of a pest. Thus, the measure of success of conservation biological control is the suppression of damaging processes, rather than the preservation of specific, natural enemies (e.g., Lockwood and Schell 1997). The material entities—nontarget predator and parasitoid species—are the means to the ends of maintaining the processes that are the essence of biological control.

NOTES

1. It is surely possible for a species to be driven to extinction with little or no impact to ecological processes (e.g., the species was very rare or another species compensated for the loss), but such cases are difficult or impossible to predict *a priori* and the moral consequences with respect to intrinsic value as well as the economic consequences with respect to the loss of option value can not be dismissed (Johnson 1991, Wilson 1988, 1992).
2. As noted in Kogan's (1998) review of IPM history and contemporary developments, the starting point for IPM was, and remains, the economic injury level (or economic threshold). This fundamental decision rule is particularly applicable to insects and to early levels of IPM integration, but the underlying

concept that the value of the reductions in pest damage should exceed the cost of a management strategy applies to all forms of IPM (including "biointense" and "ecologically-based"). Indeed, Kogan (1998) concludes with the admonition that, "no technological innovation [in IPM] will be adopted unless it contributes to producers' economic goals and meets the requisites for acceptance by society."

3. Bak and Chen (1991) hypothesize that large, interactive systems naturally develop critical states in which a minor event can lead to a catastrophic chain of events. Such systems never reach equilibrium but instead evolve from one metastable state to the next. They call this phenomenon self-organized criticality, which appears to provide a deterministic alternative to equilibrial and chaotic dynamics.

REFERENCES

Bak, P., and K. Chen. 1991. Self-organized criticality. Scientific American 249: 26–33.

Barratt, B. I. P., A. A. Evans, C. M. Ferguson, G. M. Barker, M .R. McNeill, and C. B. Phillips. 1997. Laboratory nontarget host range of the introduced parasitoids *Microctonus aethiopoides* and *M. hyperodae* (Hymenoptera: Braconidae) compared with field parasitism in New Zealand. Environ. Entomol. 26: 694–702.

Barratt, B. I. P., A. A. Evans, C. M. Ferguson, M. R. McNeill, J. R. Proffitt, and G. M. Barker. 1998. Curculionoidea (Insecta: Coleoptera) of New Zealand agricultural grassland and lucerne as potential non-target hosts of the parasitoids *Microctonus aethiopoides* Loan and *Microctonus hyperoidae* Loan (Hymenoptera: Braconidae). N.Z. J. Zool. 25: 47–63.

Berryman, A. A., and N. C. Stenseth. 1984. Behavioral catastrophes in biological systems. Behavioral Sci. 29:127–137.

Busby, F. E. 1994. Rangeland health: new methods to classify, inventory, and monitor rangelands. National Academy Press, Washington, DC, USA.

Caltagirone, L. E. 1981. Landmark examples in classical biological control. Annu. Rev. Entomol. 26:213–32.

Carson, R. 1962. Silent Spring. Houghton Mifflin, Boston.

Coppel, H. C., and J. W. Mertins. 1977. Biological insect pest suppression. Springer Verlag, New York.

DeLoach, C. J. 1976. Considerations in introducing foreign biotic agents to control native weeds of rangelands. *In* Proceedings of the 4th International Symposium, Biological Control of Weeds. Gainesville, FL, USA. p. 39–50.

Ewen, A. B., and M. K. Mukerji. 1980. Evaluation of *Nosema locustae* (Microsporida) as a control agent of grasshopper populations in Saskatchewan. Journal of Insect Pathology 35:295–303.

Ferguson, C. M., G. M. Roberts, B.I.P. Barratt, and A. A. Evans. 1994. The distribution of the parasitoid *Microctonus aethiopoides* Loan (Hymenoptera: Braconidae) in southern south island *Sitona discoideus* Gyllenhal (Coleoptera: Curculionidae) populations. Proc. 47th N. Z. Plant Protection Conf. p. 261–265.

Goeden, R. D. 1988. A capsule history of biological control of weeds. Biocontr. News Info. 9: 55–61.

Hokkanen, H. M. T., and D. Pimentel. 1984. New approach for selecting biological control agents. Can. Entomol. 116: 1109–1121.

Hokkanen, H. M. T., and D. Pimentel. 1989. New associations in biological control: theory and practice. Can. Entomol. 121: 829–840.

Howarth, F. G. 1983. Biological control: panacea or Pandora's box? Proc. Hawaiian Entomol. Soc. 24:239–244.

Howarth, F. G. 1985. Impacts of alien land arthropods and mollusks on native plants and animals in Hawaii. *In* C. P. Stone and J. M. Scott (eds.), Hawaii's terrestrial ecosystems: preservation and management. Univ. Hawaii Press, Honolulu. p. 149–179.

Howarth, F. G. 1991. Environmental impacts of classical biological control. Annu. Rev. Entomol. 36:485–509.

Howarth, F. G. 1992. Environmental impacts of species purposefully introduced for biological control of pests. Pacific Sci. 46:388–389.

Johnson, L. E. 1991. A morally deep world: An essay on moral significance and environmental ethics. Cambridge Univ. Press, New York.

Johnson, D. L., and E. Pavlikova. 1986. Reduction in consumption by grasshoppers (Orthoptera: Acrididae) infected with *Nosema locustae* Canning (Microsporidia: Nosematidae). J. Invertebrate Pathol. 48:232–238.

Kogan, M. 1998. Integrated pest management: historical perspectives and contemporary developments. Annu. Rev. Entomol. 43: 243–270.

Lewin, R. 1993. Complexity: life at the edge of chaos. Dent, London, UK.

Lockwood, J. A. 1993a. Environmental issues involved in biological control of rangeland grasshoppers (Orthoptera: Acrididae) with exotic agents. Environ. Entomol. 22:503–518.
Lockwood, J. A. 1993b. Benefits and costs of controlling rangeland grasshoppers (Orthoptera: Acrididae) with exotic organisms: search for a null hypothesis and regulatory compromise. Environ. Entomol. 22:904–914.
Lockwood, J. A. 1996. The ethics of biological control: understanding the moral implications of our most powerful ecological technology. Agric. Human Values 13:2–19.
Lockwood, J. A. 1997a. Grasshopper population dynamics: a prairie perspective. *In* S. K. Gangwere, M. C. Muralirangan, and M. Muralirangan (eds.), Bionomics of grasshoppers, katydids and their kin. CAB International, London, UK. p. 103–127.
Lockwood, J. A. 1997b. Rangeland grasshopper ecology. *In* S. K. Gangwere, M. C. Muralirangan, and M. Muralirangan (eds.), Bionomics of grasshoppers, katydids and their kin. CAB International, London, UK. p. 83–101.
Lockwood, J. A., and L. D. DeBrey. 1990a. Direct and indirect effects of *Nosema locustae* (Canning) (Microsporida: Nosematidae) on rangeland grasshoppers (Orthoptera: Acrididae). J. Econ. Entomol. 83:377–383.
Lockwood, J. A., and D. R. Lockwood. 1991. Rangeland grasshopper (Orthoptera: Acrididae) population dynamics: insights from catastrophe theory. Environ. Entomol. 20:970–980.
Lockwood, J. A., and D. R. Lockwood. 1993. Catastrophe theory: a unified paradigm for rangeland ecosystem dynamics. J. Range Mgt. 46:282–288.
Lockwood, D. R., and J. A. Lockwood. 1997. Evidence of self-organized criticality in insect populations. Complexity 2:49–58.
Lockwood, J. A., and S. P. Schell.. Decreasing economic and environmental costs through reduced area and agent insecticide treatments (RAATs) for the control of rangeland grasshoppers: Empirical results and their implications for pest management. J. Orthoptera Res. 6: 19–32.
Louda S. M., D. Kendall, J. Conner, and D. Simberloff. 1997. Ecological effects of an insect introduced for the biological control of weeds. Science 277:1088–1090.
Miller, M., and G. Aplet. 1993. Biological control: a little knowledge is a dangerous thing. Rutgers Law Rev. 45:285–334.
Office of Technology Assessment. 1993. Harmful non-indigenous species in the United States. OTA-F-565, U.S. Govt. Printing Office, Washington, D.C..
Oma, E. A., and G. B. Hewitt. 1984. Effect of *Nosema locustae* (Microsporida: Nosematidae) on food consumption in the differential grasshopper (Orthoptera: Acrididae). J. Econ. Entomol. 77:500–501.
Onsager, J. A. 1988. Assessing effectiveness of *Nosema locustae* for grasshopper control. Montana Agric. Expt. Sta., Montana AgResearch 5:12–16.
Parker, M. A. 1984. Local food depletion and the foraging behavior of a specialist grasshopper, *Hesperotettix viridis*. Ecology 65:824–35.
Perkins, J. A. 1982. Insects, experts and the insecticide crisis: the quest for new pest management strategies. Plenum, New York.
Plotkin, M. J. 1988. The outlook for new agricultural and industrial products from the tropics. *In* E. O. Wilson (ed.), Biodiversity. Ntl. Acad. Press, Washington, D.C. p. 106–116.
Saunders, P. T. 1980. An introduction to catastrophe theory. Cambridge Univ. Press, NY, USA.
Stiling, P. 1987. The frequency of density dependence in insect host-parasitoid systems. Ecology 68:844–856.
Stiling, P. 1988. Density-dependent processes and key factors in insect populations. J. Anim. Ecol. 57:581–593.
Strong, D. R. 1997. Fear no weevil? Science 277: 1058–1059.
Sussman, H. J.. and R. S. Zahler. 1976. Catastrophe theory as applied to the social and biological sciences: a critique. Synthese 37:117–216.
Sussman, H. J., and R. S. Zahler. 1978. Critique of applied catastrophe theory in behavioral sciences. Behav. Sci. 23:283–389.
United States Department of Agriculture. 1996. Grasshopper IPM user handbook. USDA, Animal and Plant Health Inspection Service, Plant Protection and Quarantine, Riverdale, Maryland.
Van Driesche, R. G., and T. S. Bellows. 1996. Biological control. Chapman and Hall, New York.
Waldrop, M. M. 1992. Complexity: The emerging science at the edge of chaos. Viking, London, UK.
Wilson, E. O. (ed.). 1988. Biodiversity. Ntl. Acad. Press, Washington, D.C.
Wilson, E. O. 1992. The diversity of life. Harvard University Press, Cambridge, Massachusetts.
Zeeman, E. C. 1976. Catastrophe theory. Scientific American 234:65–83.

3

THE FREQUENCY AND STRENGTH OF NONTARGET EFFECTS OF INVERTEBRATE BIOLOGICAL CONTROL AGENTS OF PLANT PESTS AND WEEDS

Peter Stiling
Department of Biology, University of South Florida, Tampa, Florida

Daniel Simberloff
Department of Ecology and Evolution, University of Tennessee, Knoxville, Tennessee

INTRODUCTION

In 1989, the cactus moth, *Cactoblastis cactorum*, which had been taken from South America and released around the world to control pestiferous *Opuntia* cacti on which it feeds, appeared in the Florida Keys. How it got there is uncertain. The moth was introduced into the Caribbean island of Nevis in 1957 and then moved by local authorities to other islands (Tuduri et al. 1971). It then either island hopped, as it had in Hawaii (Tuduri et al. 1971), and jumped from Cuba to the Florida Keys, or was carried inadvertently on ornamental cacti imported into Miami, Florida from pads collected in Hispaniola (Pemberton 1995). However, the moth was not noticed in Miami before it was found in the lower Keys so we tend to favor the natural dispersion explanation. The results of this invasion/introduction were rapid and serious. The moth dispersed 320 miles northward through Florida from 1989 to 1991 to reach Brevard County (Johnson and Stiling 1998). On average, from six sites throughout Florida, over 90% of *Opuntia stricta* plants with more than 10 pads exhibited feeding damage. In 1997 *Cactoblastis* was found in the Jacksonville area (Stiling, personal observation). If the moth can survive in Jacksonville, it could move across the Gulf Coast states into Texas and beyond. This prompts concern not only for native Florida cacti but also for cacti native to the rest of the North American continent (especially the *Opuntia*-rich desert southwest). Rare cacti may be particularly at risk. *Opuntia spinosissima*, whose entire U.S. population consists of 12 plants in the Florida Keys, is a preferred host of *Cactoblastis* larvae (Johnson and Stiling 1996). Individuals now exist inside cages designed to protect them from *Cactoblastis*.

It may be argued that biological control practitioners are not to blame for the *Opuntia-Cactoblastis* fiasco because they only introduced the moth into the Caribbean; the moth itself migrated to the U.S. mainland or was brought there by careless horticulturists. At the time of the moth's introduction to the Caribbean in 1957, it was not known that *Cactoblastis* was able to island hop, as it did in Hawaii. Also, it could be argued that this is a one of a kind example and similar events would be unlikely to happen today given the careful screening procedures employed in most current biological control campaigns. This led us to question whether other biological control agents of plant pests and weeds might attack more than their intended target organism (Simberloff and Stiling 1996a, b). Although there have been relatively few documented instances of damage to nontarget organisms, considering the number of releases made, we believe monitoring of most potential nontarget species is minimal, or non-existent, so the absence of evidence cannot be taken as evidence of absence. Furthermore, although many authors believe most problems occurred in the early days of biological control (e.g., Center 1995), there appears to be a recent awareness that nontarget effects may still be common. For example, the weevil *Rhinocyllus conicus* Froeh introduced to control exotic thistles has recently been shown to impact native thistles (Louda et al. 1997, Louda, this volume). The exotic coccinellid *Coccinella septempunctata* reduced the abundance of two native coccinellids, *C. transversoguttata* and *Adalia bipunctata*, 20–30 fold in South Dakota (Elliott et al. 1996, Obrycki et al., this volume). *Coccinella septempunctata* also competes with the native *C. maculata* in Iowa (Obrycki et al. 1998) and is associated with a decline in the indigenous *C. novemnotata* in the northeastern U.S. (Wheeler and Hoebeke 1995). Two species, a crayfish, *Procambarus clarkii* Girard and a mosquitofish, *Gambusia affinis* Baird and Girard, introduced into coastal regions around Los Angeles to reduce the density of mosquitoes also fed on a rare newt, *Taricha torosa* (Rathke) and may have caused local extinctions (Diamond 1996). Releases of a parasitic wasp and a fungus to control pest grasshoppers in western U.S. rangelands were stopped only after concerns were raised that these enemies might damage nontarget native grasshoppers, some of which are rare (Goodman 1993, Lockwood, this volume).

The purpose of this chapter is to gather together these isolated or anecdotal cases to systematically answer fundamental questions such as

- What is the host range of released natural enemies?

- What portion of the native biota is susceptible to nontarget effects?
- How frequent are nontarget effects of biological control agents?
- What are the strengths of the nontarget effects?

It is not our intention to suggest that "the sky is falling," despite the fact that at least one attendee of the 1997 Entomological Society of America program symposium where we first presented these data suggested this was our message. Rather we want to show that many biological control agents attack more than one host, many pest taxa are native, and nontarget effects of biological control releases are more common and more serious than we might at first think. We reach this conclusion after a synthesis of known examples of nontarget effects documented in the literature. We think these examples are relatively complete but are not the result of a systematic examination of a series of journals. We believe our chapter is particularly timely given the increasing pace of biological control releases. McEvoy (this volume) notes that while there were only four weed biological control programs per decade between 1947 and 1977, there were 24 control programs per decade between 1977 and 1997. Biological control workers are beginning to discuss more openly the pros and cons of biological control releases (e.g., Gould et al. 1996; Thomas and Willis, 1998) though conflicts still frequently arise (see Boldt 1998 and response by Louda et al. 1998, and Frank 1998 and response by Simberloff and Stiling 1998).

THE HOST RANGE OF NATURAL ENEMIES

Part of the *Cactoblastis/Opuntia* problem derives from the breadth of species attacked by *Cactoblastis.* Not only does *Cactoblastis* attack *Opuntia stricta*, its intended target, it also attacks other members of the *Opuntia* genus. In Florida, there are six species of *Opuntia.* In one study, *Cactoblastis* attacked all four *Opuntia* species (*O. stricta, O. spinosissima, O. humifusa*, and *O. triacantha)* tested (Johnson and Stiling 1996). Is this an isolated incident or do most natural enemies attack more than one host species? One of the problems in determining the host range of natural enemies is that it requires examining a huge range of potential prey or hosts to see if they are attacked. This examination requires a large amount of money and time, neither of which are available to the majority of scientists. Because of their focus on pest control, most government agencies have been concerned with gathering data on pest densities, not on nontarget effects. One of the few exceptions was the extensive rearing program of Lepidoptera in nine north-eastern states, carried out between 1915 and 1933, to find alternate hosts of parastoids of the gypsy moth and brown-tailed moths (Schaffner and Griswold 1934). During the course of the project, 11,416 collections of macrolepidoptera larvae (and a few pupae) were reared, representing 442,112 identified specimens. Sheehan (1991) summarized their database and listed host and parasitoid species. From Sheehan's work we have summarized host use patterns of these parasitoids (Table 1). Only 25 species of parasitoid use just one host while 49 species used two or more hosts. On average each species of parasitoid attacked 2.82 host species.

An extensive survey of the parasitoids of British leaf-mining Lepidoptera produced similar results (Askew and Shaw 1974). While some parasitoid genera, like *Achrysocharoides* (Eulophidae) have relatively narrow host ranges, most other gen-

Table 1. Host utilization by hymenopteran parasitoids of lepidopteran hosts in the northeastern USA (after Sheehan 1991, and Schaffner and Griswold 1934).

Number of hosts per parasitoid species	Frequency
1	25
2	16
3	21
4	3
5	2
6	2
7	0
8	1
9	0
10	2
11	0
12	0
13	1
14	1
Mean	2.82 ± 0.3 s.e.

era have much broader host ranges (Askew 1980). Inspections of the host ranges of the 10 most numerous species of polyphagous parasitoids of *Phyllonorycter* (Lepidoptera) leaf miners in Britain show that five have broad host ranges that encompass leaf mining Hymenoptera, Coleoptera, and sometimes Diptera, in addition to several other families of Lepidoptera. These five species seem to be able to develop on almost any insect that mines leaves (Askew 1994). Many of the parasitoids reared from seven genera of Nepticulidae (Lepidoptera) in northwestern Europe also have multiple hosts, and Askew (1994) concluded that most parasitoids of leaf-mining Lepidoptera are polyphagous. The average number of host families (let alone species) attacked by chalcid parasitoids in the British leafmining community is 2.97 ± 2.27 s.d. (Askew and Shaw 1986).

It could be that biological control releases involve only those enemies that are monophagous, rather than polyphagous. However, an examination of the data shows that, at least for insect parasitoids, polyphagous enemies were released in biological control campaigns more often (n = 367) than monophagous enemies (n=219) (62.6% versus 37.4%) (Stiling 1990, after data in Luck 1981). Other data reveal that, in many instances, biological control campaigns can fail if there is a lack of alternative prey or hosts for biological control agents (Clausen 1978). Stiling (1993) documented that lack of alternative prey or hosts has been attributed as a reason for failure in 16.9% of all failed biological control campaigns. Thus many biological control campaigns rely on more than the target pest to help sustain the released enemy, although it is possible that some of these alternate hosts may also be introduced pests.

THE TARGETS OF BIOLOGICAL CONTROL

While it is true that the intended targets of many biological control campaigns are exotic pests introduced from different countries, a large proportion of the potential nontarget hosts may be native insects. Barratt (this volume) noted the presence of 21 exotic and 64 native weevil species as possible hosts for introduced natural enemies of two alien pest weevils in New Zealand. In grasslands, the habitat where most releases of natural enemies of the pest weevils has occurred, there are 14 exotic and 47 native weevil species. Having a native pest as a target probably increases the risk that closely related nontarget species, such as congeners, may be attacked. Van den Bosch (1971) noted that the majority, 64%, of pest species in the United States were native but in certain regions the figure may be lower. Frank and McCoy (1993) documented that 21% of the pest insects and 25% of the pest weeds in Florida were of native origin. Perhaps more germane is the fraction of species targeted in classical biological control programs that are native versus non-native. Stiling (1990) calculated that, worldwide, 21.8% of biological control campaigns targeted native pests. An earlier worldwide review by Hall and Ehler (1979) reported only 5.7% of intended targets were native species. If the 20% figure suggested by Stiling (1990) is true, the frequency of attack of nontarget species closely related to the targeted native pests may be fairly common. Remember too that some exotic pests can have congeners in the areas of introduction of biological control agents, though we have not gathered much data on this possibility.

THE FREQUENCY OF NONTARGET EFFECTS

Having argued that (a) many enemies released in biological control campaigns attack more than one prey/host and (b) a significant number of pests targeted for biological control are native in origin and therefore likely to be associated with related species, we might expect to see much evidence of nontarget effects. However, evidence for nontarget effects is scarce, probably because monitoring of nontarget species is minimal, particularly in sites and habitats far from the point of release (Simberloff and Stiling 1996 a). Nevertheless, in recent years there has been a steady trickle of examples of nontarget effects on native organisms. For example, McFadyen (1998) lists eight recorded instances of damage to nontarget plants by weed biological control agents. The 1997 Entomological Society of America Program Symposium "Biological Control for a Small Planet: Nontarget Effects of BC" organized by the editors of this volume, Peter Follett and Jian Duan, was instrumental in providing a forum for such effects to rise to the forefront of people's minds. It was here that Les Ehler noted that fortuitous control of pests by enemies released to control other pests happens "perhaps moderately frequently," indicating that nontarget effects, albeit on other pests, may be quite common. We summarize some of the data presented in that meeting, together with some new data gathered from older sources. For example, there is some older data in an appendix of the Technical Communication of the Commonwealth Institute of Biological Control (McLeod 1962) which lists recoveries of parasites and predators from hosts/prey other than target species after releases against forest insects in Canada between 1910 and 1958. From this list we have summarized the frequency of attack of nontarget species (Table 2). The average number of species attacked is 3.95 ± 10.98 s.d. Fifteen of the 40 enemies released, 37.5%, attacked more than one host species. One species, *Compsilura concinnata* (Meigen) (Diptera: Tachinidae), introduced into North America in 1910, had over 70 recorded species of host, but is now known to attack over 200 species of Lepidoptera and Tenthredinoidea (Hymenoptera) (J. Boettner, personal communication).

Biological control workers might suggest that they would not release such a polyphagous species now, but even if we discount *Compsilura concinnata* from McLeod's list the mean number of nontarget species attacked is still 1.25 ± 6.24 s.d. Besides, release of polyphagous enemies has continued into the 1990s. The giant African snail, *Achatina fulica,* has been accidentally introduced to many parts of Asia as well as to numerous islands in the Indian and Pacific Ocean, and has recently reached the West Indies. It reached high densities and was seen as a threat to agriculture. Pressures for action became overwhelming in the face of such a highly visible invasion and caused many governments to sanction the release of predatory snails, including the New World snail, *Euglandina rosea,* beginning in the 1950s. These predatory snails wrought havoc with native snails (Civeyrel and Simberloff 1996), yet a release of *E. rosea* was made in Samoa as recently as 1992. Furthermore, release of the predaceous flatworm, *Platydemus manokwari,* a generalist that attacks snails and other soft-bodied invertebrates, was favored on Pacific islands, even after damage by *Euglandina* was known.

Recently, Hawkins and Marino (1997) provided startling data showing that at least 50 of the 313 species of parasitoids released against insect pests in the United States attack nontarget species. This study, together with studies by McLeod (1962), Turner (1985) on arthropod herbivores of weeds, and Funasaki et al. (1988) on the frequency of nontarget effects of biological control releases in Hawaii, suggest that perhaps a quarter (22.1% ± 10.6 s.d.) of biological control releases have had nontarget effects (Table 3).

A related question here is what are the taxonomic affinities of the species that tend to attack nontarget species. Perhaps the best data to address this issue come from the list of 50 insect species released in biological control campaigns which are known to attack nontarget insects (Hawkins and Marino 1977). When broken down by family, braconids and eulophids most frequently attack nontarget taxa (Table 4), but there is no statistically significant difference among the enemy taxa ($x_4^2 = 7.709$, $p = 0.173$).

We can also rephrase our question to ask how many nontarget species will be attacked for each biological control agent released. If about one fifth of all releases attack nontarget hosts (Table 3, 22.1%) and many species that attack nontarget species attack between two and three species of nontarget hosts (subtracting 1 for the target host; Table 1, 1.82, and Table 2, 2.95, mean = 2.38) then we might expect 0.221 x 2.38 = 0.526 nontarget species to be attacked for every pest species attacked by an introduced biological control agent. By this logic, release of enemies against two target species will, on average, result in attack of one nontarget species. Of course, it is possible that some of these nontarget species may themselves be pests. Our synthesis of the available data (Table 5) suggests that the overall proportion of native nontarget species attacked by at least one introduced biological control agent is 65.1% (84/129) when totaled (excluding the studies by Julien et al. [1984] and McFadyen [1998] where

Table 2. Recoveries of parasites and predators employed in biological control projects against forest insects in Canada, 1910–1958. After data in McLeod et al. 1962.

Number of species attacked	Frequency
1	25
2	4
3	4
4	0
5	2
6	2
8	1
13	1
70	1
Mean	3.95

Table 3. The overall frequency of biological control agents attacking nontarget taxa.

Type of biol. control agent	No. species released	No. attacking nontarget hosts	Area	% attacking nonnative hosts	Reference
Parasitoids	313	50	USA	16.0	Hawkins & Marino 1997
Arthropod herbivores	33	7	World	21.2	Turner 1985
All biological control agents	243	33	Hawaii	13.6	Funasaki et al. 1988
Parasitoids and predators	40	15	Canada	37.5	McLeod et al. 1962
Overall mean				22.1% ± 10.6% s.d.	

Table 4: The taxonomic distribution of insects released in biological control programs that attack nontarget hosts, n=20 for any given taxa. Other = all other taxa combined. Data extracted from Hawkins and Marino (1997).

Taxa	N	Frequency (%)
Braconidae	79	24.1
Eulophidae	22	22.7
Ichneumonidae	82	18.3
Tachinidae	63	11.1
Scoliidae	22	0
Other	40	12.5

Table 5. Attack rates of nontarget native taxa by biological control agents.

Nontarget taxa	No. of native nontarget species		Area	Responsible biol. control agent	Reference
	Examined	Attacked			
Opuntia (cacti)	6	5	Florida	*Cactoblastis*	Johnson/Stiling 1997
Curculionidae (weevils)	7	7	N.Z.	*Microctonus aethiopoides*	Barratt et al. 1997
	7	2	N.Z.	*M. hyperodae*	
Cirsium (thistles)	5	5	Midwest USA	*Rhinocyllus conicus*	Louda et al. 1997
Macrolepidoptera	71	58	NE USA	Various parasitoids	Hawkins/Marino 1997
Microlepidoptera	31	5	NE USA	Various parasitoids	Hawkins/Marino 1997
Weeds	?	5	USA	Mainly insects	Julien et al. 1984
Lepidopterans	2	2	Guam	Parasitoids	Nafus 1993

the number of nontarget taxa examined were not known) or 72.8% ± 35.5 s.d. when averaged over eight studies.

THE STRENGTH OF NONTARGET EFFECTS

If it is difficult to gauge the frequency of nontarget effects of biological control agents on nontarget taxa because of lack of data, it is even more difficult to establish the strength of such effects. Are such effects great enough to reduce population sizes or even cause extinctions of nontarget taxa? Circumstantial evidence suggests the effects can be substantial. Hawkins and Marino (1997) showed that the tachinid *Compsilura concinnata* averaged 16.7% (s.d. = 18.7) of the parasitoid individuals reared from each macrolepidopteran species in the Schaffner and Griswold (1934) study, suggesting *C. concinnata* is a non-trivial component of many parasitoid complexes. Jeff Boettner (unpublished data) put out sentinel *Cecropia* caterpillar hosts in the field and recorded 13, 26.8, and 70% mortality by *Compsilura* flies in the first, second, and third instars, respectively. Overall, *Compsilura* caused 81% total mortality rate of caterpillars. Boettner also note that while *Compsilura* has four generations per year, the intended target pest, gypsy moth, has only two generations per year so that other hosts entirely are needed to sustain *Compsilura* for two generations.

First of all we can ask how much can natural enemies reduce the density of host or prey populations? Some of the best data to answer this question come from biological control studies, because here the biological control practitioner is often interested in depression of pest populations and therefore tracks population densities through time after a release. A recent study (Williamson and Fitter 1996) showed that, worldwide, of 1128 releases of herbivores, 691 or 61.3% became established and thus might have some effect on host populations. The number of studies with a documented measure of control and thus a substantial effect on host populations, was 224 or 19.9%. For parasitoids, of 1377 releases, 417 or 30.3% became established and 146 or 10.6% exerted some measure of control. However, other data (Stiling 1990) suggest a higher rate of establishment of parasitoids (43.4%, n = 957) and a higher degree of control (21.2%). Data from Hall et al. (1980) give establishment rates of parasites and predators both as 34.0%, with success rates for parasitoids of 14% and predators 26%. The point here is that many released biological control agents have the potential to severely impact host/prey populations. Recent data from Hawkins (1994) for native insect herbivores (leaf feeders, leaf miners, gall makers, stem borers, and root feeders) also suggests high rates of attack by natural enemies, especially parasitoids, with a mean level of parasitism of about 41.4% ± 17.8 s.d. These lines of evidence suggest natural enemies released in biological control campaigns have the potential to inflict serious losses on nontarget populations because (1) natural enemies released in biological control campaigns can exert a strong degree of control on pest populations, and (2) natural enemies of native organisms can also inflict substantial mortality. However, it is also worth noting that biological control releases rarely, if ever, cause extinction of their targets, so it is probably equally unlikely that they would cause the extinction of a nontarget species.

What then of the actual data? The best data require years of painstaking field work to collect, with before-release and after-release comparisons of nontarget populations. Such data is simply not yet widely available. However, there exist some good data from highly detailed studies on specific taxa such as *Cirsium* thistles (Louda et al. 1997) and *Opuntia* cacti (Johnson and Stiling 1998). Both these studies show considerable geographic variation in attack rates. Damage to *Opuntia stricta* plants varied from 58.7% in the Florida Keys to 23.6% at Terra Ceia in central Florida. Damage to wavyleaf thistle, *Cirsium undulatum*, flowerheads by *Rhinocyllus conicus* varied from 38.7% in Mesa Verde National Park, Colorado to 77.5% in Wind Cave National Park in South Dakota.

Despite this geographical variation we can make some broad brush comparisons by taking an average damage level for nontarget organisms using all geographical locations. Thus, for *Opuntia stricta* cacti about 38.2% ± 11.8 s.d. of the pads are eaten by *Cactoblastis*, and an average of 60.4% ± 17.1 s.d. of *Cirsium undulatum* flowerheads are destroyed by *Rhinocyllus*. Furthermore, on average, 85.9% of the viable seeds produced by flowerheads of Platte thistle, *C. canescens*, were destroyed by *R. conicus*, and this native plant has been shown to be seed limited (Louda, this volume). We have gathered together data on other documented attack rates of nontarget organisms by biological control agents (Table 6). They reveal average attack rates on nontarget insects of 19.2% ± 25.6 s.d. and 39.4% ± 24.6 s.d. for plants, though it remains difficult to gauge the effects of these impacts on future population sizes.

CONCLUSIONS

Despite the views of some biological control workers (e.g., Gonzalez and Gilstrap 1992, Carruthers and Onsager 1993, Center 1995) who suggest nontarget effects are weak and infrequent and, especially, a thing of the past, our review suggests that when people take the trouble to look, nontarget effects are quite frequent and may be substantial. However, the problem may not end there. Evolution may lead to host switching of biological control agents owing to an interaction among initial genetic variation, strong selection, and ecological opportunity. Secord and Kareiva (1996) have documented at least 20 known cases of host shifts among biological control agents, some extremely unlikely to have been predictable on the basis of host specificity testing. For example, the leaf beetle *Chrysolina quadrigemina*, which was imported in the 1940s into northern California to control klamath weed, began attacking a horticultural plant, *Hypericum callycinum* L., a host shift which may have involved formation of a new host race (Ehler 1991).

Introduced species could also have indirect as well as direct effects on nontarget species. Such effects are even more imperfectly known and difficult to study. For example, prickly-pear cacti on Green Cay, San Salvador Island, in the Bahamas are a major food source (and one of only 10 plant species on the island) for rock iguanas, *Cyclura* r. *rileyi* (Hayes et al. 1995). Cacti there are under heavy attack by *Cactoblastis* and fears are great that the iguana population will be severely impacted as its food resource becomes depleted. On the other hand, gall flies released as biological control agents of introduced knapweed, *Centaurea maculosas*, in Montana, provide a bonanza for deer mice, *Peromyscus maniculatus*, permitting population increases (Pearson et al. 1999). This, in turn, disrupts small mammal community structure, which could in turn alter the structure of the predator community. Furthermore, elevated deer mouse densities may lead to an increase in the occurrence of hantavirus and its transmission to humans, given that deer mice are the primary vectors for this disease. As Pearson et al. (1999) note, even "well-behaved" biological control agents that are host-specific can have indirect effects on native communities, some of which are likely to be detrimental. Of course, as Jervis (1998) suggests, any risk of adverse environmental effects has to be weighed against those of either doing nothing or of alternative control methods. Ideally, a rigorous cost-benefit analysis needs to be carried out with every introduction under consideration, a demanding but essential task.

In the future we see a need to collect more, high quality data on nontarget effects. It is difficult, of course, to collect before and after data on nontarget effects for most existing biological control campaigns because only the "after" part is left to study. Thus, while we can discover the frequency of nontarget effects, we cannot easily know the population sizes of nontarget organisms before biological control to gauge the strength of these effects. Future campaigns, however, should be planned with an investigation of nontarget effects in mind. Low statistical power for analysis will likely be a problem because of lack of replication. Once a released enemy is established, it will likely spread into many or most areas containing the host or prey. One alternative is to maintain vulnerable nontarget taxa in greenhouse or laboratory situations and experimentally reintroduce these into the field to measure the strength of attack of biological control enemies in different habitats or sites. For example, we are currently

Table 6: Attack rates of nontarget organisms by biological control agents (n>20)

Nontarget species	Biological control agent	Site	Attack %	Reference
INSECTS				
Eutreta xanthochaeta	*Diachasmimorpha longicaudata*	Hawaii	0.7	Duan/Messing 1997
Trupanea dubautiae	*Tetrastichus giffardianus*	Hawaii		Duan/Messing 1998
	Habrocytus elevatus	Hawaii	0.5	
	Euderus metallicus	Hawaii	0.1	
Ensina sonchi	*Tetrastichus giffardianus*	Hawaii	0	
	Habrocytes elevatus	Hawaii	39	
	Euderus metallicus	Hawaii	0	
Nicaena sp.	*Microtonus aethiopoides*	N.Z.	8.6	Barratt et al. 1997
Irenimus aemulator	*M. aethiopoides*	N.Z.	61.7	Barratt et al. 1997
I. stolidus	*M. aethiopoides*	N.Z.	1.6	Barratt et al. 1997
I. aequalis	*M. hyperodae*	N.Z.	3.1	Barratt et al. 1997
Hypolimnus bolina	*Brachymeria lasus*	Guam	50.2	Nafus 1993
H. anomala	*B. lasus*	Guam	50.0	Nafus/Schreiner 1989
Agrius onvolvuli	*Trichogramma chilonis*	Guam	3.1	Nafus 1993
Macrolepidopterans	*Compsilura concinnata*	FL	16.7	Hawkins/Marino 1997
Cecropia sp.	*Compsilura concinnata*	MA	81	Boettner unpubl.
Hemileuca lucina	*Compsilura concinnata*	MA	31.6	Stamp/Bowers 1990
Coleotichus blackburniae	*Trissolcus basalis*	HI	8.9	Follett et al. (this vol.)
	Trichopoda pilipes	HI	20.2	
PLANTS				
Opuntia stricta	*Cactoblastis cactorum*	CO	38.2	Johnson/Stiling 1997
Cirsium centaureae	*Rhinocyllus conicus*	CO	45	Louda et al. 1997
C. tweedyi	*Rhinocyllus conicus*	CO	1.0	Louda et al. 1997
C. undulatum	*Rhinocyllus conicus*	CO	70	Louda et al. 1997
C. pulchellum	*Rhinocyllus conicus*	CO	24.3	Louda et al. 1997
C. canescens	*Rhinocyllus conicus*	CO	58.1	Louda et al. 1997
Mean for insects n = 14		Worldwide	19.2 ± 25.6 s.d	
Mean for plants n = 6		Worldwide	39.4 ± 24.6 s.d.	

reintroducing *Opuntia spinosissima* plants back into the wild in the Florida Keys to test which types of habitat would be best for a full-scale reintroduction to minimize the effects of *Cactoblastis.*

ACKNOWLEDGMENTS

We thank Brad Hawkins for supplying us with his data set of insects attacking nontarget taxa, and Peter Follett for the opportunity to participate in this book and the 1997 ESA symposium "Biological Control for a Small Planet: Nontarget Effects of BC" in Nashville, Tennessee. Bob Ehrig brought the *Cactoblastis-Opuntia-Cyclura* connection to our attention. Dean Pearson supplied a pre-print of his work on the indirect effects of biological control agents released to control spotted knapweed.

REFERENCES

Askew, R. R. 1986. Parasitoid communities: their size, structure, and development. *In* J. Waage and D. Greathead (eds.), Insect parasitoids. Academic Press, London, UK. p. 225–264.

Askew, R. R. 1994. Parasitoids of leaf-mining Lepidoptera: what determines their host ranges? *In* B.A. Hawkins and W. Sheehan (eds.), Parasitoid community ecology. Oxford Univ. Press, Oxford, UK. p. 177–202

Askew, R. R. 1980. The diversity of insect communities in leaf-mines and plant galls. J. Anim. Ecol. 49: 817–829.

Barratt, B. I. P., A. A. Evans, C. M. Ferguson, G. M. Barker, M. R. McNeill, and C.B. Phillips. 1997. Laboratory nontarget host range of the introduced parasitoids *Microctonus aethiopoides* and *M. hyperodae* (Hymenoptera: Braconidae) compared with field parasitism in N.Z. Environ. Entomol. 26: 694–702.

Boldt, P. E. 1998. Letter. Science. In press.

Boldt, P. E. 1997. Response of a *Rhinocyllus* researcher. Biol. Control News Info. 18(4): 100.

Carruthers, R. I., and J. A. Onsager. 1993. Perspective on the use of exotic natural enemies for biological control of pest grasshoppers (Orthoptera: Acrididae). Environ. Entomol. 22: 885–903.

Center, T. D. 1995. Review of selection criteria and ecological consequences of importing natural enemies. Edited by W. C. Kauffman and J. R. Nechols. Biodiversity and Conservation 4: 524–526.

Civeyrel, L., and D. Simberloff. 1996. A tale of two snails: is the cure worse than the disease? Biodiversity and Conservation 5: 1231–1252.

Clausen, C. P. (ed). 1978. Introduced parasites and predators of arthropod pests and weeds: a world view. U.S. Dept. Agric., Agric. Handbk. 480.

Diamond, J. M. 1996. A-bombs against amphibians. Nature 383: 386–387.

Duan, J. J., and R. H. Messing. 1996. Response of two opiine fruit fly parasitoids (Hymenoptera: Braconidae) to the lantana gall fly (Diptera: Tephritidae). Environ. Entomol. 25: 1428–1437.

Duan, J. J., and R. H. Messing. 1998. Effect of *Tetrastichus giffardianus* (Hymenoptera: Eulophidae) an nontarget flowerhead–feeding tephritids (Diptera: Tephritidae). Environ. Entomol. 27: 1022–1028.

Ehler, L. 1991. Planned introductions in biological control. *In* L. Ginzburg (ed.), assessing ecological risks of biotechnology. Butterworth-Heinemann, Boston. p. 1–39.

Elliott, N., R. Kieckhefer, and W. Kauffman. 1996. Effects of an invading coccinellid on native coccinellids in an agricultural landscape. Oecologia 105: 537–54.

Follett, P. A., M. T. Johnson, and V. P Jones. 1999. Parasitoid drift in Hawaiian pentatomoids. *In* P. A. Follett and J. J. Duan, (eds.), Nontarget effects of biological control. Kluwer Academic Publishers.

Frank, J. H. 1998. How risky is biological control? Comment. Ecology. 79: 1829–1834.

Frank, J. H., and E. D. McCoy. 1993. The introduction of insects into Florida. Florida Entomologist 76: 1–53.

Funasaki, G. Y., P.-Y. Lai, L. M. Nakahara, J. W. Beardsley, and A. K. Ota. 1988. A review of biological control introductions in Hawaii: 1890 to 1985. Proc. Hawaiian Entomol. Soc. 28: 105–160.

Gonzalez, D., and F. E. Gilstrap. 1992. Foreign exploration: assessing and prioritizing natural enemies and consequences of preintroduction studies. *In* W. C. Kauffman and J. E. Nechols (eds), Selection criteria and ecological consequences of imporing natural enemies. Entomol. Soc. Am., Lanham, Maryland. p. 51–70

Goodman, B. 1993. Research community swats grasshopper control trial. Science 260: 887.

Gould, F., G. Kennedy, and R. Kopanic. 1996. Environmental issues associated with enhancing the impact of biological control agents: a student debate. Am. Entomol. 42: 160–173.

Hajek, A. E., L. Butter, S. R. A. Walsh, J. C. Silver, F. P. Hain, F. L. Hastings, T. M. Odell, and R. D. R. Smitley. 1995. Host range of the gypsy moth (Lepidoptera: Lymantriidae) pathogen *Entomophaga maimaiga* (Zygomycetes: Entomophthorales) in the field versus laboratory. Environ. Entomol. 25: 709–721.

Hall, R. W., and L. E. Ehler. 1979. Rate of establishment of natural enemies in classical biological control. Bull. Entomol. Soc. Am. 25: 280–282.

Hall, R. W., L. E. Ehler., and B. Bisabri-Ershadi. 1980. Rate of success in classical biological control of arthropods. Bull. Entomol. Soc. Am. 26: 111–114.

Hawkins, B. A. 1994. Pattern and process in host-parasitoid interactions. Cambridge Univ. Press, Cambridge, UK.

Hawkins, B. A., and P. C. Marino. 1997. The colonization of native phytophagous insects in North America by exotic parasitoids. Oecologia 112: 566–571.
Hayes, W. K., D. M. Hayes, D. Brouhard, B. Goodge, and R. L. Carter. 1995. Population status and conservation of the endangered San Salvador rock iguana, *Cyclura* r. *rileyi*. J. Intl. Iguana Soc. 4:21–30.
Jervis, M. A. 1998. Reply from M. A. Jervis. Trends Ecol. Evolution 13: 110.
Johnson, D. M., and P. D. Stiling. 1996. Host specificity of *Cactoblastis cactorum* Berg, an exotic *Opuntia*-feeding moth, in Florida. Environ. Entomol. 25: 743–748.
Johnson, D. M., and P. D. Stiling. 1998. Distribution and dispersal of *Cactoblastis_cactorum* (Lepidoptera: Pyralidae), an exotic *Opuntia*-feeding moth, in Florida. Florida Entomologist 81: 12–22.
Julien, M. H., J. D. Kerr, and R. R. Chan. 1984. Biological control of weeds: an evaluation. Protection Ecol. 7: 3–25.
Louda, S. M., D. Kendall, J. Connor, and D. Simberloff. 1997. Ecological effects of an insect introduced for the biological control of weeds. Science 277: 1088–1090.
Louda, S. M., D. Simberloff, J. Connor, G. Boettner, D. Kendall, and A. Arnett. 1998. Insights from data on the nontarget effects of the flowerhead weevil. Biol. Control News Info. 19: 70–72.
Luck, R. F. 1981. Parasitic insects introduced as biological control agents. *In* D. Pimentel (ed), CRC Handbook of pest management in agriculture, vol. 2. CRC Press, Boca Raton, Florida. p. 125–284
McFadyen, R. E. C. 1998. Biological control of weeds. Annu. Rev. Entomol. 43: 369–393.
McLeod, J. H. 1962. A Review of the biological control attempts against insects and weeds in Canada. Technical Communication of the Commonwealth Institute of Biological Control 2, Farnham Royal, UK.
Nafus, D. M. 1993. Movement of introduced biological control agents onto nontarget butterflies, *Hypolimnas* sp. (Lepidoptera: Nymphalidae). Environ. Entomol. 22: 265–272.
Nafus, D. M., and I. H. Schreiner. 1989. Biological control activities in the Mariana Islands from 1911 to 1988. Micronesica 22: 65–106.
Obrycki, J. J., A. M. Ormond, and K. L. Giles. 1998. Interactions between an introduced and indigenous coccinellid species at different prey densities. Oecologia (in press).
Pearson, D. E., K. S. McKelvey, and L. F. Ruggiero. 1999. Indirect effects of biological control agents released to control spotted knapweed. Ecology (in press).
Pemberton, R. W. 1995. *Cactoblastis cactorum* in the United States: an immigrant biological control agent or an introduction of the nursery industry. Am. Entomologist 41: 230–232.
Schaffner, J. V., Jr., and C.C. Griswold. 1934. Macrolepideptera and their parasites reared from field collections in the northeastern part of the United States. U.S. Dept. Agric., Misc. Pub. 188, Washington D.C.
Secord, D., and P. Karieva. 1996. Perils and pitfalls in the host specificity paradigm. Bioscience 46: 448–453.
Sheehan, W. 1991. Host range patterns of hymenopteran parasitoids of exophytic lepidopteran folivores. *In* E.A. Bernays (ed.), Insect-plant interactions, vol. 3. CRC Press, Boca Raton, Florida. p. 209–248
Simberloff, D., and P. Stiling. 1996a. How risky is biological control. Ecology 77: 1965–1974.
Simberloff, D., and P. Stiling. 1996b. Risks of species introduced for biological control. Biol. Conserv. 78: 185–192.
Simberloff, D., and P. Stiling. 1998. How risky is biological control: response to J. H. Frank. Ecology 79: 1834–1836.
Stamp, N. E. and M. D. Bowers. 1990. Parasitism of New England buckmoth caterpillars (Hemileuca lucina: Saturnidae) by tachinid flies. J. Lepidopterists Soc. 44: 199–200.
Stiling, P. 1990. Calculating establishment rates of parasites in classical biological control. Am. Entomologist 36: 225–230.
Stiling, P. 1993. Why do natural enemies fail in biological control campaigns? Am. Entomologist 39: 31–37.
Thomas, M. B., and A. J. Willis. 1998. Biological control—risky but necessary? Trends Ecol. Evolution 13: 325–329.
Tuduri, J. C. G., L. F. Martorell, and S. M. Gaud. 1971. Geographical distribution and host plants of the cactus moth, *Cactoblastis cactorum* (Berg), in Puerto Rico and the United States Virgin Islands. J. Agric. Univ. Puerto Rico 58: 130–134.
Turner, C. E. 1985. Conflicting interests and biological control of weeds. *In* E. S. Delfosse (ed.), Proceedings of the Sixth International Symposium on Biological Control of Weeds; Aug 19–25 1984, Vancouver, Canada. Agriculture Canada, Ottowa, Canada. p. 203–225.
Van den Bosch, R. 1971. Biological control of insects. Annual Review of Ecology and Systematics 2: 45
Wheeler, A. G., and E. R. Hoebeke. 1995. *Coccinella novemnotata* in Northeastern North America: historical occurrence and current status (Coleoptera: Coccinellidae). Proc. Entomol. Soc. Washington 97: 701–716.
Williamson, M., and A. Fitter. 1996. The varying success of invaders. Ecology 77: 1661–1670.

4

THE IMPACT OF NONTARGET CONCERNS ON THE PRACTICE OF BIOLOGICAL CONTROL

Russell H. Messing
Department of Entomology, Kauai Agricultural Research Center, University of Hawaii, Kapaa, Hawaii

INTRODUCTION

The recent entomological literature contains a spate of articles and volumes concerning the impact of biological control agents on nontarget species. These range from thoughtful (Louda et al. 1997) to provocative (Miller and Aplet 1993) to highly critical (Howarth et al. 1997). Taken together, they set the tone for a debate which is just beginning to attract the research funding necessary to generate data of sufficient quality and quantity to try to evaluate impacts.

What has been missing in the debate so far is an assessment of the inverse perspective: *What is the impact of real (or perceived) nontarget effects on the practice of biological control?* From a strictly economic perspective, even critics of biological control have acknowledged the enormous financial benefits which accrue by protecting crops and reducing pesticide use with self-replicating, host-tracking natural enemies (Howarth in press, Lockwood 1996). And, from the viewpoint of conservation biology and preservation of biodiversity, biological control has been (although at times criticized) also acknowledged as a solution to some of our most difficult environmental problems (van Driesche and Benson, in press). However, classical biological control is coming under increasingly intense scrutiny and regulatory oversight. The degree of oversight may not always be commensurate with the degree of risk (Messing and Purcell 1999). Thus, it is useful to at least temporarily reverse the perspective and examine the ways in which concern for nontarget effects is impacting biologically-based pest management programs in both agricultural and natural settings.

TYPES OF IMPACT

For convenience sake, we may employ three categories of effects on biological control programs:

- Cases in which the estimated risk of a nontarget effect by a proposed natural enemy introduction is so great that the project is never conducted. In such cases, the interests of one segment of society outweigh the interests of those concerned with pest control, and the biological control practitioner must bow out with good grace.
- Cases in which a perceived or potential nontarget impact, based perhaps on casual observation or other non-quantitative evaluation, leads to the preclusion or abandonment of a biological control project which might otherwise prove beneficial. One might also include here projects in which a real, but minor, nontarget impact blocks a biological control program which might have substantial ecological or economic benefits.
- Cases in which a generalized apprehension about exotic organisms leads to a bureaucratic atmosphere of regulation so severe that biological control projects are either not funded nor even considered because of the obstacles involved.

These three categories are not mutually exclusive; they grade into one another to an extent that depends upon the details of each case and the perspectives of those involved. Nevertheless, categorization is a useful device to help delineate the range of impacts on biological control programs, what their foundation is, and where researchers should focus their energies.

CLEARLY DEFINED RISK

There are many examples of the first category, in which a potentially effective biological control agent is not imported or released from quarantine due to a clearly defined threat to nontarget species. The determination of this threat is one of the prime reasons for the use of quarantine facilities and host testing prior to natural enemy release. It has historically been practiced more stringently for weed biological control agents than for natural enemies of arthropods.

Field bindweed (*Convolvulus arvensis*) is a serious pest throughout the United States, particularly in arid western regions, infesting almost 2 million acres in California alone (Rosenthal 1996). A chrysomelid beetle from Italy, *Galeruca rufa*, showed considerable promise as a biological control agent, severely defoliating bindweed in laboratory and field cage tests. However, because studies in quarantine showed that the beetle had the potential to feed on sweet potato, *Ipomoea batata*, it was never released in the United States.

Another leaf feeding chrysomelid, *Chrysolina gypsophilae,* showed promise for the control of the weed dalmation toadflax, *Linaria genistifolia*, in the western United States. It is questionable whether the beetle will ever be released because of concerns about feeding on native snapdragons in the family Scrofulaceae.

Leafy spurge (*Euphorbia esula*) is an exotic weed that infests millions of acres of land in the western U.S. and Canada. It not only reduces the value of rangeland (from 50 up to 100%), but at high densities it can eliminate other grass and forb species (Pemberton 1996). Two commercially important *Euphorbia* species (poinsettia and candellia plant) and 113 native *Euphorbia* species in North America north of Mexico may be considered nontarget species at risk from leafy spurge biological control programs. Some herbivorous control agents have been shown to be host specific to members of the *Euphorbia* sub-genus *Esula*, and several of these (*Apthona* species) have been released in the field, but two promising new candidates for mesic micro-habitats (*A. ovata* and *A. venistula*) will probably not be released because of concerns regarding potential nontarget impacts. Likewise, the tingid bug *Oncochila simplex* and the torticid moth *Lobesia euphorbiana,* also showed promise as biological control agents for leafy spurge, but will not be released because of concerns about their potential host range (Pemberton 1996).

In most of these examples, there was insufficient knowledge available to realistically quantify the degree of risk to nontarget species or the extent of economic and ecological damage caused by the weeds. In some cases (i.e., leafy spurge) ecologists thought the benefits of weed control would outweigh the risk of impact to nontargets; nevertheless decisions were made to err on the side of caution.

In California, wine grapes are infested by several species of tetranychid mites, including the Willamette mite (*Eotetranychus willametti*), the Pacific mite (*Tetranychus pacificus*) and the two-spotted spider mite (*T. urticae*). Biological control of these mites by native phytoseiid predators is often disrupted by applications of sulfur to control fungal diseases. A predaceous mite (*Amblyseius aberrans*) in Italy that attacks tetranychids in wine grapes has developed genetic resistance to sulfur sprays (Frank Zalom, UC Davis, pers. comm.). Thus, it was thought that if *A. aberrans* could be introduced and established in commercial vineyards in California, it could survive the sulfur sprays and contribute to local management of *E. willametti.* However, some studies had shown that infestations by the Willamette mite have a conditioning effect on grape vines so that plants are subsequently less susceptible to damage by *T. urticae* (Karban et al. 1987). A few entomologists pointed out the possibility that successful biological control of Willamette mite could lead to more severe mite damage by *T. urticae.* Again, professional concern over the possibility of these indirect effects led to the voluntary abandonment of the project (Frank Zalom, pers. comm.).

RISK MAY BE POORLY DEFINED OR OVERSTATED

In the second category, one can find several cases in which nontarget effects which are marginal or unquantified have nevertheless prevented or delayed important biological control programs from taking place. It is this sort of case which is apt to be most controversial, where an effect considered marginal to one person may appear critical to another, and where ethical judgements and contrasting societal values come to the fore.

An example of this type concerns biological control of congressweed, *Parthenium hysterophorus*. In the area of Bangalore thousands or perhaps millions of hectares of rangeland are infested with the weed, which is not only unsuitable for cattle but also causes severe allergic dermatitis in thousands of people annually (Towers and Rao 1992). This weed is reportedly well controlled in Australia by the tortricid moth *Epiblema strenuata* (McFayden 1992). However, in India, host range testing of the moth found that it was able to feed on niger (*Guizotia abyssinica*), which is a minor oilseed crop grown in small areas. This finding led to the termination of biological control efforts using the moth.

The noxious weed Paterson's Curse (*Echium plantagineum*) is widespread on Australian rangeland, where it outcompetes pasture vegetation and is toxic to livestock. In 1985 the Australian Industries Assistance Commission estimated that successful biological control of Paterson's Curse could save Australian agriculture about $30 million per year. The proposed biological control program was given a potential benefit-cost ratio of 9:1 (CSIRO 1995). However, the Australian apiculture industry, which relies to some extent on Paterson's Curse as a honeybee forage plant, opposed any reduction in distribution and abundance of the weed. The resulting controversy (Cullen and Delfosse 1985) resulted in a delay of biological control implementation lasting 81/2 years. A similar disagreement between beekeepers and range managers in the western United States concerning biological control of yellow star thistle (*Centaurea solstitialis*) was more easily resolved when representatives of both groups agreed that the losses caused by the weed far outweighed its benefits to apiculture (Turner 1985).

Another example in which a questionable environmental concern is blocking a major biological control project concerns salt cedar, *Tamarix ramosissima.* This case represents one of the most thorough pre-release evaluations and risk/benefit analyses reported for any weed biological control project in the United States.

Salt cedar is an invasive pest causing enormous damage to riparian flora and fauna across a wide area of the western United States. Well over a million acres of habitat are infested, including many nature reserves in the Southwest. Native cottonwoods and willows, which are prime wildlife habitat, have nearly disappeared in many areas. Salt cedar is also a vigorous invader of wetlands, where it shades out many native species, reduces their reproductive potential, and contributes to the loss of biodiversity. "Many native birds and mammals are unable to utilize salt cedar, very few native insects can develop on it, and the alterations produced by salt cedar reduce habitat value and threaten the existence of many species of fish and other animals" (DeLoach 1996).

The genus *Tamarix* is of foreign origin, and no species in the family Tamaricaceae are native to the Western Hemisphere. This taxonomic isolation greatly reduces the risk that specialized biological control agents would attack native plants. *Tamarix* is an ancient genus in the Old World, so a number of host-specific insects have evolved to attack species within the genus there. It has been said that "the potential for successful control of salt cedar is as great as for any biological control project attempted anywhere in the world" (Deloach 1996).

Quarantine testing demonstrated host specificity by several herbivorous insects imported for salt cedar biological control. However, the U.S. Fish and Wildlife Service (USFW) determined that an endangered bird (the southwestern willow flycatcher, *Empidonax traillii extimus*) nests in salt cedar in a few locations in Arizona (although approximately 88% of the extant flycatcher nesting sites are dominated by native broadleaf trees or mixtures of native broadleafs and other exotic plant species: [Marshall 1996]). The USFW, feared that introduced biological control agents would immediately eliminate salt cedar stands, thereby reducing the nesting habitat of the endangered bird. Entomologists with experience in weed biological control pointed out that the insects would not eliminate salt cedar stands overnight, that there would be a gradual decline in stand vigor and other trees (which could serve as nesting sites) would re-establish in the riparian habitats (J. DeLoach, pers. comm.). However, as of this writing (August 1998) permits have still not been obtained to release the biological control agents. Thus, a questionable attempt to preserve nesting habitat for one bird species is blocking a biological control program that could potentially mitigate damage to multiple riparian ecosystems and dozens of impacted native species.

In Hawaii, tephritid fruit flies are one of the major obstacles to a more robust agricultural economy. In the 1950s several braconid parasitoids which attacked tephritids were imported from Asia, and they subsequently became established and widespread on all the major Hawaiian Islands (Clausen et al. 1965). In some situations, these parasitoids kill over 95% of the fruit flies in orchard crops (Newell and Haramoto 1968), but in other situations control is marginal. Recently, following the promising results of several pilot tests (Wong et al. 1991, Sivinski et al. 1996), a joint team of USDA and University scientists proposed to mass-rear one of the parasitoids and augmentatively release it to reduce Oriental fruit fly populations in the weedy strawberry guava that dominates parts of the island of Kauai (Wong et al. 1992). In over 40 years of widespread establishment in Hawaii, the parasitoid (*Diachasmimorpha longicaudata*) had not been recorded from any native species, but had been recorded attacking several beneficial tephritids introduced for weed biological control (Bess and Haramoto 1972). Under the mien of a concern regarding nontarget effects, an environmental review group threatened legal action to block the augmentation project, and succeeded in postponing part of the program for over two years (Minutes: 2/19/92 meeting: Environmental Action Group of the Hawaii Fruit Fly Committee).

DIFFUSE DREAD OF EXOTICS

Regarding the third category (i.e., general fear of exotic organisms), there are several notable statements in the literature that exemplify the attitudes of apprehension and mistrust which have led to a tightening of regulatory control. For example, Howarth (1991) enjoined that "biological control may be an inappropriate strategy for some third world countries." Asquith (1994) stated that the entire practice of biological control is "a permanent solution to a temporary problem" (implying that introduced biological control agents will pollute ecosystems long after the agriculture in a region is abandoned). And Lockwood (1999) used the phrase "Necessary evils: murder, drugs, and biological control" in a discussion of the ethics of importing natural enemies.

The extensive regulatory oversight enacted in response to statements such as these is best illustrated by the recent situation in the state of Hawaii (Messing and Purcell 1999), where the introduction of biological control agents and the issuance of release permits has slowed to a trickle (Figure 1). For example, the average number of parasitoids introduced into Hawaii has fallen from 3.8/yr between 1900 and 1979, to 2.3/yr between 1980 and 1989, to 0.8/yr between 1990 and 1998. This slowdown has been due in part to decreased funding and a reduction in the political and economic power of agriculture in the state, but a major contributing factor has been a significant increase in the complexity of regulatory bureaucracy overseeing introductions of all exotic species, including host-specific natural enemies.

For example, a few years ago an application was filed to introduce an aphidiid parasitoid (*Aphidius colemani*) into Hawaii to help control the melon aphid (*Aphis gossypii*), a major pest of several important agricultural crops. Throughout the world, all species in the family Aphidiidae (or sub-family Aphidiinae) have been well documented to attack aphids, and only aphids (Mackauer and Stary 1967, Stary 1970). There are no endemic aphids in Hawaii, nor are there any beneficial aphids imported for biological control of weeds or any other purpose. Yet, the application to import this parasitoid took over two years to make it through regulatory channels.

In another case, an application by the author to import *Psyttalia insignipennis*, an opiine braconid parasitoid of the Mediterranean fruit fly from Reunion Island into Hawaii has been pending (as of this writing, August 1998) for over four years and still has not even made it to the public-hearing phase of the regulatory process. This is despite the fact that a strong body of evidence from recent studies has provided data regarding the safety of opiine braconids in Hawaiian ecosystems (Duan and Messing 1996a, b, 1997; Duan et al. 1996, 1997a, b; Purcell et al. 1997).

Hawaii Department of Agriculture entomologists introduced another aphidiid, *Lysiphlebus ambiguous*, to control the yellow sugarcane aphid (*Sipha flava*), a pest of sugarcane and forage grasses. Although there are neither endemic nor beneficial aphids in Hawaii, and thus no logical "nontarget" aphids to use for testing in quarantine, host testing was considered a requirement. Entomologists thus felt compelled to conduct host range tests with psyllids, even though an aphidiid has never been recorded to successfully attack a psyllid anywhere in the world for as long as host records have been kept (Mackauer and Stary 1967). Three species of alien (pest) psyllids and two species of native psyllids were used in the "nontarget" testing and no successful parasitism occurred. The waste of time, money, and quarantine space, and the drop in research morale that this engenders is detrimental to the practice of biological control in a state where many exotic pests threaten both native and agricultural ecosystems.

Projects on biological control of weeds in Hawaii using plant pathogens have also suffered delays due to local opposition and the conservative nature of the regulatory system. The noxious weed banana poka (*Passiflora tripartita*) is an invader of high elevation native forests, and has been declared "the major threat to the existence of this unique forest ecosystem" (Markin and Pemberton 1996). A proposal to introduce the fungus *Septoria passiflorae* into Hawaii to control banana poka was thoroughly analyzed according to State regulatory policy, with three committee levels of critical review, two of which were open to public comment. There were no objections to the release of the pathogen at any level in the State of Hawaii. At the Federal level, four plant pathologists on a USDA-APHIS-BATS review team all approved the pathogen for release, based on its host specificity and the fact that there are no native Passifloraceae in Hawaii. However, a local office of the U.S. Fish and Wildlife Ser-

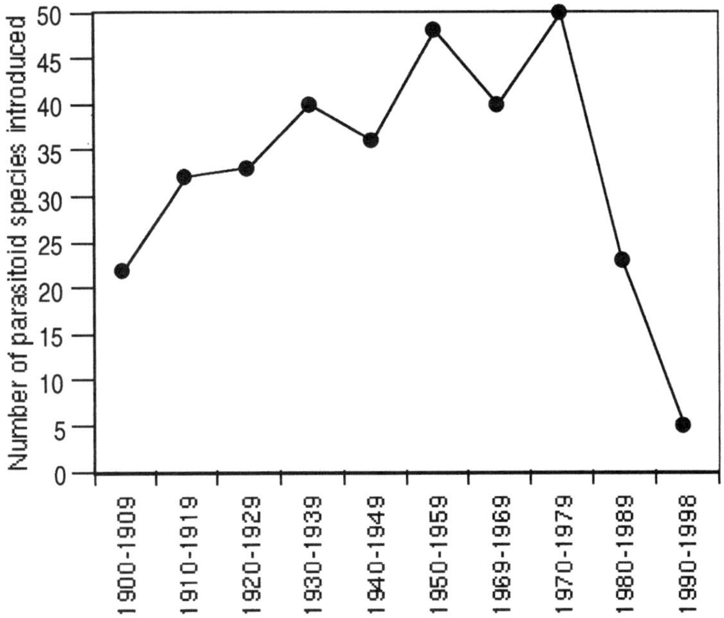

Figure 1. Introductions of biological control agents into Hawaii, 1900–1998.

vice (USFW) objected to importation of the pathogen. This objection challenged the concept of host specificity in fungal pathogens, in contradiction to the unanimous opinion of state and federal advisors with expertise in plant pathology. In response, the deciding regulatory agency (USDA-APHIS) requested that host testing must be conducted with other genera close to *Passiflora* within the Passifloraceae, despite the fact that an environmental assessment had already documented that no other genera of Passifloraceae exist in Hawaii. All the correspondence and requests and rebuttals caused significant delays and added substantially to the cost of conducting the project.

That these types of delay have significant impact on the practice of biological control in the state is evidenced by the fact that scientifically sound research and implementation proposals have been denied funding because grant review panels concluded that release permits were unlikely to be obtainable within the time frame of the grant cycle (Dr. Kenneth Rohrbach, University of Hawaii at Manoa, College of Tropical Agriculture and Human Resources, pers. comm.).

It is not only classical biological control which suffers from this regulatory constraint. A growing number of effective natural enemies are available from commercial insectaries in the continental U.S. for augmentative release (i.e., *Encarsia formosa* for greenhouse whitefly; predaceous phytoseiids for tetranychid mites). Many of these species already are established in Hawaii. However, farmers are prohibited from bringing these agents into the state without import permits and quarantine evaluation of each shipment, which is impractical. In some cases where attempts were made to obtain import permits for these agents, applications were denied for fear of taxonomic misidentification at the insectary of origin, or contamination with hyperparasites.

In other states besides Hawaii the practice of biological control is also being impeded by regulatory constraint. In Oregon, increased levels of risk analysis, lengthier testing protocols, time delays in the permit process, and the increase in power which uninformed parties have to block permits have been cited by biological control scientists as factors involved in the slowdown of introductions of weed biological control agents (E. Coombs, pers. comm.).

At the federal level, apprehension about nontarget impacts of arthropod biological control agents has shifted the overall position of research and regulatory administrators in the United States toward a defensive posture. This has most recently become manifest following a policy change in USDA-APHIS-PPQ, which absolved itself of responsibility for regulation of "non-plant pests" (e.g., natural enemies of arthropods). Without APHIS-PPQ oversight, those most involved in the actual introduction and release of arthropod natural enemies (USDA-ARS scientists, or other scientists funded by USDA grants) must now take responsibility for meeting all legal requirements emanating from the Endangered Species Act and the National Environmental Policy Act. This has resulted in a slow-down of new natural enemy introductions and a stay on the release of at least one new parasitoid species (J. Coulson, pers. comm.).

RESPONSIBILITIES AND PRIORITIES

Every year approximately 17 alien arthropods are accidentally introduced into Hawaii, either as contaminants on imported plant material, in tourists' baggage, or by other means (Beardsley 1979). None of these arthropods are screened, examined, or even identified until it is too late, and many of them go on to become significant pests of agricultural, urban, and natural ecosystems. The situation is similar in Florida (Frank and McCoy 1992) and the trend is also true in much of the mainland United States (OTA 1993), as both human population densities and human and cargo air travel continue to increase. It seems as though a disproportionate amount of time and energy have been directed toward discussion, evaluation, and regulation of deliberate (carefully screened) species introductions when these intellectual and financial resources could have a much greater impact if directed toward the accidental introductions that have much more serious ecological consequences.

Applied entomologists undertake a serious responsibility when importing nonindigenous species for biological pest control, and they recognize the need to weigh potential risks to nontarget species. Thorough review of exotic species proposed for introduction will inevitably lead to some degree of slowdown in the rate of classical biological control. As Ehler (this volume) points out, there may be some compensation in the fact that this level of review will also likely improve our understanding of natural enemy-host relationships, and may lead to greater predictability and higher success rates for biological control programs. However, there will always be some cases in which divergent scientific perspectives or personal values lead to differences of opinion in evaluating risk, and in evaluating the benefit side of the equation as well. Establishment of clear, coherent, streamlined regulations at the national level should provide a framework to address these differences. Such regulations should help allay some of the diffuse fears of exotic species introductions which tend to exaggerate the estimation of risk, while the incorporation of strict time-lines can help alleviate excessive delays in implementation programs. The drafters of these regulations should bear in mind that biological control is among the most cost-effective and environmentally sound methods of pest control available, both for the agriculture, which is the foundation of our nation's wealth, and also for our natural ecosystems which are beleaguered by exotic pest species.

ACKNOWLEDGEMENTS

Ernst Neering of the Dept. of Theoretical Production Ecology, Wageningen University, provided information about biological control of congressweed. I thank Jack DeLoach, USDA-ARS, for information on the proposed biological control of salt cedar. Juli Gould, Frank Zalom, Les Ehler, and Bob Nowierski also shared with me examples of biological control projects impacted by concern over nontarget effects. I also appreciate the help of Patrick Conant and Peter Follett in compiling data on biological control introductions into Hawaii.

REFERENCES

Asquith, A. 1995. Alien species and the extinction crisis of Hawaii's invertebrates. Endangered Species Update 12: 6–11.

Beardsley, J. W. 1979. New immigrant insects in Hawaii: 1962 through 1976. Proc. Hawaiian Entomol. Soc. 23: 35–44.

Bess, H. A., and F. H. Haramoto. 1972. Biological control of pamakani, *Eupatorium adenophorum*, in Hawaii by a tephritid gall fly, *Procecidochares utilis*, 3: status of weed, fly and parasites of the fly in 1966–1971 versus 1950–1957. Proc. Hawaiian Entomol. Soc. 21: 165–178.

Clausen, C. P., D. W. Clancy, and Q. C. Chock. 1965. Biological control of the oriental fruit fly and other fruit flies in Hawaii. U. S. Dept. Agric. Tech. Bull. 1322.

CSIRO. 1995. Entomology, Report of research 1993–95. <http://www.ento.csiro.au/research/>.

Cullen, J.M., and E. S. Delfosse. 1985. *Echium plantagineum* in Australia: catalyst for conflict and change. *In* E. Delfosse (ed.), Proc. VI Int. Symposium on Biological Control of Weeds. Vancouver, Canada. p. 249–292.

DeLoach, C. J. 1996. Salt cedar biological control: methodology, exploration, laboratory trials, proposals for field releases, and expected environmental effects. Proc. Salt Cedar Management and Riparian Restoration Workshop, Las Vegas, Sept. 1996.

Duan, J. J., and K. Joshi, M. Ahmad, and R. H. Messing. 1997. Evaluation of the impact of the fruit fly parasitoid *Diachasmimorpha longicaudata* (Hymenoptera: Braconidae) on a nontarget tephritid, *Eutreta xanthochaeta*. Biol. Control 8: 58–64.

Duan, J. J., and R. H. Messing. 1996. Response of two opiine fruit fly parasitoids (Hymenoptera: Braconidae) to the lantana gall fly (Diptera: Tephritidae). Environ. Entomol. 25: 1428–1437.

Duan, J. J., and R. H. Messing. 1996. Risk analysis and decision-making in biological control—a case study with fruit fly parasitoids. J. Agric. Human Values 13: 1–10.

Duan, J. J., and R. H. Messing. 1997. Effect of two opiine parasitoids (Hymenoptera: Braconidae) introduced for fruit fly control on a native Hawaiian tephritid *Trupanea dubautiae* (Diptera: Tephritidae). Biol. Control 8: 177–184.

Duan, J. J., M. F. Purcell, and R. H. Messing. 1996. Parasitoids of nontarget tephritid flies in Hawaii: Implications for biological control of fruit fly pests. Entomophaga 41: 245–256.

Duan, J. J., M. F. Purcell, and R. H. Messing. 1997. Ovipositional responses of three opiine fruit fly parasitoids to gall forming tephritids. Biol Control 9: 81–88.

Frank, J. H., and E. D. McCoy. 1992. The immigration of insects to Florida, with a tabulation of records

published since 1970. Florida Entomologist 75: 1–28.
Howarth, F. G. 1983. Classical biological control: panacea or pandora's box. Proc. Hawaiian Entomol. Soc. 24: 239–244.
Howarth, F. G. 1985. Impacts of alien land arthropods and mollusks on native plants and animals in Hawaii. *In* Stone and Scotts (eds.), Hawaii's terrestrial ecosystems: preservation and management. Univ. Hawaii Press, Honolulu.
Howarth, F. G. 1991. Environmental impacts of classical biological control. Annu. Rev. Entomol. 36: 485–509.
Howarth, F. G. Environmental issues concerning the importation of non–indigenous biological control agents. *In* M. F. Purcell, F. G. Howarth, and J. Lockwood, (eds.),Balancing nature: assessing the impact of importing exotic biological control agents: an international perspective. Thomas Say Publications in Entomol. In press.
Howarth, F. G., G. Nishida, and A. Asquith. 1997. Insects of Hawaii. *In* Our living resources, a report to the nation on the distribution, abundance, and health of U.S. plants animals, and ecosystems. U.S. Dept. of the Interior, Natl. Biol. Svc. p. 365–372
Karban, R., R. Adamchak, and W. C. Schnathorst. 1987. Induced resistance and interspecific competition between spider mites and a vascular wilt fungus. Science 235: 678–680.
Lockwood, J. A. 1996. The ethics of biological control: understanding the moral implications of our most powerful ecological technology. Agric. Human Values 13: 2–19.
Lockwood, J. A. 1999. The ethics of "classical" biological control and the value of place. *In* M. Purcell, F. Howarth, and J. Lockwood, (eds.), Importation and legislation issues affecting foreign exploration for natural enemies. Thomas Say Publications in Entomol. In press.
Louda, S. M., D. Kendall, J. Connor, and D. Simberloff. 1997. Ecological effects of an insect introduced for the biological control of weeds. Science 277: 1088–1090.
Mackauer, M. and P. Stary. 1967. Hymenoptera: Ichneumonoidea. World Aphidiidae. *In* V. Delucchi and G. Remaudiere (eds.), Index of entomophagous insects. Paris, France.
Markin, G. P., and R. W. Pemberton. 1996. Case history: banana poka. *In* L. Andres, R. Goeden, G. Jackson, and J. Beardsley (eds.), Biological control in the U.S. western region. DANR Publications, Univ. California. Chap. 81.
Marshall, R. M. 1996. Status, distribution, and current threats to the endangered Southwestern willow flycatcher. Proc. Salt Cedar Management and Riparian Restoration Workshop, Las Vegas, Sept. 1996.
McFadyen, R.C. 1992. Biological control against parthenium weed in Australia. Crop Protection 11: 400–407.
Messing, R. H., and M. F. Purcell. Biological control on the ropes: bureaucracy stifles science in Hawaii. *In* M. F. Purcell, F. G. Howarth, and J. Lockwood (eds.), Thomas Say Publications in Entomol. In press.
Miller, M., and G. Aplet. 1993. Biological control: a little knowledge is a dangerous thing. Rutgers Law Rev. 45: 285–314.
Newell, I. M. and F. H. Haramoto. 1968. Biotic factors influencing populations of *Dacus dorsalis* in Hawaii. Proc. Hawaiian Entomol. Soc. 20: 81–139.
OTA (Office of Technology Assessment). 1993. Harmful non-indigenous species in the United States. Report no. OTA-F-565. U.S. Govt. Printing Office, Washington D.C.
Pemberton, R. W. 1996. Case histories: Leafy spurge. *In* L. Andres, R. Goeden, G. Jackson, and J. Beardsley (eds.), Biological control in the U.S. western region. DANR Publications, Univ. California. Chap. 76
Purcell, M. F., Duan, J. J., and R. H. Messing. 1997. Response of three hymenopteran parasitoids introduced for fruit fly control to a gall-forming tephritid, *Procecidochares alani*. Biol. Control 9: 193–200.
Rosenthal, S. S. 1996. Case histories: field bindweed. *In* L. Andres, R. Goeden, G. Jackson, and J. Beardsley (eds.), Biological control in the U.S. western region. DANR Publications, Univ. California. Chap. 75.
Simberloff, D. 1992. Conservation of pristine habitats and unintended effects of biological control. *In* W. C. Kaufman and J. Nichols (eds.), Selection criteria and ecological consequences of importing natural enemies. Thomas Say Publications in Entomol.
Sivinski, J. M., C. O. Calkins, R. Baranowski, D. Harris, J. Brambila, J. Diaz, R. E. Burns, T. Holler, and G. Dodson. 1996. Suppression of a Caribbean fruit fly population through augmented releases of the parasitoid *Diachasmimorpha longicaudata*. Biol. Control 6: 177–185.
Stary, P. 1970. Biology of aphid parasites with respect to integrated control. Series Entomologica 6. Dr. W. Junk, The Hague, Netherlands. 643 p.
Towers, G. H. N., and P. V. S. Rao. 1992. Impact of the pan-tropical weed, *Parthenium hysterophorus* L. on human affairs. Proc. 1st International Weed Control Congress, vol. 1: 134–138. Weed Science Society

of Victoria, Melbourne, Australia.
Turner, C. E. 1985. Conflicting interests and biological control of weeds. *In* E. S. Delfosse, (ed.), Proc. 6th Intl. Symp. Biol. Control Weeds, Vancouver, BC, Canada.
van Driesche, R. G., and J. Benson. Invading nature: releasing exotic species to protect natural systems.
Wong, T. T. Y., M. F. Purcell, R. H. Messing, and C. G. Jackson. 1992. Draft environmental assessment for the evaluation of parasitoid and sterile fly releases for management of the Oriental fruit fly, *Bactrocera dorsalis*. U.S. Dept. Agric, Agric. Res. Svc. 18 p.
Wong, T.T.Y., M. M. Ramadan, D. O. McInnis, N. Mochizuki, J. A. Nishimoto and J. C. Herr. 1991. Augmentative releases of *Diachasmimorpha tryoni* to suppress a Mediterranean fruit fly population in Kula, Maui, Hawaii. Biol. Control 1: 2–7.

PART 2

PARASITOIDS AND PREDATORS

5

PREDICTING THE RISK FROM BIOLOGICAL CONTROL AGENT INTRODUCTIONS: A NEW ZEALAND APPROACH

Barbara I. P. Barratt and Colin M. Ferguson
New Zealand Pastoral Agriculture Research Institute Ltd., Invermay Agricultural Centre, Mosgiel, New Zealand

Stephen L. Goldson and Craig M. Phillips
New Zealand Pastoral Agriculture Research Institute Ltd., Canterbury Agriculture and Science Centre, New Zealand

Donald J. Hannah
Environmental Risk Management Authority, Wellington, New Zealand

INTRODUCTION

Since the publication of *Silent Spring* (Carson 1963) there has been increasing demand to reduce the amount of agricultural pesticides used. This need has been further supported by the development of increasing levels of pesticide resistance, and the market advantages for products from "sustainable" agricultural systems. Classical biological control, where a predator, parasite, or pathogen is imported to control a weed or pest, is one obvious alternative to pesticides and has often been promoted as environmentally safe (e.g., DeBach and Rosen 1991).

There have been some spectacular successes in classical biological control that have been well publicized (e.g., Caltagirone 1981). Biological control has in many instances proved to be a highly effective form of agricultural pest management and a valuable component of integrated pest management systems. However, the failure rate has been high, and in a survey of the literature, Ehler (1990) estimated that both the rate of establishment and success of biological control agents released to control arthropod pests was lower than that achieved for weed biological control. It was suggested by Ehler (1990) that this was attributable to the more detailed evaluation and host-range testing required to obtain approval to release phytophagous insects, which might threaten crops, than has been the case for entomophagous insects. Therefore the investment in weed biological control programs, the effort to ensure successful releases, and monitoring post-release impacts on the target pests has been greater.

When comparing biological and chemical control methods, it is clear that in comparison with the latter, biological control is self-perpetuating, self-dispersing, and irreversible. These attributes are some of the features that are seen as advantages of biological control; however, they are also factors that have alerted researchers to the potential environmental risks of such introductions. Regulation of biological control agent introductions is required in the public interest because of its irreversibility (Harris 1990, Waage 1997) and the potential for biological control agents to disperse to habitats other than those where they were released (Van Lenteren 1997).

ENVIRONMENTAL SAFETY AND BIOLOGICAL CONTROL

The environmental safety of biological control, earlier taken for granted, is now a topic for debate. Many authors (e.g., Howarth 1991, Howarth and Ramsay 1991, Cullen 1993, Samways 1994, Simberloff and Stiling 1996a, 1996b) have pointed out that lack of evidence for adverse environmental impacts of biological control introductions can be attributed more to lack of study of effects than the absence of such impacts. Similarly, Waage (1997) pointed out that, while there is little evidence of negative effects from biological control agent introductions, pre-release testing, when carried out, has rarely included nontarget indigenous species. Lockwood (1993, 1996) expressed concerns that nontarget effects of biological control could lead to extinctions and also that the failure rate of biological control programs is high and rarely explained. He attributed some of these outcomes to lack of adequate legislative requirements for quarantine and post-release studies of biological control agents for insect pests.

While there are few documented cases, there is circumstantial evidence to suggest that biological control agents have the capacity to cause species extinctions, particularly in island communities. In Fiji, the introduction of the tachinid *Bessa remota* (Aldrich) to control the coconut moth, *Levuana iridescens* Bethune-Baker, is thought to have caused the extinction of its host (Howarth 1991). Similarly, the citrus psylla, *Trioza erytreae* (Del Guericio) apparently was exterminated from Reunion Island by the eulophid parasitoid *Tetrastichus dryi* Waterston (Aubert and Quilici 1983). In both

Table 1. Summary of results from laboratory host range tests and field monitoring of nontarget weevil populations (updated from Barratt et al. 1997).

	Parasitoid species	
	M. aethiopoides	*M. hyperodae*
Laboratory tests		
No. nontarget species parasitized (no. tested)	12 (13)	7 (30)*
No. nontarget genera parasitized (no. tested)	9 (11)	3 (21)*
No. tribes parasitized (no. tested)	4 (5)	1 (4)*
No. subfamilies parasitized (no. tested)	2 (2)	1 (2)*
Mean % parasitism of nontarget hosts (target host)	58 (62)	13 (61)
Mean % showing immune response in nontarget hosts (target host)	12 (0)	32 (2)
Field monitoring		
No. nontarget species parasitized (no. examined)	14 (48)	2 (48)
No. genera parasitized (no. examined)	8 (23)	2 (18)
No. tribes parasitized (no. examined)	4 (8)	1 (7)
No. subfamilies parasitized (no. examined)	2 (2)	1 (2)
Mean % parasitism recorded (range)	23 (0–71)	1.8 (0–3)
No. sites where nontarget parasitism was found (total sites sampled)	17 (33)	1(33)

*Includes (Goldson et al. 1992) data.

cases, the parasitoids maintained their populations, despite the decline of the target hosts, on alternative species.

Since 1874, over 240 biological control agents have been released in New Zealand for weed and pest control, of which about 31% have established (Cameron et al. 1993). Although no known extinctions have resulted from biological control introductions in New Zealand, concerns have been expressed about the lack of pre-release assessment of impact on the native fauna (Roberts 1986, Longworth 1987). It has been suggested that some introduced biological control agents have adversely affected the native fauna. Roberts (1986) cites the example of the release in the 1960s of the tachinid parasitoid *Trigonospila brevifacies* (Hardy) for control of light brown apple moth, *Epiphyas postvittana* (Walker). This fly is now found parasitizing a wide range of Lepidoptera, including native species, with anecdotal evidence to suggest an associated decline in the numbers of some native leafrollers. Similarly, the red admiral, *Bassaris gonerilla* (F.), is considered less numerous since the introduction of parasitoids to control white butterfly, *Pieris rapae* (L.) (Gibbs 1980). Recent research has provided more definitive examples of nontarget effects of biological control agents in New Zealand (Ferguson et al. 1994, Barratt et al. 1995, 1997). Ideas for practical means to improve the environmental safety of biological control have been proposed (Goldson and Phillips 1990, 1993, Goldson et al. 1998, Barratt et al. 1998).

Pre-release risk assessment of weed biological control agents poses fewer problems than those attempted for insects. In New Zealand there are approximately 2000 species of native plants, of which about 95% are described. This allows a "centrifugal phylogenetic strategy" (Wapshere 1974) to be adopted whereby, in quarantine, a sequence of plants is exposed to the potential biological control agent, from those most closely related to those successively more distantly related to the target weed, to identify the potential host range. In contrast with plants, there are estimated to be well over 20,000 species of native insects in New Zealand, of which about half are estimated to be described (Emberson 1994). Rigorous host-range testing of biological control agents for insect pests is therefore constrained by poor knowledge of the fauna, this limitation having been termed a "taxonomic impediment" (Howarth and Ramsay 1991). Other safeguards are also more easily implemented for phytophagous compared with entomophagous biological control agents. For example potential nontarget host plants are often transported to the country of origin of a proposed weed biological control agent for field testing to evaluate nontarget effects. Seed storage is possible in the case of rare and endangered plant species that could be at risk.

In the USA, procedures for host specificity testing of weed biological control agents have been stringent. At the same time, natural enemies of insect pests have required no such testing. The prevailing notion has been that insect natural enemies have evolved so closely with their hosts that their host range is restricted to one or a few host species (Knutson and Coulson 1997).

The environmental impact of biological control agents can be defined as any measurable effect on a nontarget species resulting from the introduction of an exotic organism. Moller (1989) pointed out that these could be extremely complex and emphasized the need to monitor effects on other species in the food web, spanning three or more trophic levels. An indirect effect of an introduced organism could be increased competition for a host-prey with native natural enemies, or habitat modification (Van Lenteren 1997).

Clearly the risks associated with the introduction of a biological control agent have to be weighed against the economic benefits but also balanced against the environmental cost of doing nothing. In the case studies described below, the target species are pests of exotic agricultural plants and unlikely to establish extensively in native vegetation, therefore posing little direct risk to native species in their natural habitat. However, this is a serious consideration in many instances where an exotic plant or insect might threaten the existence of native species or habitats in the absence of control, and where biological control might be the only option.

REGULATORY LEGISLATION

International

In 1995, the FAO Council ratified an international Code of Conduct for the Import and Release of Biological Control Agents with the intention of providing a set of standards and guidelines for "best practice" biological control agent introduction. Schulten (1997) has summarized the stages of development, main elements, and status of the code. It is recommended in the code that proposed importers of biological control agents provide information on agricultural and environmental nontarget effects (FAO 1996). It also recommends that relevant international conventions, such as the Convention on Biological Diversity, be observed. While some countries find the code too prescriptive (e.g., Van Halteren 1997), others such as New Zealand have developed legislation that goes beyond the recommendations of the code in relation to environmental safety.

New Zealand

Legislation relating to the importation of new organisms, along with hazardous substances, has recently been reformed in New Zealand. The aim of the Hazardous Substances and New Organisms Act 1996 (HSNO) is to provide a comprehensive and consistent approach to the management of these importations, superseding the Animals Act 1970 (which previously covered biological control agents) and the Plants Act 1967, as well as the Explosives Act 1957, the Dangerous Goods Act 1974, the Toxic Substances Act 1979, and the Pesticides Act 1979. Under HSNO, an independent statutory body known as the Environmental Risk Management Authority (ERMA New Zealand) has been established, which is responsible for making decisions on the manufacture, importation, and/or release of hazardous substances, new organisms, and genetically modified organisms in New Zealand.

The methods by which ERMA New Zealand will process applications under HSNO have been subject to a public consultation process (ERMA New Zealand 1997a, 1997b), which included nation-wide meetings with interest groups and the public. The procedures and protocols for the processing of applications were then finalized, and HSNO was enforced in 1998.

Before the implementation of HSNO, applications to import proposed biological control agents were considered by the Ministry of Agriculture and Fisheries (MAF) Regulatory Authority, a responsibility which is now assumed by ERMA New Zealand, although importations will still have to meet requirements of Import Health Standards approved by MAF under the provisions of the Biosecurity Act 1993. The HSNO legislation is focused on the health and safety of people and the environment, stating that ERMA New Zealand must consider the following principles with every application:

- safeguarding the life-supporting capacity of air, water, and ecosystems
- maintaining and enhancing the capacity of people and communities to provide for their own economic, social, and cultural well being and for the reasonable foreseeable needs of future generations and, where applicable:
  - the sustainability of all native and valued introduced flora and fauna
  - the intrinsic value of ecosystems
  - public health
  - the relationship of Maori and their culture and traditions with their ancestral lands, water, sites, *waahi tapu* (historic places of sacred cultural significance), valued flora and fauna and other *taonga* (treasures)
  - economic and related values
  - New Zealand's international obligations.

HSNO incorporates a "precautionary approach" stating that "All persons exercising functions, powers, and duties under this Act...shall take into account the need for caution in managing adverse effects where there is scientific and technical uncertainty about those effects."

NEW ZEALAND TEST CASES FOR BIOLOGICAL CONTROL AGENT HOST RANGE PREDICTION AND IMPACT ASSESSMENT

The biological control agents *Microctonus aethiopoides* Loan and *Microctonus hyperodae* Loan (Hymenoptera: Braconidae) have been introduced into New Zealand to control the forage pests *Sitona discoideus* Gyllenhal and *Listronotus bonariensis* (Kuschel) (Coleoptera: Curculionidae), respectively. They are both solitary, koinobiont endoparasitoids of the adult stage of the target weevil species. These two parasitoids were selected as "test cases" to investigate the predictability of quarantine testing by carrying out pre-release assessments and laboratory host range tests to predict host range, and then to assess the accuracy of the predictions by investigating realized field host range post-release.

Microctonus aethiopoides was released in New Zealand in 1982 from Australia (Stufkens et al. 1987) where it had been introduced from the Mediterranean region to control *S. discoideus*. Surveys throughout New Zealand have shown that *M. aethiopoides* is well established in *S. discoideus* populations in lucerne-growing areas (Stufkens et al. 1987, Ferguson et al. 1994, Barratt et al. 1998, Barker and Addison unpublished), where it has been shown to suppress *S. discoideus* populations (Goldson et al. 1993). *M. aethiopoides* was, however, released with limited host range testing in quarantine, which revealed no evidence of attack on nontarget species (M. Stufkens, personal communication).

Microctonus hyperodae of South American origin was originally released in 1991 at several sites throughout New Zealand where it has established successfully (Goldson et al. 1994a, 1994b). Spread from the release sites has been slow, particularly at the more southerly sites (Ferguson et al. 1997). *M. hyperodae* was released after extensive quarantine testing (Goldson et al. 1992), which indicated that the parasitoid was oligophagous, but the authors predicted that one or two native weevil species could be parasitized in the field.

These two *Microctonus* spp. were used to compare pre-release predictions with post-release realized field parasitism as follows:

1. identify potentially "at risk" nontarget hosts of *M. aethiopoides* and *M. hyperodae* species by examining phylogenetic, ecological, and behavioral affinities of the native fauna with the intended target hosts, *S. discoideus* and *L. bonariensis*, respectively
2. determine to what extent species identified above are found in mixed populations with the target hosts, and therefore in direct contact with parasitoid populations
3. conduct laboratory host range tests (in conditions similar to those in quarantine) on a sub-set of potential "at risk" species using the information collected in 1–2 as a guide
4. predict field host range using data from 1–3
5. investigate realized field host range by sampling and collecting nontarget species from within and outside of the normal environment of the target hosts
6. compare laboratory defined potential parasitism with realized parasitism in the field.

PHYLOGENETIC, ECOLOGICAL AND BEHAVIORAL AFFINITIES OF THE NEW ZEALAND NATIVE FAUNA WITH THE TARGET HOSTS

Phylogeny

The target hosts, *S. discoideus* and *L. bonariensis* are both in the weevil subfamily Brachycerinae (broad-nosed weevils), and the tribes Entimini and Rhytirhinini, respectively (Kuschel 1990, 1995). This subfamily and both tribes are well represented in New Zealand by a large number of native species, especially the Entimini, many of which inhabit grassland environments (Barratt et al. 1998). It could be assumed, therefore, that the native Brachycerinae could be possible targets for parasitism by *Microctonus* species in New Zealand. Elsewhere *Microctonus* species are known to utilize a wide range of host sub-families within the Curculionidae, and even within the Coleoptera, including carabids, cerambycids, tenebrionids, and chrysomelids (Shaw 1985, 1988). *M. aethiopoides* has been recorded parasitizing chrysomelid species (Smith 1953).

The New Zealand Rhytirhinini include alpine and sub-alpine grassland native genera such as *Gromilus*, *Liparogetus*, *Steriphus* and *Nestrius*, as well as introduced genera such as *Listronotus* and *Listroderes*. The Entimini include large numbers of largely undescribed species of native grassland and sub-alpine genera such as *Irenimus*, *Catoptes*, *Nicaeana*, *Brachyolus*, and *Nonnotus* (Barratt and Kuschel 1996) which are in the same size range as *L. bonariensis* and *S. discoideus*. Endemic Brachycerinae, which are largely flightless in contrast to the introduced genera, are of particular significance in New Zealand because of the extensive speciation that has occurred, which is scarcely documented, and their often limited distributional range (Barratt and Kuschel 1996).

Ecology and Phenology

Little is known about the biology and ecology of broad-nosed weevils in New Zealand. The adults feed on seedling plants (Barratt et al. 1992) and mature plant foliage of a broad range of plants species (Bremner 1988). Field observations suggest that they also feed on pollen of flowering plants, and adult survival studies in the laboratory (unpublished data) have indicated that pollen may represent an important source of protein for oviposition as it does for *L. bonariensis* (Evans and Barratt 1995). Larvae of the native species are soil dwelling and feed on root material (Bremner 1988; May 1993).

The phenology of adults of a number of native weevil species has been studied to determine whether seasonal abundance and activity patterns are similar to those of the introduced target hosts, or whether native hosts might be available to parasitoids at times when the target hosts are scarce, placing the native hosts particularly at risk. Figure 1 shows a simplified plot of adult seasonal abundance of *S. discoideus*, *L. bonariensis*, and three native brachycerines identified as potential nontarget hosts. It shows that there are periods when *S. discoideus* is unavailable to *M. aethiopoides* in late spring when one generation dies and before the new one emerges, and during late summer and early autumn, when the majority of the *S. discoideus* population is aestivating, and unavailable as hosts. Most *M. aethiopoides* which parasitize newly emerged *S. discoideus* at this time enter aestivation along with the host, although it has been found that a small percentage are able to continue development (Goldson et al. 1990). The native Entimini shown are moderately abundant in spring, and could represent alternative hosts for *M. aethiopoides* between *S. discoideus* generations.

Listronotus bonariensis is least abundant in pastures in spring after the overwintering generation has largely died and before the new generation adults emerge. This is a period when some native species are at their peak of abundance e.g. *Steriphus variabilis* which therefore could be particularly vulnerable to parasitism attempts by *M. hyperodae*.

Behavior

For a parasitoid to successfully exploit a host, the activity patterns of both must bring about contact at times appropriate for parasitism to occur. This was explored in laboratory experiments which investigated the circadian patterns of oviposition of *M.*

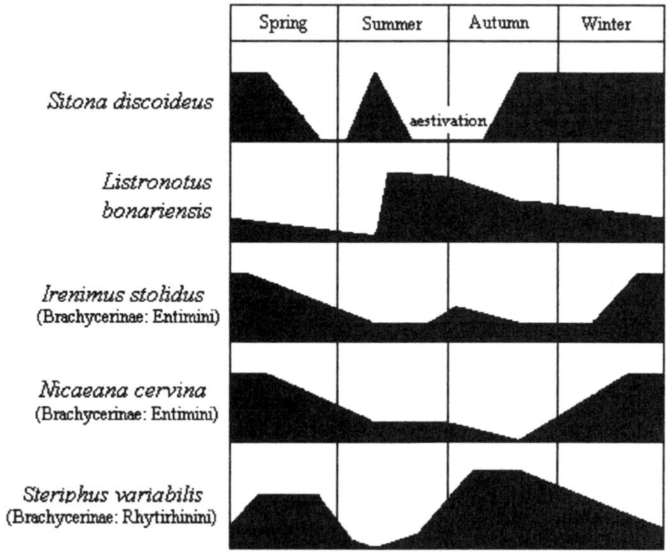

Figure 1. Seasonality and relative abundance of adult weevils of the two target and examples of three potential nontarget hosts.

aethiopoides and *M. hyperodae* in their respective target hosts *S. discoideus* and *L. bonariensis*, and then related to target and nontarget host activity patterns. Successful oviposition by *M. aethiopoides* in *S. discoideus* (which is a diurnal feeder) occurred more frequently during light than during dark periods, but *M. hyperodae* achieved higher parasitism levels in *L. bonariensis* during hours of darkness (Armstrong et al. 1996). However, when *M. aethiopoides* was exposed to *L. bonariensis* which feeds during dark periods, but oviposits throughout the 24 hour period (Barratt et al. 1995), parasitoid oviposition occurred equally during light and dark periods (Couch et al. 1997). This suggests that the activity pattern of the host may be more important in regulating ovipositional activity by *M. aethiopoides* than any endogenous rhythm of its own (Couch et al. 1997). Observations by Loan and Holdaway (1961) suggested that *M. aethiopoides* is reliant on host movement for successful stalking and oviposition. Couch (1997) found that the activity patterns of two native brachycerine weevil species, as measured by feeding and oviposition, were less clearly defined than those of the target hosts, but both tended to feed during light periods. The results of these studies indicate that *M. aethiopoides* is an opportunistic parasitoid that hunts for hosts throughout the 24 hour period, probably limited only by the activity patterns of the available hosts. For *M. hyperodae*, it was not determined whether its apparent nocturnal activity is an endogenous rhythm or whether it also is dependent upon the activity pattern of its host. If the former is true, the impact that it might have on diurnally active weevils may be reduced.

HABITAT OVERLAP BETWEEN TARGET AND POTENTIAL NONTARGET HOSTS

To determine which potential nontarget species coexist with the target species, and might therefore be particularly at risk of attack, a survey of the weevil fauna of pasture, lucerne, and grazed native grasslands up to 1000m altitude was carried out, particularly in lucerne-growing areas where *M. aethiopoides* is known to have established, and areas where *M. hyperodae* has been released (Goldson *et al.* 1994b). Over 150 sites were sampled from which 62 native and 23 introduced species of Curculionoidea were found (Barratt et al. 1998). About 75 of the species found were in the subfamily Brachycerinae, and over 80% of these were in the tribe Entimini. Only nine species of rhytirhinines, of which five were native, were found in these environments.

The survey showed that New Zealand has a diverse range of Curculionoidea, including many native species that occur within its agricultural environments. As expected, native species diversity was higher in less intensively modified grassland environments, and the proportion of exotic weevil species was lower. Semi-permanent grasslands clearly represent an environment that is favorable to survival of a component of New Zealand's weevil biodiversity. The survey indicated that there are a large number of species, particularly 41 native species of weevils in the tribe Entimini, which have potential to become targets for the introduced parasitoid, *M. aethiopoides*, but far fewer Rhytirhinini are available for *M. hyperodae*.

Limited taxonomic and ecological knowledge is a critical problem in studies of nontarget host range, particularly in New Zealand where the fauna is so poorly known and incompletely described. This precludes an exhaustive study of potential hosts and objective selection of species for laboratory testing, and hence to some extent compromises prediction of impacts.

LABORATORY HOST RANGE TESTS

A series of nontarget weevil species were exposed to the parasitoid species in laboratory cage tests. Given the limitations outlined above, species selection was based on those identified as being 'at risk', particularly in relation to taxonomy, ecology, and habitat overlap, and influenced to some extent by availability of native species in sufficient numbers for replicated experiments. Methodology was based on protocols developed by Goldson et al. (1992) for *M. hyperodae* and standardized so that the nontarget host ranges of *M. hyperodae* and *M. aethiopoides* could be compared (Barratt et al. 1996). The standard procedure adopted was to expose five replicate cages of 20 test weevils to three female parasitoids for 48 hours after which the weevils were maintained until parasitoids emerged. Surviving weevils and those that died during the experiment were dissected for evidence of unsuccessful parasitism.

Microctonus aethiopoides

The results of these laboratory tests are summarized in Table 1. In total, 12 weevil species were exposed to *M. aethiopoides*, of which seven were native to New Zealand. *M. aethiopoides* oviposited in all the native species to which it was exposed with percent parasitism averaging 58% over all species, which was similar to the 62% parasitism recorded in tests carried out in parallel with the target host *S. discoideus* (Barratt et al. 1997) . Parasitoid development times from when adult parasitoids were removed from test cages to when parasitoid prepupae emerged from the test species also were similar to that for *S. discoideus*. About 80% of the pre-pupal parasitoids that emerged from native weevils developed successfully to the adult stage. The proportion of the parasitoids found during dissection which were melanized, encapsulated, or showing other signs of a host immune response (rejection by the host) in nontarget species was about 12% of total parasitism, whereas no host immunity was detected with *S. discoideus* (Table 1).

The host species attacked by *M. aethiopoides* ranged in size from 3-12 mm long, compared with *S. discoideus*, which is about 4 mm long, suggesting that in the laboratory, host size outside the normal range was not an obstacle to oviposition attempts by *M. aethiopoides*. There is some evidence to suggest, however, that some native species may be too small to support successful parasitoid development (unpublished data).

M. aethiopoides oviposited successfully in the weed biological control agent *Rhinocyllus conicus* (Froelich) (Curculioninae: Cleonini) and *L. bonariensis*, both to a level of about 40%, and to a lesser extent in *Rhinonchus australis* Oke (Curculioninae: Ceutorhynchini). Oviposition in another weed biological control agent, *Trichosirocalus horridus* Panzer (Curculioninae: Ceutorhynchini) resulted in melanized and encapsulated parasitoid larvae. Limited tests undertaken with non-cuculionid Coleoptera, including chrysomelids, has detected no oviposition by *M. aethiopoides*.

Microctonus hyperodae

While *M. hyperodae* was in quarantine, studies were undertaken to determine its likely host range in New Zealand (Goldson et al. 1990, 1992). *M. hyperodae* oviposited in only a small proportion of the native species to which it was exposed, and in all but one species parasitoid development was impeded, with up to 50% of the larvae becoming encapsulated. The quarantine investigation concluded that in view of the relatively oligophagous behavior of *M. hyperodae*, and the potential benefit to be derived from successful biological control of *L. bonariensis*, the parasitoid should be recommended as suitable for release (Goldson et al. 1992). Benefits were based on the economic value of lost pasture production and animal health problems associated with changes in pasture quality as a result of damage from *L. bonariensis*, estimated at NZ$78-251million per annum (Prestidge et al. 1991).

Results from the quarantine investigations, and those from further laboratory studies carried out post-release (Barratt et al. 1997), are summarized in Table 1. In total, 30 nontarget species of weevils were exposed to *M. hyperodae*, of which 20 were New Zealand native species. *M. hyperodae* oviposited in seven species, all of which were native. In the tests carried out post-release (Barratt et al. 1997), parasitism in those species which were attacked averaged 13% in native species compared with 61% in *L. bonariensis*. The mean development time for *M. hyperodae* was similar in all weevil species, with no differences between development times in *L. bonariensis* and native species (Barratt et al. 1997) although longer development times were noted in the pre-release tests (Goldson et al. 1992). The proportion of the parasitoids found during dissection which were exhibiting signs of a host immune response in native weevils was about 32% of total parasitism compared with 2% in *L. bonariensis* (Barratt et al. 1997).

SUMMARY OF PREDICTED FIELD HOST RANGE

M. aethiopoides

From the phlyogenetic analysis, it was anticipated that potential nontarget hosts for *M. aethiopoides* were likely to be adult weevils in the subfamily Brachycerinae and tribe Entimini, although host range testing indicated that a wider range of taxa could be utilized successfully, for example members of the Curculioninae. Like the target host, native brachycerines are essentially ground surface dwelling and feed on foliage mainly within 1m of the ground, many are a similar size to *S. discoideus*, and they are largely diurnally active, so it was assumed that they would be encountered by *M. aethiopoides*. The weevil fauna in lucerne, in which *S. discoideus* is most commonly found, were expected to be most at risk, although weevils feeding on other vegetation types were attacked in laboratory host range tests. *M. aethiopoides* oviposited in over 90% of the species to which it was exposed in laboratory tests indicating that it is polyphagous. In agricultural environments the number of brachycerine species which could be at risk of attack is large (at least 61spp.). The target host is very mobile, especially after autumn dispersal flights, and is frequently found at high altitude in areas of natural vegetation, and so the parasitoid could be transported to these areas.

Microctonus hyperodae

Comparative analysis of phylogenetic, ecological, and behavioral affinities of *L. bonariensis* with nontarget species, along with the data from habitat overlap surveys and laboratory host range tests, supported the conclusions of (Goldson et al. 1992) that *M. hyperodae* was oligophagous and likely to parasitize very few nontarget species in the field. Taxa at risk were considered likely to be brachycerines, particularly in the tribe Rhytirhinini, however, among those species available for laboratory testing, no oviposition was detected. In total, 23% of species exposed to *M. hyperodae* were attacked, but parasitoid development in nontarget species was less successful than that achieved in the target host. Grassland-dwelling species were predicted to be most at risk, and while this is a potentially large number for brachycerines, the number of native rhytirhinines were relatively few. *L. bonariensis* is crepuscular/nocturnal, but little is known about the circadian rhythms of native rhytirhinines. *L. bonariensis* is sometimes found at high altitude and in natural vegetation, and larvae have been found in association with adults suggesting that, unlike *S. discoideus* which depends upon lucerne, *L. bonariensis* could complete development in such areas, and possibly introduce *M. hyperodae* to these environments.

ACTUAL FIELD HOST RANGE

To compare predicted with realized host ranges of *M. aethiopoides* and *M. hyperodae*, a total of 48 species of weevils were collected from 33 sites (lucerne, pasture and grazed natural grassland) throughout New Zealand, and examined for parasitism by rearing and/or dissection (Table 1). Ten New Zealand native species and four introduced nontarget species were found to be parasitized by *M. aethiopoides* in the field, approximately 30% of the species examined, with levels of parasitism up to 70% (Barratt et al. 1997). At over half of the sites sampled, nontarget species were found parasitized by *M. aethiopoides* (Table 1).

M. hyperodae has been recovered from only two nontarget weevil species in the field, the native species *Irenimus aequalis* (Broun), and an exotic species recently discovered in New Zealand, *Sitona lepidus* Gyllenhal (Barratt et al. 1997). Both were found at a North Island site, where populations densities of *M. hyperodae* are very high, and in each instance, only a single parasitized host individual has been detected to date.

Goldson et al. (1992) postulated that *M. hyperodae* might be ecologically separated from some native species, which might escape contact with the parasitoid in alpine environments. However, both *L. bonariensis* and *S. discoideus* frequently occur at high altitude in native vegetation (e.g. Patrick et al. 1993) and could potentially take parasitoids to these environments. For example, *L. bonariensis* parasitized by *M. aethiopoides* has been found at 1650m on Coronet Peak, Otago.

Most nontarget species found parasitized by *M. aethiopoides* in the field were apterous except for two native species, *Nonnotus albicans* Broun, which has been found up to 850 m altitude, and *S. variabilis*, and four exotic species. These species could also be involved in transporting *M. aethiopoides*.

COMPARISON BETWEEN POTENTIAL AND REALIZED FIELD PARASITISM

Predictions about the likely host range of *M. aethiopoides* and *M. hyperodae* which were made after the laboratory investigations were complete, were generally borne out by what was found in the field. *M. aethiopoides* has proved to be polyphagous, developing successfully with quite high levels of parasitism in a diversity of nontarget taxa in a range of habitats. *M. hyperodae* has to date proved perhaps even more oligophagous than was anticipated with only two confirmed species, and indeed individuals having been discovered parasitized in the field. Significantly, one of these, the native species, *I. aequalis*, was predicted from the quarantine investigation to be a likely host for *M. hyperodae* (Goldson et al. 1992). In this comparison between the two parasitoid species, allowance must be made for the fact that *M. hyperodae* is currently less widely distributed throughout New Zealand than *M. aethiopoides*, and that it has been present in the environment for only seven years, compared with 16 years for *M. aethiopoides*.

SCIENCE UNDERPINNING THE NEEDS OF PROPOSED IMPORTERS AND REGULATORS

In New Zealand, ERMA New Zealand has made it clear that the provision of information to support applications to introduce new organisms into New Zealand is the responsibility of the applicant (ERMA New Zealand 1997), and that the outcome of the application will be to a large extent contingent on the quality of the information provided. In order to streamline the application process, ERMA New Zealand recommends that applicants consult with them directly before formal submission of a proposal, at which time they are given guidance on the nature of the information that will be required to process the application.

In the case of biological control agents, one of the most difficult tasks faced by the applicant will be supplying adequate information on potential impacts of the new organism on the environment. While the regulators will have to accept that this will always be an area of some scientific uncertainty, and the final decision based on an assessment of risks, costs and benefits, research can assist in reducing uncertainty. Case studies where predicted and realized host range are compared can assist considerably in evaluating the quality of predictions that can be made a result of containment investigations (Cullen 1997). The examples given above for the two *Microctonus* spp. go some way towards providing this sort of information, but have not provided definitive quantitative information on actual impacts of the parasitoids on nontarget weevil populations. It is not possible to predict at this stage the impact of the two *Microctonus* spp. in the long-term, although further population modeling will aspire to achieve this. Further research will also endeavor to determine the impact of *M. aethiopoides* on native weevil population dynamics and hence provide much needed information about the significance of the parasitism levels recorded in field populations. However, the exercise has provided a framework of methodologies that will facilitate similar studies in the future. To be generally applicable, such case studies need to be repeated for a number of representative host/parasitoid models, for example those involving different taxa, developmental stage attacked, those living in different environments, using different host searching mechanisms etc. The target/nontarget host relative abundancies and phenologies are also likely to be important. Provision of such data to regulators would constitute a valuable resource for decision support.

Clearly the intensity of research which has contributed to the case studies discussed above cannot be supported for all biological control introductions. However, post-release investigations of biological control agent introductions, especially in relation to impacts on nontarget species, have generally been sparse, if indeed carried out at all. The large investment in biological control programs should warrant some level of post-release assessment of impact at least on the target, if not also nontarget species. Again, a database of such information would represent a valuable resource for regulators in future decision-making.

Cage testing is often criticized as being artificial and for over-emphasizing host range (e.g. Sands 1993), especially for entomophagous biological control agents. There are many examples in the literature of parasitoids that attack species in a cage, when this has never been found to occur in the field. Current research is investigating the scope for improving the range of containment host range testing methodologies available, which would benefit importers and regulators. By separating host/parasitoid compatibility into its behavioral and physiological components, it may be possible in some cases to improve the interpretation of conventional cage testing. For example, behavioral acceptance of a host by a parasitoid is being measured using a pathogen marker on the parasitoid's ovipositor to indicate, by means of the rapid death of the test host, an oviposition attempt. Assessment of physiological permissiveness may provide a conservative measure of parasitoid success that should apply even if behavioral changes occur post-release. Methods of measuring physiological compatibility are being investigated by studying parasitoid immunosuppression of the host.

CONCLUSIONS

The case studies described in this contribution have shown that for the two braconids studied, pre-release investigations including host specificity testing were generally predictive of post-release field host range. Information from a wider series of host/parasitoid models could enhance the precision of decision making by regulators, who will invariably be faced with some degree of scientific uncertainty regarding proposed biological control introductions. Clearly it will never be possible to determine absolute environmental impacts since such studies provide only a 'snap shot' assessment in what is a dynamic process, driven by complex ecological factors such as parasitoid adaptation to new hosts, and interactions and impacts further along the food chain. However, if the decision support system for biological control introduction proposals can capitalize on feedback from previous experiences, it should be possible not only to substantially reduce the level of uncertainty in many cases, and the risk of adverse environmental impacts, but also to improve the predictability of biological control success and the overall effectiveness of pest management.

ACKNOWLEDGMENTS

We thank Adrian Evans, Mark McNeill, John Proffitt, Rory Logan, Gary Barker, Paul Addison for assistance with this research, and The New Zealand Foundation for Research, Science and Technology for funding.

REFERENCES

Armstrong, S. M., B. I. P. Barratt, and A. A. Evans. 1996. Circadian pattern of oviposition in the parasitoids *Microctonus aethiopoides* Loan and *M. hyperodae* Loan (Hymenoptera: Braconidae), in relation to host activity. *In* Proc., 49th N.Z. Plant Protection Soc. Conf., August 13–15, 1996. N.Z. Plant Protection Soc. Inc. p. 280–284.

Aubert, B., and S. Quilici. 1983. Nouvel équilibre biologique observe à la Réunion sur les populations de psyllides après l'introduction et l'éstablissement d'hymenopteres chalcidiens. Fruits 38: 771–780.

Barratt, B. I. P., C. M. Ferguson. P. A. Jones, and P. D. Johnstone. 1992. The effect of insects on oversown white clover establishment and yield in New Zealand tussock grassland. N.Z. J. Agric. Res. 35: 63–73.

Barratt, B. I. P., A. A. Evans, and C. M. Ferguson. 1995. Circadian patterns of oviposition and feeding in *Listronotus bonariensis* (Kuschel) (Coleoptera: Curculionidae) in the laboratory. *In* Proc., 48th N.Z. Plant Protection Soc. Conf., August 8–10, 1995. N.Z. Plant Protection Soc. Inc. p. 219–223.

Barratt, B. I. P., C. M. Ferguson, and A. A. Evans. 1995. Assessing the fate of introduced biological control agents in natural and agricultural ecosystems, *In* S. Field (ed.), Proc., Asia Pacific Agri-industry Community Conf., August 9–11, 1995. Australian Inst. Agric. Sci. Occas. Public. 89. p. 8.

Barratt, B. I. P., A. A. Evans, and P. D. Johnstone. 1996. Effect of the ratios of *Listronotus bonariensis* and *Sitona discoideus* (Coleoptera: Curculionidae) to their respective parasitoids *Microctonus hyperodae* and *Microctonus aethiopoides* (Hymenoptera: Braconidae), on parasitism, host oviposition and feeding in the laboratory. Bull. Entomol. Res. 86: 101–108.

Barratt, B. I. P., and G. Kuschel. 1996. Broad–nosed weevils (Curculionidae: Brachycerinae: Entimini) of the Lammermoor and Rock and Pillar Ranges in Otago, with descriptions of four new species of *Irenimus*. N.Z. J. Zool. 23: 359–374.

Barratt, B. I. P., A. A. Evans, C. M. Ferguson, G. M. Barker, M. R. McNeill, and C. B. Phillips. 1997. Laboratory nontarget host range of the introduced parasitoids *Microctonus aethiopoides* and *Microctonus hyperodae* (Hymenoptera: Braconidae) compared with field parasitism in New Zealand. Environ. Entomol. 26: 694–702.

Barratt, B. I. P., A. A. Evans, C. M. Ferguson, M. R. McNeill, J. R. Proffitt, and G. M. Barker. 1998. Curculionoidea (Insecta: Coleoptera) of agricultural grassland and lucerne as potential nontarget hosts of the parasitoids *Microctonus aethiopoides* Loan and *Microctonus hyperodae* Loan (Hymenoptera: Braconidae). N.Z. J. Zool. 25: 47–63.

Barratt, B.I.P., C.M. Ferguson, M.R. McNeill, and S.L. Goldson. Parasitoid host specificity testing to predict host range. *In* T. M Withers, L. Barton-Browne, and J. N. Stanley (eds.), Recommendations for host specificity testing procedures in Australasia—towards improved assays for biological control agents. CRC for Tropical Pest Management, Brisbane, Australia. In press.

Caltagirone, L. E. 1981. Landmark examples in classical biological control. Annu. Rev. Entomol. 26: 213–232.

Cameron. p. J., R. L. Hill, J. Bain, and W. P. Thomas. 1993. Analysis of importations for biological control of insect pests and weeds in New Zealand. Biol. Control Sci. Technol. 3: 387–404.

Carson, R. 1963. Silent Spring. Hamish Hamilton, London, UK.

Couch, K. M., A. S. Cresswell, B. I. P. Barratt, and A. A. Evans. 1997. Implications of host weevil circadian activity for parasitism by *Microctonus aethiopoides* (Hymenoptera: Braconidae). *In* Proc., 50th N.Z. Plant Protection Soc. Conf., August 18–21, 1997. N.Z. Plant Protection Soc. Inc. p. 227–231.

Cullen, J. M. 1993. Opportunities and challenges in biological control. *In* S. A. Corey, D. J. Dall, and W. M. Milne (eds.), Pest control and sustainable agriculture. CSIRO Info. Svcs., Melbourne, Australia. p. 44–50.

Cullen, J. M. 1997. Biological control and impacts on nontarget species. p. 195–201. *In* Proc., 50th N.Z. Plant Protection Soc. Conf., August18–21, 1997. N.Z. Plant Protection Soc. Inc.

DeBach. p., and D. Rosen. 1991. Biological control by natural enemies. 2nd ed. Cambridge Univ. Press, Cambridge, UK.

Ehler, L. E. 1990. Environmental impact of introduced biological control agents: implications for agricultural biotechnology. *In* J. J. Marois and G. Breuning (eds.), Risk assessment in agricultural biotechnology. California Div. Agric. Natural Resources, Oakland, California. p. 85–96.

Emberson, R. M. 1994. Taxonomic impediments to the development of sustainable practices in conservation and production. *In* Proc., 43rd Annu. Conf. Entomol. Soc. N.Z., May 15–18, 1994. Entomological Society of New Zealand. p. 71–78.

ERMA, New Zealand. 1997a. A proposed methodology for the consideration of applications for hazardous substances and new organisms under the HSNO Act 1996; Consultation document. ERMA New Zealand, Wellington.

ERMA, New Zealand. 1997b. Considering applications: proposed procedures and information requirements for the consideration of applications for the introduction of hazardous substances and new organisms under the HSNO Act 1996; Consultation document no. 2. ERMA New Zealand, Wellington.

Evans, A. A., and B. I. P. Barratt. 1995. Effect of a ryegrass diet supplemented with pollen on *Listronotus bonariensis* (Kuschel) (Coleoptera: Curculionidae) fecundity, feeding and survival. *In* Proc., 48th N.Z. Plant Protection Soc. Conf., 8–19 August 1995. N.Z. Plant Protection Soc. Inc. p. 242–244.

Ferguson, C. M., A. A. Evans, B. I. P. Barratt, and C. B. Phillips. 1997. Establishment and dispersal of *Microctonus hyperodae* Loan (Hymenoptera: Braconidae) in Otago and Southland. *In* Proc., 50th N. Z. Plant Protection Soc. Conf., August 18–21, 1997. N.Z. Plant Protection Soc. Inc. p. 41–46.

Ferguson, C. M., G. M. Roberts, B. I. P. Barratt and A. A. Evans. 1994. The distribution of the parasitoid *Microctonus aethiopoides* Loan (Hymenoptera: Braconidae) in southern South Island *Sitona discoideus* Gyllenhal (Coleoptera: Curculionidae) populations. *In* Proc., 47th N.Z. Plant Prot. Soc. Conf., August 9–11, 1994. N.Z. Plant Protection Soc. Inc. p. 261–265.

Gibbs, G. W. 1980. New Zealand butterflies: identification and natural history. Collins, Auckland, NZ.

Goldson, S. L., and C. B. Phillips. 1990. Biological control in pasture and lucerne and the requirements for further responsible introduction of entomophagous insects. Bull. Entomol. Soc. N. Z. 10: 63–74.

Goldson, S. L., and C. B. Phillips. 1993. Science and biological control. *In* R. A. Prestidge (ed.), Proc., 6th Australian Grassl. Invert. Ecol. Conf., AgResearch, Hamilton, New Zealand. p. 347–354.

Goldson, S. L., J. R. Proffitt, and M. R. McNeill. 1990. Seasonal biology and ecology in New Zealand of *Microctonus aethiopoides* (Hymenoptera: Braconidae), a parasitoid of *Sitona* sp. (Coleoptera: Curculionidae), with special emphasis on atypical behavior. J. Appl. Ecol. 27: 703–722.

Goldson, S. L., M. R. McNeill, C. B. Phillips, and J. R. Proffitt. 1992. Host specificity testing and suitability of the parasitoid *Microctonus hyperodae* (Hym.: Braconidae, Euphorinae) as a biological control agent of *Listronotus bonariensis* (Col.: Curculionidae) in New Zealand. Entomophaga 37: 483–498.

Goldson, S. L., J. R. Proffitt, and K. E. Muscroft-Taylor. 1993. The economic value of achieving biological control of *Sitona discoideus*. *In* D. M. Suckling and A. J. Popay (eds.), Plant protection: costs, benefits and trade implications. N.Z. Plant Protection Soc. Inc., New Zealand. p. 45–60.

Goldson, S. L., G. M. Barker, B. I. P. Barratt, and N. D. Barlow. 1994a. Progress in the biological control of Argentine stem weevil and comment on its potential. *In* Proc., N.Z. Grassl. Assoc. Conf. p. 39–42.

Goldson, S. L., G. M. Barker, B. I. P. Barratt, and A. J. Popay. 1994b. The establishment of an Argentine stem weevil parasitoid at its release sites. p. 274–276. *In* Proc., 47th N.Z. Plant Protection Soc. Conf., 9–11 August 1994. N.Z. Plant Protection Soc. Inc.

Goldson, S.L., B.I.P. Barratt, N.D. Barlow and C.B. Phillips. 1998. What is a safe biological control agent? *In* M. Zalucki, R. Drew, and G. White (eds.), Pest management—future challenges: Proc. 6th Australasian Appl. Entomol. Res. Conf., vol. 1. Brisbane, Australia, September 29 – October 2, 1998. The Cooperative Research Centre for Tropical Pest Management, Brisbane, Australia. p. 530–538.

Harris. p. 1990. Environmental impact of introduced biological control agents. *In* M. Mackauer, L. E. Ehler, and J. Roland (eds.), Critical issues in biological control. Intercept, Andover, Hampshire, UK. p. 289–300.

Howarth, F. G. 1991. Environmental impacts of classical biological control. Annu. Rev. Entomol. 36: 489–509.

Howarth, F. G. and G. W. Ramsay. 1991. The conservation of island insects and their habitats. *In* N. M. Collins and J. A. Thomas (eds.), The conservation of insects and their habitat. Academic Press, London, UK. p. 71–107.

Knutson, L. and J. R. Coulson. 1997. Procedures and policies in the USA regarding precautions in the introduction of classical biological control agents. *In* I. M. Smith (ed.), EPPO/CABI workshop on safety and efficacy of biological control in Europe. Blackwell Science Ltd., Oxford, UK. p. 133–142.

Kuschel, G. 1995. A phylogenetic classification of Curculionoidea to family and subfamily level. *In* R. S.

Anderson and C. H. C. Lyal (eds.), Biology and phylogeny of Curculionoidea. Proc., XVIII International Congress of Entomology, July 1988. Mem. Entomol. Soc. Washington. p. 5–36.
Loan, C., and F. G. Holdaway. 1961. *Microctonus aethiops* (Nees) auctt. and *Perilitis rutilis*(Nees) auctt. (Hymenoptera: Braconidae), European parasites of *Sitona* weevils (Coleoptera: Curculionidae). Can. Entomol. 93: 1057–1079
Lockwood, J. A. 1993. Environmental issues involved in biological control of rangeland grasshoppers (Orthoptera: Acrididae) with exotic agents. Environ. Entomol. 22: 504–518.
Lockwood, J. A. 1996. The ethics of biological control: Understanding the moral implications of our most powerful ecological technology. Agric. Human Values 13: 2–19.
Longworth, J. F. 1987. Biological control in New Zealand: policy and procedures. N.Z. Entomol. 10: 1–7.
May, B. M. 1993. Larvae of Curculionoidea (Insecta: Coleoptera): a systematic overview. Fauna of New Zealand no. 28. Manaaki Whenua Press, Lincoln, Canterbury, New Zealand.
Moller, H. 1989. Towards constructive ecological engineering; the biological control of pests for the restoration of mainland habitats. *In* D. A. Norton (ed.), Management of New Zealand's natural estate. N.Z. Ecol. Soc. p. 89–93.
Prestidge, R. A., G. M. Barker, and R. P. Pottinger. 1991. The economic cost of Argentine stem weevil in New Zealand. *In* Proc., 44th N.Z. Weed and Pest Control Conf. August 13–15, 1991. N.Z.Weed Pest Contr. Soc. Inc. p. 165–170.
Roberts, L. I. N. 1986. The practice of biological control – implications for conservation, science and the community. The Weta, Entomol. Soc. N.Z. 9: 76–84.
Samways, M. J. 1994. Insect Conservation Biology. ed. Chapman and Hall, London, UK.
Sands, D. P. A. 1993. Effects of confinement on parasitoid-host interactions: interpretation and assessment for biological control of arthropod pests. *In* S. A. Corey, D. J. Dall, and W. M. Milne (eds.), Pest Control in Sustainable Agriculture, CSIRO, Canberra, Australia. p. 196–199.
Schulten, G. G. M. 1997. The FAO Code of Conduct for the import and release of exotic biological. *In* I. M. Smith (ed.), EPPO/CABI workshop on safety and efficacy of biological control in Europe. Blackwell Science Ltd., Oxford, UK. p. 29–36.
Shaw, S. R. 1985. A phylogenetic study of the subfamilies Meteorinae and Euphorinae. Entomography 3: 277–370.
Shaw, S. R. 1988. Euphorine phylogeny: the evolution of diversity in host–utilization by parasitoid wasps (Hymenoptera: Braconidae). Ecol. Entomol. 13: 323–335.
Simberloff, D., and P. Stiling. 1996a. How risky is biological control? Ecology 77: 1965–1974.
Simberloff, D., and P. Stiling. 1996b. Risks of species introduced for biological control. Biol. Conserv. 78: 185–192.
Smith, O. J. 1953. Species, distribution, and host records of the braconid genera *Microctonus* and *Perilitus* (Hymenoptera: Braconidae). Ohio J. Sci. 53: 173–178.
Stufkens, M. W., J. A. Farrell, and S. L. Goldson. 1987. Establishment of *Microctonus aethiopoides*, a parasitoid of the sitona weevil in New Zealand. *In* Proc., 40th N.Z. Plant Protection Soc. Conf., August 11–13, 1987. N.Z. Plant Protection Soc. Inc. p. 31–32.
Van Halteren. P. 1997. A code of conduct for the import and release of exotic biological control agents for Europe?. *In* I. M. Smith (ed.), EPPO/CABI workshop on safety and efficacy of biological control in Europe. Blackwell Science Ltd., Oxford, UK. p. 45–48.
Van Lenteren, J. C. 1997. Benefits and risks of introducing exotic macro–biological control agents into Europe. *In* I. M. Smith (ed.), EPPO/CABI workshop on safety and efficacy of biological control in Europe. Blackwell Science Ltd., Oxford, UK. p. 15–27.
Waage, J. 1997. Global developments in biological control and the implications for Europe. *In* I. M. Smith (ed.), EPPO/CABI workshop on safety and efficacy of biological control in Europe. Blackwell Science Ltd., Oxford, UK. p. 5–13.
Wapshere, A. J. 1974. A strategy for evaluating the safety of organisms for biological weed control. Ann. Appl. Biol. 77: 201–211.

6

PARASITOID DRIFT IN HAWAIIAN PENTATOMOIDS

Peter A. Follett
USDA-ARS Pacific Basin Agricultural Research Center, Hilo, Hawaii

M. Tracy Johnson and Vincent P. Jones
Department of Entomology, University of Hawaii at Manoa, Honolulu, Hawaii

INTRODUCTION

Following the accidental introduction of the pest *Nezara viridula* (L.) (Heteroptera: Pentatomidae), the southern green stink bug, into Hawaii in 1961, several parasitoids were imported and released in an effort at biological control (Davis 1964). Two parasitoids established: *Trissolcus basalis* Woolaston (Hymenoptera: Scelionidae), an egg parasitoid, and *Trichopoda pilipes* (F.) (Hymenoptera: Tachinidae), a parasitoid of adults and late-stage nymphs. By 1963, propagation facilities for these two parasitoids had been established on Oahu, Hawaii, Maui, and Kauai and releases were being made on all the major Hawaiian Islands (Davis 1964).

Most of the introduced parasitoids were propagated on *Nezara viridula*, but the use of other bugs to increase production was explored, including use of the native Hawaiian koa bug, *Coleotichus blackburniae* White (Heteroptera: Scutelleridae), and the harlequin bug, *Murgantia histrionica* (Hahn) (Heteroptera: Pentatomidae), an alien minor-pest stink bug. *Trichopoda pilipes* attacked koa bug and harlequin bug adults readily under insectary conditions but recovery of puparia was poor. *Trissolcus basalis* attacked koa bug eggs in the insectary and parasitism averaged 96 percent (Davis 1964).

Field surveys at the time these two parasitoids were being released against *N. viridula* showed high levels of parasitism of this bug in agricultural areas (Davis 1964). On Oahu in 1963, surveys indicated parasitism by *Trissolcus basalis* averaged 95 percent, and parasitism by *Trichopoda pilipes* ranged from 35 to 100 percent (Davis 1964). However, on the island of Hawaii, parasitism by *Trissolcus basalis* was lower (35–50 percent). Nishida (1966) noted that at some locations mortality caused by the big-headed ant, *Pheidole megacephala*, an accidentally introduced predator, was as high as 80–90 percent. Based on the high rates of parasitism, the *Nezara* biological control program was declared a success and terminated in 1965.

In the years following the introduction of the parasitoids, the native koa bug reportedly declined and the harlequin bug disappeared altogether (Howarth 1991). Also, certain native predaceous stink bugs (*Oechalia* spp.) and a native rhopalid *Ithamar hawaiiensis,* which were common once at lower elevations, have not been collected in the years since the parasitoids were introduced. During this same period, *N. viridula* has remained ubiquitous. Although several studies have been conducted on *N. viridula* and its parasitoids in Hawaii since the initial biological control program (Ganesalingam 1965, Nishida 1966, Shahjahan 1966, Mitchell and Mau 1971, Jones 1995), reports of nontarget attack by *Trichopoda pilipes* and *Trissolcus basalis* have been sparse and anecodotal (Gagne 1983, Funasaki et al. 1988). No data have been collected systematically to document the decline of pentatomoid (stink bugs and their relatives) populations, and no field data have been gathered to document rates of parasitism by *T. pilipes* and *T. basalis* of nontarget species. It was known at the time of their release that *T. pilipes* and *T. basalis* could attack nontarget native pentatomoids in the laboratory, but it was not considered an important concern in the field.

Stricter regulations now govern the release of biological control agents in Hawaii, yet the problem remains of how to properly assess the potential impact of released organisms. Uniform host specificity protocols are not in place, in part, because there is a lack of case studies to guide the design of pre- and post-release tests for nontarget effects. We have initiated a retrospective analysis of nontarget effects in the Hawaiian pentatomoid system, with an emphasis on the koa bug, to develop an understanding of the patterns and processes involved in "parasitoid drift" after biological control introductions. In this chapter, we first discuss parasitoid host use in general terms as a backdrop to looking in depth at the biological control program against *Nezara* and the vulnerability of the Hawaiian pentatomoids. We then discuss a series of museum, laboratory, and field studies that focus on the Hawaiian koa bug and patterns of attack by the introduced *Nezara* parasitoids. Finally, we examine the pros and cons of conducting retrospective studies of biological control programs to generate case histories that will support the planning and evaluation of future programs to reduce risk.

PARASITOID HOST USE

Most parasitoid species are restricted to attacking relatively few hosts that either share similiar life history traits or exist in a common habitat, and they usually will attack hosts within a specific range of sizes or ages (Hawkins 1994, Strand and Obrycki 1996). One or more of several patterns may appear when an introduced parasitoid begins to attack a nontarget species including host switching, host range expansion, and host shifts. Host switching is characterized by the parasitoid concentrating on the most common prey species (Murdoch 1969, Cornell 1976, Cornell and Pimentel 1978, Orr et al. 1986). This may arise with visually orienting parasitoids if a search image is formed, and, in this case, attack rates are often density dependent (e.g., Orr et al. 1986). Conditions that favor switching include a patchy distribution of prey in time and space relative to the parasitoid, and the use of sensory detection systems that work at a distance (Cornell 1976). Host range expansion occurs when a parasitoid begins to attack a new species, regardless of density, in addition to the primary host. The term host shift commonly is used to describe a change in host plant preference by a herbivore. Host shift is equally appropriate for describing a change in parasitoid preference for one host species over another, especially if the new host is preferred or, in the case of biological control introductions, if the introduced species establishes on a nontarget host and not the target (Hawkins and Marino 1997).

Any of these three overlapping patterns may operate when an introduced biological control agent begins attacking a nontarget species. We propose the term "drift" to encompass all three processes. Drift may require a genetic change (adaptation) in the biological control species after introduction to a new environment (Roderick 1992), or the species may be preadapted to attack a nontarget species. For example, preadaptation operates when a parasitoid uses a kairomone shared by the original and novel host. Drift onto a nontarget species also may result from ecological association—a species may be attacked simply because it shares a host plant with, or is in close proximity to, the primary host.

Introduced parasitoids can negatively effect target and nontarget hosts in several ways. It may kill the host outright before it can reproduce. Alternatively, there are parasitoids that do not always kill their host before the host reproduces. These parasitoids, which reduce but do not eliminate the reproductive output of the parasitized hosts, can be called "sublethal" or "nonlethal" parasitoids (DeVries 1984, English-Loeb et al. 1990). In this case, the loss of reproductive fitness by the host depends on the timing of parasitization. Other direct trophic interactions also may be involved in nontarget effects, including death through feeding or probing behavior. Indirect effects, such as interference competition or broader community level effects, also may be important, although they often are difficult to measure (Elliot et al. 1996, Lockwood, this volume). Refuges from parasitism are a major factor in determining the consequences of parasitoid attack on a host population. For hosts with sufficiently large or numerous refuges, parasitoids will be unable to exploit fully the host population and appreciably depress its density (Hawkins et al. 1993). Refuges can result from poor host searching behavior, poor synchronization with the host, differential response of host and parasitoid to climatic gradients, and other factors.

NEZARA VIRIDULA

Nezara viridula is one of the most important pentatomid pests of agricultural crops worldwide. It is common in tropical, subtropical, and temperate regions of the Americas, Africa, Australasia, and Europe, and occurs between the latitudes 45°N and 45°S (Todd 1989). *N. viridula* is highly polyphagous, with a host range that includes over 30 families of dicotyledenous plants and a number of monocots. Agricultural hosts include legumes, vegetable crops, and citrus fruits. In Hawaii, *Nezara* is a pest on mangoes, corn, tomatoes, soybeans, and green beans; it also is a major pest of macadamia nut, the largest tree crop grown in the state.

As *N. viridula* spread throughout the world and emerged as a serious pest, numerous efforts were made to establish natural enemies into newly invaded areas. Over 50 parasitoids on *N. viridula* have been recorded worldwide (Jones 1988). The most successful biological control agents have been the scelionid egg parasitoid, *Trissolcus basalis*, and two tachinid flies, *Trichopoda pilipes* and *T. pennipes*, that attack the adult and late-stage nymphs of *Nezara*. In fact, biological control of *N. viridula* has been considered a landmark example of classical biological control of an introduced pest (Caltigirone 1981), although recent studies have cast doubt on this assessment (Clarke 1990, 1992a, b; Jones 1995).

Nezara eggs are laid in clusters of 50–75. Nymphs and adults typically feed on tender growing shoots and developing fruits or pods of their hosts. In addition to the defensive scent gland secretion that is characteristic of stink bugs, *N. viridula* males elicit a sex or aggregation pheromone that is attractive to adult females and late stage nymphs (Mitchell and Mau 1969; Harris and Todd 1980). This pheromone acts as a host-finding cue (kairomone) for the tachinid *Trichopoda pilipes* (Mitchell and Mau 1971). An adhesive secretion applied to the egg chorion by *N. viridula* during oviposition is used by *Trissolcus basalis* as a host recognition kairomone (Bin et al. 1993).

THE NATIVE KOA BUG

Coleotichus blackburniae is endemic to Hawaii and is oligophagous, feeding principally on the seeds and pods of two native hosts—*Acacia koa*, the dominant tree in native forests, and the shrub *Dodonea viscosa*—and on a few imported species of *Acacia* (mainly *A. confusa* Merrill) in lowland areas (Zimmerman 1948). The koa bug is the only native herbivorous scutellerid and there are no native herbivorous pentatomids. With its large size (18 mm) and it's brilliant iridescent green, red, blue and sometimes yellow markings it is considered one of the most spectacular Hawaiian insects. The koa bug does not feed on any agricultural crops, ornamentals, or other economically important plants.

Koa bug eggs are deposited in clusters of approximately 25. Both adults and nymphs are frequently gregarious, lying closely packed when at rest (Zimmerman 1948). In the early 1900s, the koa bug could be found from sea level to 2500 m or more. Today its distribution is more fragmented, occurring on its native host plants primarily at elevations above 800 m and on *A. confusa* below 800 m. According to Howarth (1991) the koa bug declined sharply in the late 1960s and early 1970s, at the same time the *Nezara* parasites were building up. Whether the koa bug emits an aggregation pheromone which *Trichopoda pilipes* could use as a kairomone, and whether its adhesive secretion applied to eggs during oviposition is attractive to *Trissolcus basalis*, are both untested.

PARASITISM OF HAWAIIAN PENTATOMOIDS

Evidence from Museum Specimens

Evidence for nontarget attack by *Nezara* parasitoids has been accumulating in museum collections. Museum specimens of pentatomoids offer two pieces of information of potential application to the question of nontarget effects: a record of the relative abundance of various pentatomoid hosts over time, and for *Trichopoda pilipes*, a record of the presence and rate of parasitism. Claims of a decline in koa bug populations (Howarth et al. 1995) were based on specimens deposited in collections at the Bishop Museum and University of Hawaii at Manoa. Adult koa bugs are common in these collections from 1900 to 1978, but relatively few specimens have been deposited since 1978. Whether the decline in collections represents a true decline in nature is suspect. First, the time frame for the alleged decline of koa bugs lags far behind the build-up of parasitoids and high rates of *N. viridula* attack reported in the mid-1960s (Davis 1964). Secondly, no systematic surveys were conducted on all the islands to support evidence of a decline from collections. The main bias in the available information is that it focuses on one island, Oahu. As the population center of Hawaii, Oahu has undergone the greatest degree of urbanization and natural habitat degradation which may have contributed to the decline of koa bug on this island. A global decline in koa bug might not have been assumed if collections had been made from other islands over this period. For example, entomology students at the University of Hawaii at Hilo on the island of Hawaii have been collecting koa bugs from the introduced host *Acacia confusa* in the town of Hilo consistently throughout the 1980s and 1990s, although at low numbers. Our extensive survey for the past two years of koa bug habitats, particularly of the island of Hawaii, has produced hundreds of individuals at a range of elevations. This suggests that museum evidence for a decline in koa bug may have been the result of limited sampling effort.

One piece of hard evidence of nontarget parasitism is the remnant eggshells of *Trichopoda pilipes* attached to museum specimens of Heteroptera collected since the *Nezara* biological control program was initiated. *Trichopoda* females glue eggs firmly to the cuticle of the host. Eggs hatch in 1–2 days and upon hatching the larva penetrates the bottom of the eggshell and burrows into the body wall. The eggshell remains glued to the body after the larva has hatched. Using the presence of eggs attached to the cuticle as an estimate of percent parasitism is probably valid. Using dissection, Harris and Todd (1981) found that only 16.8% (n = 1000) of *N. viridula* adults carrying old tachinid eggs were not parasitized, which was offset by the finding that 16.6% of adults without tachinid eggs were parasitized. (Parasitized adults without tachinid eggshells were probably parasitized as late stage nymphs and eggshells were shed during the molt to adult.)

We conducted a survey of pentatomoids collected from 1965 to 1995 in the five principal museum collections in Hawaii to examine the range of potential hosts on which *T. pilipes* will oviposit (Table 1). Egg remnants were most commonly found on the target species, *Nezara viridula* (52/302=17.2%) and on the native koa bug (9/107 = 8.4%). Eggshells were also found on several alien nontarget pentatomids (*Brochymena quadripustulatus* [F.], *Plautia stali* Scott, and *Thyanta custator accerra* McAtee), but none was found on the native predaceous *Oechalia* spp. (the only native Hawaiian pentatomids). Specimens of several other alien bugs were inspected, including a rhopalid (*Jadera haematoloma* [Herrich-Schaeffer]; n = 30) and a reduviid (*Haemataloccha rubescens* Distant, n = 10), but no *T. pilipes* eggs were found.

Parasitization of the koa bug and other Hawaiian pentatomoids by *T. pilipes* is an example of preadapted host range expansion. When *T. pilipes* initially was imported into quarantine in Hawaii they attacked and completed development on koa bug adults in the laboratory (Davis 1964). *Trichopoda* spp. are known to attack a wide variety of stink bugs using male aggregation pheromones as a host finding cue (Mitchell and Mau 1971, Aldrich 1988). Scutellerids (such as the koa bug) generally resemble stink bugs in the arrangement and chemistry of their metathoracic glands and larval dorsal abdominal glands (Aldrich 1988), therefore, the attraction of *T. pilipes* to a presumptive pheromone from the koa bug seems likely.

Field Parasitism

Over the past several years, we have initiated long-term studies in a range of habitats reflecting the distribution of the *N. viridula* and the koa bug to study patterns of parasitism by *Trichopoda pilipes* and *Trissolcus basalis*. For three years, we have made repeated collections of adults from a koa bug population at a high elevation (Top Strip Rd., 2070 m elevation) site in Hawaii Volcanoes National Park (HAVO) on the island of Hawaii. For the past year we also have been collecting koa bugs of all ages from a mid-elevation (Hilina Pali, 670 m elevation) site in the park. HAVO extends from sea level to 4000 m and provides some of the best remaining natural habitat in the islands for the koa bug. At the high elevation site, adult parasitism by *Trichopoda pilipes* over the 3-year period ranged from 0 to 25%, with an average parasitism of 4.7% (Fig. 1; n = 15 collection dates, percent parasitism weighted by sample size), half the rate found in museum collections. This level of parasitism would not normally lead to a rapid decline in a host population (Hawkins et al. 1993). In contrast, parasitism by *T. pilipes* at the mid-elevation site (Hilina Pali Rd.) was considerably higher, ranging from 15 to 71%, with an average parasitism of 20.2% (Fig. 2; n = 8 collections, percent parasitism weighted by sample size). Collections at the mid-elevation site included adults and third, fourth, and fifth instars, and parasitism by the tachinid occured in all these stages.

From these data alone we would predict a greater impact of *T. pilpes* on the koa bug population at the mid-elevation site compared with the high elevation site. However, parasitism rates alone do not tell the whole story. Life history or demographic traits are equally as important. Our population census data indicate koa bugs may have three to four generations per year at the mid-elevation site but only one generation at the high elevation site. Shorter generation time of the koa bug at the mid-elevation site may offset the higher rate of parasitism and result in a larger, more stable local population. In fact, population densities during the summer months ap-

Table 1. *Trichopoda* eggs on Hawaiian museum specimens of pentatomoids collected between 1965 and 1995[1].

Taxon	N	Parasitized	%
ALIEN PENTATOMIDAE			
Nezara viridula	302	52	17.2
Plautia stali	160	7	4.4
Thyanta custator accerra	58	3	5.2
Brochymena quadripustulatus	62	1	1.6
Eysarcoris ventralis	3	0	0.0
NATIVE PENTATOMIDAE			
Oechalia pacifica	64	0	0.0
O. virigula	12	0	0.0
O. virescens	9	0	0.0
O. grisea	4	0	0.0
O. patreulis	4	0	0.0
O. hirtipes	3	0	0.0
NATIVE SCUTELLERIDAE			
Coleotichus blackburniae	107	9	8.4

[1]Museum specimens from Univ. Hawaii-Manoa, Univ. Hawaii-Hilo, Bishop Museum (Honolulu), Hawaii Volcanoes National Park, and Hawaii Department of Agriculture.

pear to be higher at the mid-elevation site. Therefore, a demographic approach is needed to measure the relative impact of parasitism by *T. pilipes* on the two populations. We now have established 15 koa bug sites across three islands and have begun monitoring levels of parasitism by *T. pilipes*.

T. pilipes is a nonlethal or sublethal parasitoid. Studies with *N. viridula* have shown that loss of reproductive fitness by the host depends on the timing of parasitization (Shajahan 1966). When the fly larva is developing inside an adult host, the adult remains active—feeding, mating and ovipositing—and may experience only partial loss of reproductive fitness. However, when third, fourth, or fifth instars are parasitized, later reproduction, if they reach the adult stage, will be zero (A. Taylor, personal communication). Separate estimates of parasitism must be made in nymphs and adults because *T. pilipes* eggs attached to the cuticle are lost during molts, and attack of nymphs will have a greater cost to reproduction than attack of adults. Little was known previously about *T. pilipes* attack of koa bug immatures. On two of the collection dates shown in Fig. 1 at low population densities, fifth stage nymphs were collected with adults, and three out of 39 total nymphs collected had tachinid eggs attached. On the final three dates shown in Fig. 2, when koa bug densities were high and no adults were present, parasitism of third, fourth, and fifth instars was substantial (Table 2). Dissections of nymphs and adults would be the most accurate method to determine

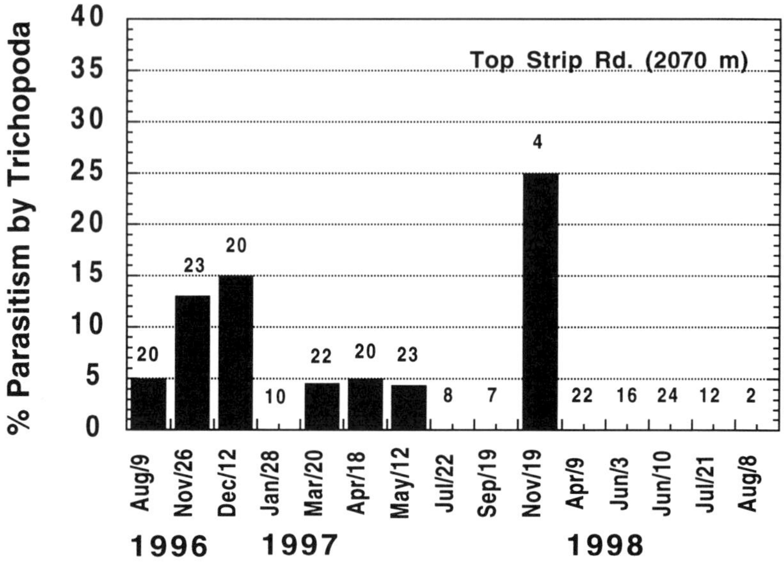

Figure 1. Parasitism by *Trichopoda pilipes* of adult koa bugs on *Dodonea viscosa* at Top Strip Rd. (2070 m), Hawaii Volcanoes National Park. Numbers over bars are the number of koa bugs observed.

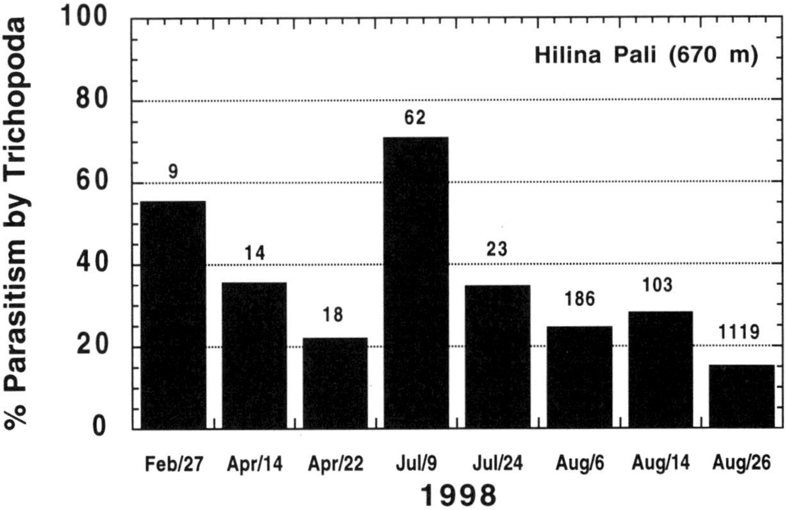

Figure 2. Parasitism by *Trichopoda pilipes* of adult koa bugs on *Dodonea viscosa* at Hilina Pali (670 m), Hawaii Volcanoes National Park. Numbers over bars are the number of koa bugs observed.

Table 2. *Trichopoda* parasitism of third, fourth and fifth stage koa bugs on *Dodonea viscosa* at Hilina Pali (670 m), Hawaii Volcanoes National Park.

Date	Nymphal stage		
	3	4	5
06 Aug 1998			
n	8	17	131
%parasitism	12.5	23.5	30.5
14 Aug 1998			
n	2	28	73
% parasitism	50.0	21.4	30.1
26 Aug 1998			
n	30	394	695
% parasitism	0.0	2.8	23.0

parasitism in the different life stages, but we prefer to sample nondestructively because the size of native koa bug populations often is small. In light of the frequency of attack of nymphs by *Trichopoda pilipes*, parasitism rates determined from adult specimens in museums clearly are underestimates of total parasitism.

T. pilipes also exhibits superparasitism which affects the success of the parasitoid and possibly the host. A single fly will lay only one egg at a time, so superparasitism represents multiple attacks. In one case, 237 eggs were found on a single field-collected *N. viridula* in Hawaii (Mitchell 1964). Although many larvae many enter the bug host only one will emerge. Shajahan (1966) showed that the rate of successful *T. pilipes* pupation was 78% when one larva attacked its *N. viridula* host and 38% when 10 larvae attacked the host. In our inspection of museum specimens, the mean number of eggs on parasitized adult *N. viridula* and koa bugs was 6.5 (range 1–43, n = 50) and 2.0 (range 1–3, n = 9), respectively. It is not known how superparasitism affects host reproductive output.

Parasitoid drift in Hawaiian pentatomoids may be sex-biased and chemically mediated. *T. pilipes* attraction to the male aggregation pheromone of *N. viridula* may lead the fly to oviposit on alternate host stink bugs that emit a similar pheromone, or to oviposit on other insects residing on the same host plant as *N. viridula* (Aldrich et al. 1987). It is not known at this time whether the native Hawaiian koa bug emits such a pheromone, but *T. pilipes* attack in Hawaii appears to be strongly male-biased in koa bug as it is with *N. viridula* (Mitchell and Mau 1971). Eleven parasitized adult koa bugs were collected from the high-elevation site in HAVO between August 1996 and August 1998; of these, ten were males and one was female while the overall sex ratio of collected adults was approximately 1:1 (n = 233). Another possibility is that parasitoid drift is the result of ecological association. In Hawaii, *T. pilipes* occasionally oviposits on hosts other than pentatomoids that are found in pentatomoid habitats or on pentatomoid host plants. For example, *T. pilipes* eggs were found on the Australian mantis, *Tenodera australasiae* (Hardy 1981), and the endemic mirid, *Hyalopeplus pellucidus* (Gagne 1983), and eggs have been observed on grasshoppers and cockroaches on the *Nezara* host plant *Crotalaria* (R. Mau, personal communication). *N. viridula* is polyphagous but rarely feeds on the koa bug's host plants (*Acacia* spp. and *Dodonea viscosa*), so host switching (or host range expansion) originally was probably mediated by a koa bug aggregation pheromone rather than by host plant cues (Vet and Dicke 1992). However, *T. pilipes* typically flies to *D. viscosa* shrubs with ripe berries then conducts a within-shrub search, while avoiding nonfruiting plants. This behavior suggests plant cues are an important component to host finding.

Establishing past nontarget effects of the egg parasitioid *Trissolcus basalis* poses a greater problem because it has left no footprint. *T. basalis* is known to attack a broad range of pentatomid eggs (Jones 1988). However, except for *N. viridula*, nothing is known from Hawaii about the susceptibility of pentatomoid eggs to attack by *T. basalis*, or the prevalence of attack in the field. Between August 1997 and July 1998 we looked for koa bug egg parasitism in koa bug habitats at different elevations (Table 3). We either observed koa bug eggs that had been laid naturally, or transferred fresh eggs onto koa bug host plants. Parasitism at two low elevation sites (50 m) in Hilo on the island of Hawaii ranged from 0 to 50% (Table 3), which suggests that koa bug eggs are attacked readily by *T. basalis* when they are found. No parasitism was observed at the higher elevation (850–2070 m) sites in HAVO, indicating spatial or temporal refuges from egg parasitism by *T. basalis* may exist. This result has been corroborated by additional studies in which sentinel egg masses of *N. viridula* were placed along elevational transects between 50–2070 m; *T. basalis* attack of *N. viridula* eggs was observed at low-elevation sites in Hilo, but not at higher elevations (Johnson and Follett, unpublished data). Refuge theory predicts that hosts that have a substantial portion of their population in refuges will be less exploitable than hosts with small refuges (Hawkins et al 1993). The high-elevation refuge from *T. basalis* attack should prevent global extinction of koa bug by this parasitoid. As with *Trichopoda pilipes*, host range expansion in *Trissolcus basalis* in Hawaii may be chemically mediated if the adhesive secretion the koa bug applies to its eggs has kairomonal properties similar to that of *N. viridula* (Bin et al. 1993, Strand and Vinson 1993).

An accidentally introduced parasitoid, *Anastatus* sp., also attacks koa bug eggs in Hawaii (Table 3), and predation is common at some higher elevations (three collections, HAVO July 1998, Table 3). Predators collected at our study sites included coccinellids, vespids, and ants. Ants, especially *Pheidole megacephala*, previously have been reported as key natural enemies in macadamia orchards in Hawaii (Jones 1995). Life table studies are in progress to separate the relative importance of parasitoids and various predators at different elevations.

Laboratory Host Preference Study

A common method to screen for nontarget effects for biocontrol agents is to expose the agent to a range of potential hosts in cages in a quarantine laboratory. For simplicity and due to space limitations, confining a parasitoid and its host in small cages is the

Table 3. Natural enemy attack of koa bug eggs determined by placement of eggs or observation of naturally occurring eggs.

Site	Date	Elev. (m)	Host[1]	No. egg masses	No. eggs	Normal eclosion		*Trissolcus basalis*		*Anastatus* sp.		Predation[2]	
						% egg masses[3]	% eggs[4]	% egg masses attacked	% eggs attacked	% egg masses attacked	% eggs attacked	% egg masses attacked	% eggs attacked
Hilo-A	Aug97	30	*Ac*	11	333	45.0	46.9	45.5	48.1	0	0	0	0
Hilo-A	Aug97	30	*Ac*	13	414	100	94.7	0	0	0	0	0	0
HAVO[5]	Sep97	2070	*Dv*	4	77	—[6]	—	0	0	0	0	0	0
Hilo-B	Sep97	30	*Ac*	4	78	—[6]	—	0	0	0	0	0	0
Hilo-B	Apr98	30	*Ac*	4	151	100	97.4	50.0	16.7	0	0	0	0
Hilo-B	Jun98	30	*Ac*	7	228	71.4	64.0	28.6	15.4	28.6	14.9	14.3	0.4
HAVO	Jul98	1180	*Ak*	12	380	100	79.2	0	0	0	0	75.0	9.2
HAVO	Jul98	880	*Ak*	33	1047	75.8	38.3	0	0	6.1	1.3	90.9	33.6
HAVO	Jul98	850	*Ak*	17	521	100	71.4	0	0	0	0	58.8	7.9

[1] Host plants: *Dv = Dodonea viscosa, Ac = Acacia confusa, Ak = Acacia koa.*
[2] Evidence of punctures or grazing; possible predators include ants, coccinellids, and vespids.
[3] Percent egg masses with at least one normally hatching egg.
[4] Percent of total eggs hatching normally.
[5] Hawaii Volcanoes National Park.
[6] Eggs were refrigerated at 4.5°C before use, which affected normal hatch.

norm. This sort of forced interaction may result in unrealistically high rates of host acceptance and parasitism (e.g., Duan and Messing 1997a). Nevertheless, confined cage studies can be used to demonstrate parasitoid host acceptance and development on novel hosts, and in some cases can demonstrate preference for one host species over another (e.g., Barratt et al. 1997).

We conducted host choice tests in a 50x50x50 cm wooden cage fitted with cloth sleeves, parasitoid screen and a plexiglas ceiling. To test *Trissolcus basalis* preference for *N. viridula* and koa bugs eggs on different hosts, one 1–2 day old egg mass of each bug species was pinned to the underside of separate leaves on 20 cm sprigs of *Acacia confusa* (a koa bug host) and *Amaranthus* sp. (a *Nezara* host). For each test, a sprig of each of the two plant species with egg masses attached was placed in the cage with 10 inexperienced 2–5-day-old *T. basalis* females. After 24 hours, leaves with eggs were removed and held in petri dishes for parasitoid emergence. The test was replicated four times.

Untransformed data were subjected to analysis of variance (ANOVA) after the Shapiro-Wilks W test for normality was performed (SAS Institute 1994). Two-way ANOVA with interaction was used to test for the effects of insect species, plant species, and insect by plant interaction on percent parasitism. ANOVA indicated significant effects of insect species ($F = 17.6$; df = 1; $P = 0.01$), plant species ($F = 33.9$; df = 1; $P = 0.004$) and the insect by plant interaction ($F = 8.2$; df = 1; $P = 0.05$). *N. viridula* eggs were parasitized at a higher rate than koa bug eggs on both plant species, and *Acacia confusa* was a preferred foraging substrate for *T. basalis* compared to *Amaranthus* sp. (Fig. 3). The lower parasitism of eggs placed on *Amaranthus* compared with *Acacia confusa* is counterintuitive since *Amaranthus* is a highly preferred host for *N. viridula.*

Several factors are known to influence the outcome of choice tests for host preference/specificity including the degree of host deprivation; parasitoid age and egg load; densities of parasitoids and hosts; length of testing time; and the size and complexity of the testing arena (e.g. Duan and Messing 1997b). An important question is whether parasitoid behavior in the laboratory under confined conditions and forced interactions with a host correlates well with field behavior and ecological opportunity. As suggested by Howarth (1991) and Secord and Karieva (1996) it would be beneficial to conduct more field studies in the geographic region of origin of the biological control agents. One of the problems with this is that one would not be permitted to release elsewhere a nontarget insect species that is endemic to the potential new home of the biological control agent (such as the koa bug). In the case of *Trissolcus basalis*, studies have shown that frozen *N. viridula* and koa bug eggs (that will not hatch) are as attractive as fresh eggs, so field releases are safe and this limitation can be avoided. The responsiveness of *T. basalis* to frozen eggs allows us to conduct comparative studies with pest and native pentatomoids in nature preserves in Hawaii that normally would not allow the release of a pest species.

RETROSPECTIVE STUDIES

A wealth of information on the patterns and processes of nontarget effects remains unexcavated in the outcomes of previous biological control introductions. Retrospec-

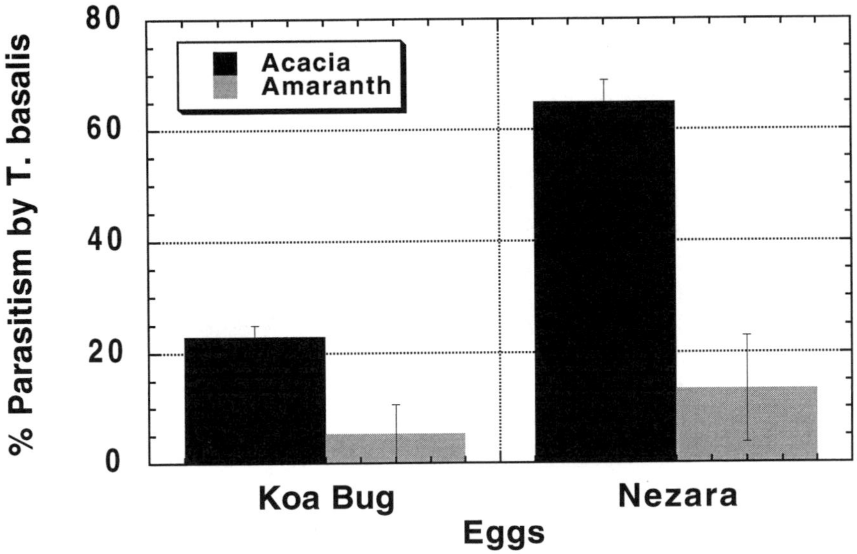

Figure 3. Host choice test for *Trissolcus basalis* attacking *Nezara viridula* and koa bug eggs on two host plants, *Acacia confusa* and *Amaranthus* sp.

tive studies that examine the outcomes of previous biological control programs relative to environmental effects would generate case histories to support the design of host testing protocols and pre- and post-release monitoring procedures for vulnerable nontarget species or habitats (Follett et al., in press). A first step in retrospective studies is to determine the distribution of the introduced parasitoid, its intended hosts, and potential nontarget hosts. Over the geographical range of the parasitoid, the suite of hosts may vary and studies should therefore be replicated in space and time. The evaluation of nontarget effects ideally should consider the complete set of mortality agents and examine the relative influence of each on nontarget population dynamics. Life tables provide a useful framework to evaluate nontarget mortality after parasitoid release, as they allow us to examine the ecological role a natural enemy plays in a given system (Carey 1993, Bellows et al. 1992). In some systems, more sensitive and data-hungry approaches such as constructing food webs (Memmott, this volume) or system process analysis (Lockwood, this volume) may be more revealing.

Although retrospective studies may offer some insights, there are several problems to consider. The nontarget species may now be absent from all or part of its range, which poses obvious problems for conducting studies to establish a causative relationship between nontarget decline and the biological control agent. For example, the harlequin bug was a minor pest in Hawaii of several crops until it rapidly declined about the time the *Nezara* parasitoids were introduced. Some believe the harlequin bug was driven to extinction by *Trichopoda* (F. Howarth, personal communication), but its decline happened so rapidly that data were not collected. Whether *Trichopoda* or *Trissolcus basalis* or some other factor caused its disappearance will remain a mystery.

Another problem is that the nontarget species and its natural enemy may be so rare that it is difficult to conduct experiments with statistical rigor, particularly if laboratory rearing is difficult or impractical. If a sufficient number of individuals or populations are present, a historical problem arises—current rates of parasitism in the field may not accurately reflect the parasitoid's previous parasitism rate and its original potential to destabilize nontarget populations. Also, rarity often implies that the geographical distribution of the nontarget species is now fragmented or collapsed relative to its original state, and hence, the impact of the parasitoid may have been greatest over the part of the nontarget's range where it no longer survives (Kruess and Tscharntke 1994). On the other hand, if another factor, such as habitat loss (Onstad and McManus 1996, Hopper 1995), caused the population decline in the first place, parasitism may be more or less intense where the nontarget presently survives compared with its original state. The time frame for the retrospective study may be misleading in short-term studies, as nontarget effects may be delayed (Louda et al. 1997). Also, post-colonization evolution of the nontarget host or parasitoid may lead to false conclusions about the host/parasitoid interaction (Roderick 1992). In Hawaii, an important additional factor is the accidental introduction of alien keystone species after the time the biological control agent is introduced. Ants, predatory wasps, and other inadvertently introduced parasitoids can have devastating impacts on native insect species that outweigh the importance of purposely-introduced species (Gambino et al. 1990, Cole et al. 1992, Howarth et al. 1995, Jones 1995). If the actions of these other natural enemies are in flux, their current impact may not reflect their relative importance at an earlier time. Using hindsight, we can only speculate on the fate of populations and suggest probable scenarios of regulating factors.

CONCLUSIONS

The Hawaiian pentatomoids are popularly cited by scientists in Hawaii as an example in which biological control introductions have lead to the decline of native species (Howarth 1983, Gagne and Howarth 1985, Howarth 1991, Howarth et al. 1995). *Trichopoda pilipes* is cited most often as the culprit, probably because it is a generalist that attacks adult bugs (the most frequently collected life stage) and the eggs it attaches to the host cuticle leave a permanent and easily visible record of attack. In contrast, eggs of pentatomoids are less commonly collected or recognized in the field, so egg parasitism by *T. basalis* of Heteroptera other than *Nezara viridula* has received little attention (Follett et al., in press). Data on attack of bugs by generalist predators such as ants and wasps are rarer still.

Our work with the Hawaiian pentatomoid system is at an early stage, but some patterns have emerged. The koa bug is attacked in nature by the introduced *Nezara* parasitoids, *Trichopoda pilipes* and *Trissolcus basalis*. Parasitism rates of koa bug appear to be a function of elevation. Above 800 m, where much of the remaining pristine habitat of the koa bug exists, attack by *Trichopoda pilipes* is generally low and *Trissolcus basalis* appears to be absent. At lower elevations, parasitism rates of koa bug by these two parasitoids can be substantial, and other natural enemies such as ants can be important mortality factors as well. Hence, elevation gradients create refuges that may be critical in reducing the global impact of *T. pilipes* and *T. basalis* on the koa bug.

Did *Nezara* parasitoids cause koa bug populations to decline, fragment, or collapse? We cannot answer the question definitively without having tracked parasitism rates before and after introduction of the parasitoids. This is a shortcoming of most retrospective analyses of biological control introductions. If we assume that present day patterns of parasitoid/host interactions are congruent with the past, we can infer the level of nontarget impact. The more modest question becomes, Do *Trichopoda pilipes* and *Trissolcus basalis* attack the koa bug in nature and are they substantial, persistent mortality factors? *T. pilipes* and *T. basalis* are permanently established in koa bug habitats, which typically are distinct from *N. viridula* habitats. Future life table studies will separate the importance of different sources of koa bug mortality and estimate the relative impact of the parasitoids on the koa bug in a wide range of habitats. Studies of purposely introduced biological control agents should not overshadow studies of other natural enemies; the invasion of koa bug habitats by alien keystone predators such as ants poses perhaps the greatest risk to the long-term stability of koa bug populations. Continuing habitat degradation could compound the negative effects of enemy attack.

The Hawaiian pentatomoids is an ideal system to test hypotheses about parasitoid drift because it provides a useful array of ecological and biological features for comparative studies: both predatory and herbivorous feeders occur; native and alien species coexist; variation in size, color, and various biological attributes exists among species; some bug species are polyphagous and others oligophagous; and many species are found at a range of elevations (sea level to 2000 m or more) and on the dry and wet sides of the islands. From an experimentalist's viewpoint, Hawaii is a unique laboratory for studying the effects of biological control introductions because releases were often made (as it was with *Nezara* parasitoids) on different islands, allowing for replicated field studies. With the list of arthropod biological control introductions into Hawaii numbering close to 700 over the past 100 years (Funasaki et al. 1988, Follett et al., in press), there are many other systems that lend themselves to retrospective analysis.

ACKNOWLEDGMENTS

We appreciate of the efforts of Gwen Ochikubo who maintained our *Trissolcus basalis* and *Nezara viridula* colonies during the early stages of our work. Andrew D. Taylor graciously supplied *Trichopoda pilipes* and a back-up colony of *N. viridula*. David Foote provided logistical support at Hawaii Volcanoes National Park.

REFERENCES

Aldrich, J. R. 1988. Chemical ecology of the Heteroptera. Annu. Rev. Entomol. 33: 211–238.

Aldrich, J. R., J. E. Oliver, W. R. Lusby, and J. P. Kochansky. 1987. Pheromone strains of the cosmopolitan pest, *Nezara viridula* (Heteroptera: Pentatomidae). J. Exper. Zool. 244: 171–175.

Bellows, T. S. , R. G. Van Driesche, and J. S. Elkinton. 1992. Life–table construction and analysis in the evaluation of natural enemies. Annu. Rev. Entomol. 37: 587–614.

Barratt, B.I.P., A. A. Evans, C. M. Ferguson, G. M. Barker, M. R. McNeill, and C. B. Phillips. 1997. Laboratory nontarget host range of the introduced parasitoids *Microctonus aethiopodes* and *Microctonus hyperodae* (Hymenoptera: Braconidae) compared with field parasitism in New Zealand. Environ. Entomol. 26: 694–702.

Bin, F., S. B. Vinson, M. R. Stand, S. Colazza, and W. A. Jones. 1993. Source of an egg kairomone for *Trissolcus basalis*, a parasitoid of *Nezara viridula*. Physiol. Entomol. 18: 7–15.

Caltigirone, L. E. 1981. Landmark examples in classical biological control. Annu. Rev. Entomol. 26: 213–232.

Carey, J. R. 1993. Applied demography for biologists. Oxford University Press, New York.

Clarke, A. R. 1990. The control of *Nezara viridula* L. with introduced egg parasitoids in Australia. A review of a "landmark" example of classical biological control. Aust. J. Agric. Res. 41: 1127–1146.

Clarke, A. R. 1992a. Current distribution and pest status of *Nezara viridula* (L.) (Hemiptera: Pentatomidae) in Australia. J. Aust. Entomol. Soc. 31: 289–297.

Clarke, A. R. 1992b. Investigations into the biological control of *Nezara viridula* (L.) by the egg parasitoid *Trissolcus basalis* (Wollaston) in Eastern Australia. Unpublished Ph.D. dissertation. Univ. Queensland, Brisbane, Australia.

Cole, F. R., A. C. Medeiros, L. L. Loope, and W. W. Zuehlke. 1992. Effects of the Argentine ant on arthropod fauna of Hawaiian high-elevation shrubland. Ecology 73: 1313–1322.

Cornell, H. V. 1976. Search strategies and the adaptive significance of switching in some general predators. Amer. Nat. 110: 317–320.

Cornell, H. V., and D. Pimentel. 1978. Switching in the parasitoid *Nasonia vitripennis* and its effect on host competition. Ecology 59: 297–308.

Davis, C. J. 1964. The introduction, propagation, liberation, and establishment of parasites to control *Nezara viridula* variety *smaragdula* (Fabricius) in Hawaii (Heteroptera: Pentatomidae). Proc. Hawaiian Entomol. Soc. 18 (3): 369–375.

DeVries, P. J. 1984. Butterflies and Tachinidae: does the parasite always kill the host? J. Nat. Hist. 18: 323–326.

Duan, J. J., and R. H. Messing. 1997a. Effects of two opiine parasitoids (Hymenoptera: Braconidae) introduced for fruit fly control on a non-target native Hawaiian tephritid, *Trupanea dubautiae* (Diptera: Tephritidae). Biol. Control 8: 177–184.

Duan, J. J., and R. H. Messing. 1997b. Biological control of fruit flies in Hawaii: factors affecting non–target risk analysis. Agric. Human Values 14: 227–236.

Elliott, N. C., R. W. Kieckhefer, and W. C. Kauffman. 1996. Effects of an invading coccinellid on native coccinellids in an agricultural landscape. Oecologia 105: 537–544.

English–Loeb, G. M., R. Karban, and A. K. Brody. 1990. Arctiid larvae survive attack by a tachinid parasitoid and produce viable offspring. Ecol. Entomol. 15: 361–362.

Follett, P. A., J. J. Duan, R. H. Messing, and V. P. Jones. In press. Parasitoid drift after biological control introductions: re-examining Pandora' s box. Am. Entomologist.

Funasaki, G. Y., P-Y Lai, L. M. Nakahara, J. W. Beardsley, and A. K. Ota. 1988. A review of biological control introductions in Hawaii: 1890–1985. Proc. Hawaiian Entomol. Soc. 28: 105–160.

Gagne, W. C. 1983. Notes and exhibitions. Proc. Hawaiian Entomol. Soc. 24: 191.

Gagne, W. C., and F. G. Howarth. 1985. Conservation status of endemic Hawaiian Lepidoptera. pp. 74–84, *In* Proc. 3rd Congr. European Lepidopterology Cambridge. 1982. Karlsruhe: Societas Europaea Lepidopterologica.

Gambino, P., A. C. Medeiros, and L. L. Loope. 1990. Invasion and colonization of upper elevations on East Maui (Hawaii) by *Paravespula pensylvanica* (Hymenoptera: Vespidae). Ann. Entomol. Soc. Am. 83: 1088–1095.

Ganesalingam, V. K. 1965. Some environmental factors influencing parasitization of the eggs of *Nezara viridula* Linnaeus (Pentatomidae) by the egg parasite *Telenomus basalis* Wollaston (Hymenoptera: Scelionidae). M.S. Thesis, Univ. Hawaii at Manoa, Honolulu.

Gould, F., G. G. Kennedy, and R. Kopanic. 1996. Environmental issues associated with enhancing the impacts of biological control agents. Am. Entomol. 42: 160–173.

Hardy, D. E. 1981. Insects of Hawaii, Diptera: Cyclorrapha IV. 14: 420–425.

Harris, V. E., and J. W. Todd. 1980. Male-mediated aggregation of male, female and 5th instar southern green stink bugs and concomitant attraction of a tachinid parasite, *Trichopoda pennipes*. Entomol. Exp. Appl. 27: 117–126.

Harris, V. E., and J. W. Todd. 1981. Validity of estimating percentage parasitization of *Nezara viridula* populations by *Trichopoda pennipes* using parasite-egg presence on the host cuticle as the indicator. J. Georgia Entomol. Soc. 16: 505–510.

Hawkins, B. A. 1994. Pattern and process in host-parasitoid interactions. Cambridge Univ. Press, Cambridge, UK.
Hawkins, B. A., M. B. Thomas, and M. E. Hochberg. 1993. Refuge theory and biological control. Science 262: 1429–1432.
Hawkins, B. A., and P. C. Marino. 1997. The colonization of native phytophagous insects in North America by exotic parasitoids. Oecologia 112: 566–571.
Hopper, K. R. 1995. Potential impacts on threatened and endangered insect species in the United States from introductions of parasitic Hymenoptera for the control of insect pests. *In* H. M. T. Hokkanen and J. M. Lynch (eds.), Biological control: benefits and risks. Cambridge Univ. Press, Cambridge, UK.
Howarth, F. G. 1991. Environmental impacts of classical biological control. Annu. Rev. Entomol. 36: 485–509.
Howarth, F. W. 1983. Classical biocontrol: panacea or pandora's box? Proc. Hawaiian Entomol. Soc. 24: 239–244.
Howarth, F. G., G. Nishida, and A. Asquith. 1995. Insects of Hawaii. *In* E. T. Laroe et al. (eds.), Our living resources, U.S. Dept. Interior, Ntl. Biol. Svc., Washington, D.C. p. 365–368.
Jones, V. P. 1995. Reassessment of the role of predators and *Trissolcus basalis* in biological control of southern green stink bug (Hemiptera: Pentatomidae) in Hawaii. Biol. Control 5: 566–572.
Jones, W. A. 1988. World review of the parasitoids of the southern green stink bug, *Nezara viridula* (L.) (Heteroptera: Pentatomidae). Ann. Entomol. Soc. Am. 81: 262–273.
Kruess, A., and T. Tscharntke. 1994. Habitat fragmentation, species loss, and biological control. Science 264: 1581–1584.
Louda, S. M., D. Kendall, J. Connor, and D. Simberloff. 1997. Ecological effects of an insect introduced for the biological control of weeds. Science 277: 1088–1090.
Mitchell, W. C. 1964. Note, Proc. Hawaiian Entomol. Soc. 18: 345.
Mitchell, W. C., and R. F. L. Mau. 1971. Response of the female southern green stinkbug and its parasite, *Trichopoda pennipes*, to male stink bug pheromones. J. Econ. Entomol. 64: 856–859.
Murdoch, W. W. 1969. Switching in general predators: experiments on predator specificity and stability of prey populations. Ecol. Monogr. 39: 335–354.
Nishida, T. 1966. Behavior and mortality of the southern green stink bug *Nezara viridula* in Hawaii. Res. Pop. Ecol. 8: 78–88.
Onstad, D. W., and M. L. McManus 1996. Risks of host range expansion by parasites of insects. Bioscience 46: 430–435.
Orr, D. B., J. S. Russin, D. J. Boethel, and W. A. Jones. 1986. Stink bug (Hemiptera: Pentatomidae) egg parasitism in Louisiana soybeans. Environ. Entomol. 15: 1250–1254.
Roderick, G. K. 1992. Post-colonization evolution of natural enemies. *In* W. C. Kauffman and J. E. Nechols (eds.), Thomas Say Proceedings: Selection criteria and ecological consequences of importing natural enemies. Entomol. Soc. Am., Lanham, Maryland.
SAS Institiute. 1998. JMP user's guide, version 3, SAS Institute, Inc., Cary, North Carolina.
Secord, D., and P. Kareiva. 1996. Perils and pitfalls in the host specificity paradigm. BioScience 46: 436–447.
Shahjahan, M. 1966. Some aspects of the biology of *Trichopoda pennipes* Fabricius (Diptera, Tachinidae), a parasite of *Nezara viridula* (Fabricius) in Hawaii. M.S. Thesis, Univ. Hawaii at Manoa, Honolulu.
Strand, M. R., and J. J. Obrycki. 1996. Host specificity of insect parasitoids and predators. BioScience 46: 422–429.
Strand, M. R., and S. B. Vinson. 1983. Factors affecting host recognition and acceptance in the egg parasitoid *Telenomus heliothidis* (Hymenoptera: Scelionidae). Environ. Entomol. 12: 1114–1119.
Todd, J. W. 1989. Ecology and behavior of *Nezara viridula*. Annu. Rev. Entomol. 34: 273–292.
Vet, L. E., and M. Dicke. 1992. Ecology of infochemical use by natural enemies in a tritrophic context. Annu. Rev. Entomol. 37: 141–172.
Zimmerman, E. C. 1948. Insects of Hawaii: Heteroptera (vol. 3). Univ. Hawaii Press, Honolulu, USA.

7

EVALUATING NONTARGET EFFECTS OF CLASSICAL BIOLOGICAL CONTROL: FRUIT FLY PARASITOIDS IN HAWAII AS A CASE STUDY

Jian J. Duan and Russell H. Messing
Department of Entomology, Kauai Agricultural Research Center, University of Hawaii, Kapaa, Hawaii

INTRODUCTION

Classical biological control is a method of pest management that employs the deliberate introduction of exotic living organisms to suppress pest species. The ecological basis of classical biological control rests on the principle that many invasive pest populations are limited in their native habitat by the action of upper trophic level organisms (often termed natural enemies), or in some cases, by competitors within the same trophic level (Van Driesche and Bellows 1996). Many species achieve pest status when they are moved (accidentally or intentionally) to a new location without their natural enemies, and where local natural enemies are ineffective in suppressing them. The anticipated outcome of a biological control introduction is the establishment of a new balance with both the pests and natural enemies coexisting at low densities.

Although throughout its history it has been viewed as an environmentally safe alternative to chemical pest control, classical biological control is not free of risk. There are generally two categories of risk associated with biological control introductions: (1) accidental introductions of unwanted species, and (2) introduction of agents that also attack nontarget species of economic or ecological value. Van Driesche and Benson (1999) recently reviewed some methods for mitigating these two types of risk. In general, the first type can be managed by the means of "safe-guard" quarantine procedures and by studying the agents' biology in a quarantine laboratory before release. The second type, however, is more difficult to manage because the same agent may pose variable risks in different ecological contexts and because it is not possible to be completely certain from laboratory host-specificity studies what ecological interactions will occur once the agent is released into a new community. Thus, the heart of the matter of nontarget impacts of biological control, as Lockwood (1997) put it, is "a reflection of the fundamental challenge facing ecology."

Classical weed biological control has long sought to reduce nontarget risks by screening the herbivorous agents (arthropods and pathogens) based on host-specificity testing (e.g., Wapshere 1974, Harris 1990). It currently has a fairly defined system in place for managing direct nontarget risks of the concerned agents, although it still has limitations for predicting indirect nontarget impacts (see review in Marohasy 1998). Classical biological control of arthropod pests, in contrast, does not have a comparable system as yet. This is in part because predaceous arthropod agents have been historically considered "low-risk," and host specificity testing with those agents has been generally not required, nor used to estimate potential nontarget risks. Currently, however, critics of biological control in general (Lockwood 1996, Simberloff and Stiling 1996) and on island ecosystems in particular (e.g, Howarth 1991, Asquith 1995), have increasingly voiced concerns about the potential impacts of introduced arthropod predators and parasitoids on nontarget invertebrates of ecological and economic value.

In this chapter, we will use biological control of fruit fly (Diptera: Tephritidae) pests in Hawaii as a case study to discuss the development of approaches for evaluating nontarget impacts of entomophagous insect introductions. First, we briefly review the history of biological control of tephritid fruit fly pests in Hawaii, and the current concerns regarding its potential nontarget impacts. Secondly, based on our current knowledge and findings from recent nontarget studies, we examine what impacts the previously introduced fruit fly parasitoids may have, or have had, on the majority of nonpest tephritids. Finally, we discuss some of the important ecological issues relating to the development of efficient protocols for estimating nontarget risks of insect parasitoid introductions.

CLASSICAL BIOLOGICAL CONTROL OF TEPHRITID PESTS IN HAWAII: A BRIEF HISTORY AND NONTARGET CONCERNS

Beginning in the late 1800s, four exotic species of tephritid pests invaded Hawaii, causing severe damage to fruit and vegetable crops and also posing threats to the southern U.S. mainland. These tephritids included melon fly (*Bactrocera cucurbitae*), Mediterranean fruit fly (medfly, *Ceratitis capitata*), Oriental fruit fly (*B. dorsalis*), and solanum fly (*B. latifrons*). Extensive classical biological control programs were conducted in Hawaii against the first three species between 1913 and 1950 (see reviews in Clausen 1978, Wharton 1989). These resulted in the importation and release of over 30 species of exotic hymenopteran parasitoids from six families: Braconidae, Eulophidae, Pteromalidae, Chalcididae, Diapriidae, and Eucoilidae (Silvestri 1914;

Table 1. Tephritid flies endemic to Hawaii and their host-plant associations.

Genus	No. species*	Host plants	Association with host plant
Neotephritis	2	*Argyroxiphium* sp.	Stem miners
Phaeogramma	3	*Bidens* sp.	Stem gall former
Trupanea	21	*Argyroxyphium* sp. *Artemisia* sp. *Bidens* sp. *Dubautia* sp.	Flowerhead or seed feeder (19 spp.) and stem gall former (2 spp.)

* Individual species names can be found in Hardy and Delfinado (1980).

Bess et al. 1961; Clausen 1956, 1978; Funasaki et al. 1988). These parasitoids came from many foreign locations, including India, Southeast Asia, Africa, and Australia. Six opiines (Family Braconidae, subfamily Opiinae) and one eulophid (*Tetrastichus giffardianus*, Family Eulophidae) became widely established in Hawaii and contribute significantly to the reduction of pest populations (Newell and Haramoto 1968, Wong et al. 1984, Wong and Ramadan 1987). For example, it has been demonstrated that the egg parasitoid *Fopius arisanus* (imported to Hawaii from Asia) kills over 90% of the Oriental fruit fly eggs in some crops (Newell and Haramoto 1968).

Classical biological control programs against alien fruit fly pests conducted earlier in the century paid scant attention to the possible environmental impacts of introduced natural enemies (parasitoids). At the time when the parasitoids were introduced, host-specificity tests were conducted only with targeted frugivorous tephritid pests.

Hardy and Delfinado (1980) described 33 species of nonpest tephritids in Hawaii, including 26 Hawaiian endemic species (Table 1), and five deliberately introduced and two inadventently introduced weed control agents (Table 2). Some endemic species (in the genus *Phaeogramma*) are candidates for listing as endangered species by the U.S. Fish and Wildlife Service (1996). Clearly, future fruit fly biological control programs need to address the potential impacts of introduced parasitoids on these nontarget tephritid species.

NONTARGET IMPACTS OF PREVIOUSLY INTRODUCED FRUIT FLY PARASITOIDS

An important step towards resolving the current controversy about the nontarget risks of arthropod biological control introduction is to investigate what impacts previously introduced natural enemies have had on nontarget species in regions where the natural enemies have been successfully established after their release (e.g., Hawkins and Marino 1997). From such retrospective studies, we may learn a great deal about the interaction of introduced biological control agents with ecological communities and thus gain insight into the development of sound protocols for analyzing nontarget risks in future biological control introductions. The introduction and establishment of exotic fruit fly parasitoids in Hawaii in the first half of the century provides us with an excellent opportunity to conduct retrospective analyses of nontarget impacts of previous introductions.

Table 2. Beneficial tephritids established for biological weed control in Hawaii.

Tephritid species	Year first released or recorded	Country of origin	Target weed	Target-weed family	Association with host plant
Deliberately introduced species					
Eutreta xanthochaeta (lantanagall fly)	1902	Mexico	*Lantana camara*	Verbenaceae	stem gall maker
Procecidochares utilis (Maui pamakani gall fly)	1944	Mexico	*Ageratina adenophorum*	Asteraceae	stem gall maker
Acinia picturata	1959	Guatemala	*Pluchea odorata*	Asteraceae	flowerhead feeder
Tetraeuaresta obscuriventris	1961	Fiji	*Elephantopus mollis*	Asteraceae	flowerhead feeder
Procecidochares alani (Hamakua pamakani gall fly)	1973	Mexico	*Ageratina riparia*	Asteraceae	stem gall maker
Adventive species					
Dioxyna sororcula	1966	?	*Bidens pilosa*	Asteraceae	Flowerhead feeder
Ensina sonchi	1968	Europe or Asia	*Sonchus oleraceus*	Asteraceae	Flowerhead feeder

To determine the impacts of previously introduced fruit fly parasitoids, we first conducted field surveys to determine the *status quo* with the extant parasitoid guilds attacking nontarget tephritids. Based on findings of these surveys, we then conducted experiments to examine the degree to which previously introduced parasitoids have impacted populations of associated nontarget hosts.

Parasitoid Guilds Presently Attacking Nonpest Tephritids

Nonpest tephritids in Hawaii are all vegetative feeders, and can be classified according to microhabitat as gall formers, flowerhead feeders, and stem miners. Surveys are not yet complete, but for those species that have been studied, available data reveal some useful patterns regarding parasitism of these nontarget tephritids:

(a) Seven endemic flowerhead-feeders in the genus *Trupanea* and one endemic gall-former (*Phaeograma lortnocoibon*) were recently surveyed on the island of Kauai and Maui (John Herr, unpublished data, Duan et al. 1996, Duan and Messing 1998; Table 3). While no deliberately introduced fruit fly parasitoids were recovered from any of these flies, two adventive parasitoids, *Habrocytus elevatus* (Hymenoptera: Pteromalidae) and *Euderus metallicus* (Hymenoptera: Eluphidae) were frequently recovered from field-collected flowerheads infested with *Trupanea* flies. *E. metallicus* and a torymid, *Torymus advenus* (Torymidae) were recovered from stem galls formed by *P. lortnocoibon* on the endemic composite, *Bidens cosmoides*.

(b) Extensive surveys on the deliberately introduced weed control tephritids (three stem gall forming species) were conducted by Bess and Haramoto (1959, 1972), Wong et al. (1991), and Duan et al. (1996) on several major Hawaiian islands. Those surveys revealed that deliberately introduced beneficial tephritids were frequently attacked by a robust guild of hymenopterous parasitoids from several families (Table 3). The two deliberately introduced opiine braconids *Diachasmimorpha tryoni* and *D. longicaudata* were frequently recovered from field-collected stem galls formed by the lantana gall fly, *Eutreta xanthochaeta*, on *Lantana camara*, and by the euporatorium gall fly, *Procecidochares utilis*, on *Ageratina* (= Eupatorium) *adenophorum*. On the islands of Kauai and Maui, *D. tryoni* was, in fact, the most abundant parasitoids attacking *E. xanthochaeta* (Wong et al. 1991, Duan et al. 1996)

(c) A recent survey conducted on the island of Kauai by Duan and Messing (1998) showed that while no deliberately introduced fruit fly parasitoids were recovered from field-collected *Ensina sonchi* larvae feeding on the flowerheads of the introduced weed, *S. oleraceus, H. elevatus* was frequently recovered from field-collected fly larvae.

These survey data suggest that gall forming species are more likely to be attacked by deliberately introduced, opiine fruit fly larval parasitoids than flowerhead-feeders. Thus, future introduction of related opiine fruit fly larval parasitoids against frugivorous tephritid pests should consider at least the potential impact on this group of nontarget tephritids.

The Degree of Impact on Populations of Associated Nontarget Hosts

The magnitude of effect of a parasitoid (or predator) on nontarget species can range from trivial to significant. Realistic estimation of the degree of attack that is likely to

Table 3. Extant parasitoid guilds associated with nonpest tephritids on the major Hawaiian islands.

Nontarget tephritids	Survey location	Associated parasitoid species[a]	References
Endemic species			
Phaeogramma lortnocoibon	Kauai	*Euderus metallicus, Torymus advenus*	Duan et al. (1996)
Trupanea dubautiae	Kauai	*Habrocytus elevatus* *Euderus metallicus*	Duan et al. (1996) Duan and Messing (1998)
Trupanea spp. (incl. *T. artemisiae, T. cratericola, T. beardsleyi, T. denotata, T. joycei, T. limpidapex*)	Maui	*Habrocytus elevatus* *Euderus metallicus, Bracon teryi* *Eurytoma tephritidis, Cyrtogaster fluscitarsis*	John Herr (unpublished)
Introduced species			
Procecidochares utilis	Maui, Oahu, Molokai, Lanai, Hawaii	*Bracon terryi* *Eupelmus cushmani, Eurytoma tephritidis* *Diachasmimorpha tryoni, D. longicaudata*	Bess and Haramoto (1959, 1972)
Procecidochares alani	Hawaii island	*Eurytoma tephritidis, Bracon terryi* *Eupelmus cushmani, Habrocytus elevatus* *Torymus advenus, Eupelmus allynii*	Duan et al. (1996)
Eutreta xanthochaeta	Maui, Kauai	*Diachasmimorpha tryoni* *D. longicaudata, Bracon terryi* *Torymus advenus, Eupelmus* sp.	Wong et al. (1991), Duan et al. (1996)

occur in the field following the release of an introduced parasitoid is often complicated, and usually must rely upon extensive ecological studies under natural conditions. Only with a thorough knowledge of the ecology of concerned species and communities, can we design tests to accurately predict the degree of attack that might occur. Many ecological factors (such as proximity to nontarget hosts, presence of alternate hosts, parasitoids' physiological and experiential states) affect the degree of impact of a parasitoid, and these cannot be duplicated easily in the laboratory.

Although the previously mentioned parasitoid-guild survey data are appropriate to evaluate the composition of parasitoid guilds and relative abundance of each parasitoid species associated with given tephritid species, the ecological role of the parasitoids in the population dynamics of nontarget hosts cannot be fully determined in these faunal studies. Surveys do not provide information on host population levels, or on host and parasitoid mortality caused by predators, competitors, and natural death; thus the parasitoid-host relationship (e.g., whether or not the impact is critical, or density dependent) remains obscure.

Recently, using a recruitment method similar to that described by Van Driesche and Bellows (1988), we constructed detailed budgets of population losses of lantana gall fly (*E. xanthochaeta*) larvae due to major mortality factors, including parasitoids, predators, competitors, and natural death, during a three month period on the island of Kauai (Duan et al. 1998). The recruitment protocol involved direct measurement of total host populations entering a defined larval stage, as well as losses in that stage due to specified mortality factors. Beginning in April 1995, we sampled weekly three patches (5.6–10 m^2) of lantana plants growing in each of the following three distinct ecological habitats: (a) lowland pasture at an elevation of approximately 124 m, (b) mid-elevation sugarcane fields at an elevation of approximately 500 m, and (c) upland forests at an elevation of approximately 1000 m. On each sampling date, we recorded the number of recently formed immature galls and losses of previously recruited mature galls caused by parasitoids, predators, competitors, and natural abortions. Over the entire study period and across all sites, a total of 1695 lantana stem galls formed by *E. xanthochaeta* were observed. Among the recruited galls, 42–71% were lost from the combined effects of predation (presumably by birds and rodents), damage by a competing herbivore, the lantana tortricid moth (*Crocidosema lantana*), and the death of resident larvae inside the gall. Although *D. tryoni* is the most abundant parasitoid species attacking lantana gall flies on Kauai (accounting for >86% of total parasitism), it caused only about 10% indispensable mortality of gall fly populations in the field (Duan et al., 1998). Regression analysis indicated that the percentage of parasitism by *D. tryoni* had little relationship to the density of immature *E. xanthochaeta* galls growing in patches of lantana plants, and varied significantly with the location of these patches. This indicates that the effect of *D. tryoni* on *E. xanthochaeta* populations is not density-dependent, rather it is habitat-related.

To the best of our knowledge, this is the only study to quantify the impact of a deliberately introduced parasitoid on a nontarget host population. Although this type of ecological study is *post hoc*, it provides insights into the degree of potential risk from future implementation of similar classical biological control programs and may also improve our ability to design relevant tests of host associations of related parasitoid species before their release. However, we wish to point out that this type of nontarget impact study is often labor-intensive, and may be cost-prohibitive if many nontarget species potentially are involved in the system.

Indirect Nontarget Effects

Potential nontarget impacts of introduced species generally are classified (Simberloff, 1992) as either (1) direct effects from immediate trophic interactions (i.e., direct victim/exploiter relationships, as discussed above), or (2) indirect effects from ecological interactions other than immediate victim/exploiter relationships, such as inducing changes in habitat, competing for the same resources, sharing the same natural enemies, etc. Detection of indirect nontarget impact is often more complicated and requires considerable knowledge about the fauna in natural settings. A retrospective analysis of previously introduced parasitoids can provide insight on the potential for indirect effects.

While examples of habitat changes induced by introduced parasitic or predaceous arthropods have not been documented, competitive exclusion has been noted in insect parasitoids (e.g, Ehler and Hall 1982, Heinz and Nelson 1996) and predators (e.g., Elliott et al. 1996). For classical biological control of tephritid pests, competitive displacement between introduced braconid parasitoids has been noted in Hawaii. In the early 1900s, a number of opiine larval parasitoids were introduced into Hawaii for control of medflies. Willard and Manson (1937) documented that parasitism of medflies by *Psyttalia humilis*, was reduced from 31.5 to 4.6% from 1914 to 1920, during which time parasitism of medflies by *D. tryoni and D. fullawayi* (released in 1915 and 1917 respectively), increased from 0.3 to 33.7% (Figure 1). In the late 1940s the egg-larval parasitoid *Fopius arisanus* was successfully introduced into Hawaii for control of medflies and Oriental fruit flies. Following the successful establishment of *F. arisanus* on medflies in major host plants (coffee and peaches), the abundance of earlier-introduced opiine larval parasitoids (e.g., *P. humilis*, *D. fullawayi*, and *D. tryoni*) was reduced sharply (Willard and Mason 1937, Haramoto and Bess 1970, Wong and Ramadan 1987, Vargas et al. 1995). In recent years, *P. humilis* and *D. fullawayi* rarely have been collected from medflies in the field and the abundance of *D. tryoni* attacking *C. capitata* on some crops (e.g., coffee and peach) in Hawaii has been very low (< 0.5% parasitism, Wong and Ramadan 1987, Vargas et al. 1995). However, *D. tryoni* has apparently become the most abundant parasitoid attacking the nontarget tephritid, *E. xanthochaeta* (Wong et al. 1991, Duan et al. 1996). Whether the increased abundance of *D. tryoni* on the nontarget host (*E. xanthochaeta*) is related to the competitive pressure from *F. arisanus* on the primary host (*C. capitata*) remains to be investigated.

Currently, there are no data indicating that the introduced opiine fruit fly parasitoids have any impact on Hawaiian endemic insect communities through indirect ecological interactions. However, the potential indirect impact of a new biological control introduction on the previously-introduced, effective biological control agents should be evaluated in future fruit fly biological control programs.

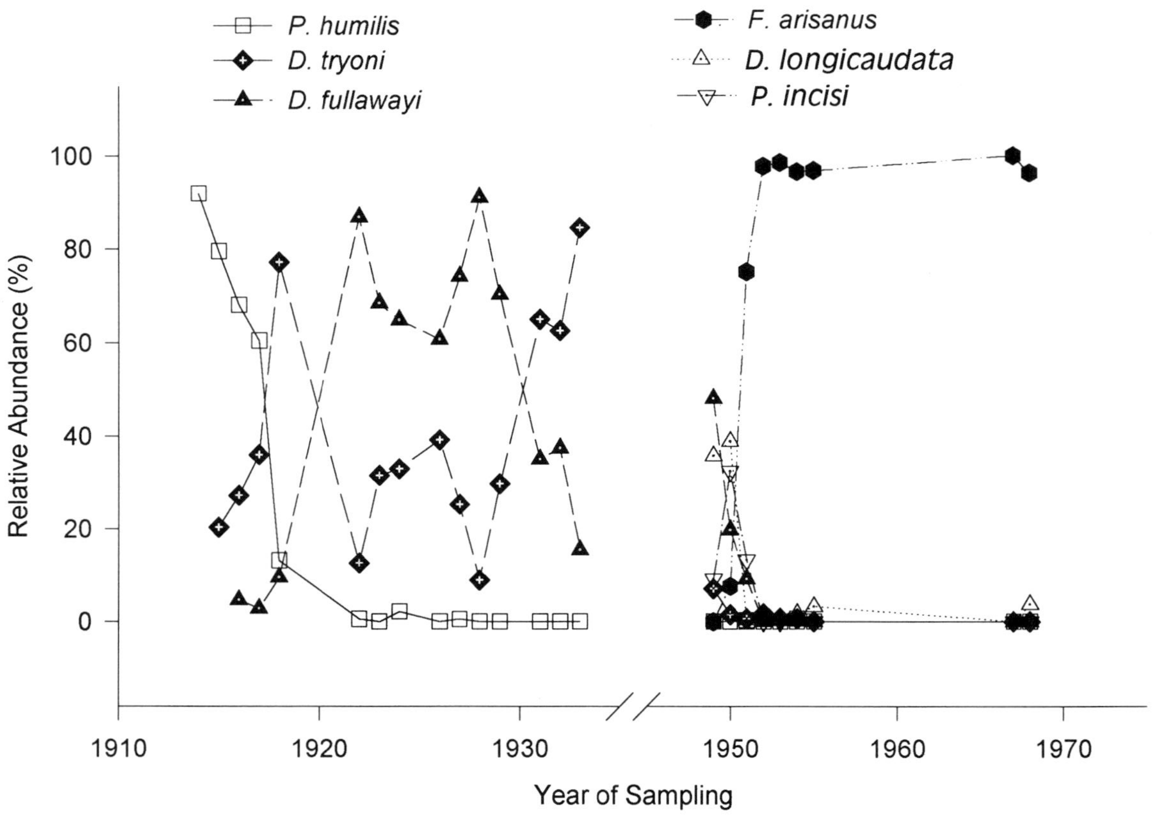

Figure 1. Opiine parasitoids successively established in Hawaii for control of medflies (*C. capitata*) provide an illustration of competitive interactions among introduced biological control agents. Data presented here are from Willard and Mason (1937) and Haramoto and Bess (1970).

ESTIMATING NONTARGET IMPACTS OF FUTURE FRUIT FLY PARASITOID INTRODUCTIONS

Pre-Release Evaluation of Nontarget Risks

Evaluation of nontarget risks of introduced biological control agents prior to their field release is a critical step in classical biological control programs. The extent of the evaluation should depend on our knowledge about the community into which the agents are being introduced, as well as the biology, behavior, and ecology of the species concerned. For classical weed control programs, risk evaluation prior to the release of herbivorous arthropod agents has focused on safety testing with a list of valuable plant species that are taxonomically or phylogenetically related to the targeted host plant species, aiming at estimating potential host ranges of the agents (see review in Zwölfer and Harris 1971; Wapshere 1974). This approach has proven effective in predicting direct nontarget risks in the majority of weed biological control programs, although it still has some limitations with regard to estimating indirect ecological impacts (see Louda et al. 1997).

As pointed out by Van Driesche and Hoddle (1997), testing of parasitoids and predators with a long list of nontarget arthropods related to targeted arthropod pests may prove costly and unrealistic. This is mainly because host ranges of insect parasitoids and predators are often less phylogenetically or taxonomically constrained than those of herbivorous insects, and collecting and rearing many species of arthropods for safety tests can be much more labor-intensive and difficult to accomplish with limited resources. In cases where an arthropod species is rare or endangered, safety testing may be extremely difficult, if not impossible. To date, relatively few projects of arthropod biological control have attempted to conduct safety testing of parasitoids and predators before release (but see Fied and Darby 1991, Barratt et al. 1997).

Besides the logistical difficulties, it also has been well documented that results of safety tests conducted in the laboratory are subject to the influence of many experimental factors (see review in Duan and Messing 1997a). For example, physiologically suitable tephritid fly hosts often will not be recognized by opiine parasitoids if removed from their food substrates. On the other hand, non-host species may be attacked by the parasitoids when placed in a true host's substrate (Nishida 1956; Duan and Messing 1996, 1997b). In nature, the usual microhabitats that are attractive to *D. longicaudata* (and other opiine fruit fly larval parasitoids) are decayed-fruit (due to infestation by fruit fly larvae). In a field-cage test, we found that *D. longicaudata* visited significantly more nontarget microhabitats (stem galls) in the absence of the parasitoids' normal rearing hosts than in their presence (Duan et al. 1997a). Also, we demonstrated that when gravid female *D. longicaudata* were confined with *E. xanthochaeta* galls in small cages for 24 hours, 57% of the exposed larvae were parasitized (Duan and Messing 1996), whereas under both large field-cage and open field conditions, less than 1% of fly larvae were attacked by the parasitoid, even at higher parasitoid/host ratios and longer exposure periods (Duan et al. 1997a). Together, these findings point to the need for developing ecologically relevant methods for screening potential nontarget effects of newly imported parasitoids.

Future fruit fly biological control programs should encourage investigations into the host association of introduced parasitoids with representive species of ecologically distinct groups of non-frugivorous tephritids. However, for some groups of parasitoids such as opiine braconids that are well studied, our current knowledge suggest that host-specificity tests could be restricted to nontarget tephritids that develop as larvae in microhabitats that, to some extent, are fruit-like, such as stem galls (Duan et al., 1997b). This conclusion is supported by our knowledge on patterns of host utilization by this group of parasitoids. First, known host ranges of this group of opiine fruit fly larval parasitoids are exclusively restricted to dipterans in the family Tephritidae that live in plant substrates as larvae. Polyphagy can vary from several to dozens of host species and from a single to several genera (Fisher 1971, 1973, 1977; Wharton and Gilstrap 1983; Wharton 1997). Secondly, known host ranges of fruit fly larval parasitoids that have been used in biological control programs consist mainly of fruit-inhabiting species, plus a very few non-frugivorous (mainly gall-making) tephritids (see Duan and Messing 1997a). Thirdly, studies on host finding and acceptance behavior of the fruit fly parasitoids indicate that host microhabitat selection plays a central role in the process of host selection (Nishida 1956, Messing et al. 1996). Visual and olfactory stimuli (e.g., fruit shape, size, color, and odors) have been shown to play a central role in host-microhabitat recognition by opiines attacking fruit-inhabiting tephritid hosts (Greany et al. 1977, Vargas et al. 1991, Leyva et al. 1991, Messing and Jang 1992, Baustista and Harris 1996, Messing et al. 1996). Natural microhabitats recognized by fruit fly parasitoids include host-infested fruit (often associated with abundant olfactory stimuli resulting from chemical changes due to infestation) and fruit-shaped galls. Parasitoids raised in the laboratory recognize screened-dishes containing host fruit pulp or decaying wheat-based artificial diet of the host larvae as potential-host habitats (Messing et al. 1996). Finally, even in the family Tephritidae, some taxa of non-frugivorous tephritids may not be at risk from attack. For example, we recently studied responses of four introduced fruit fly parasitoids to an endemic flowerhead-feeding fly (*Trupanea dubautiae*), infesting an endemic composite plant. All four species were unable to recognize *T. dubautiae*-infested flowerheads as potential host habitats, and showed little ovipositional activity (Duan and Messing 1997b, 1998, also unpublished data). To the extent that other endemic Hawaiian *Trupanea* inhabiting flowerheads of composite plants have life history traits and infestation ecology similar to *T. dubautiae* (Hardy and Delfinado 1980), these flowerhead feeders may be at minimal risk from introduced opiine fruit fly parasitoids.

Post-Release Monitoring of Nontarget Impacts

Evaluation of nontarget impacts prior to release of introduced species is often done without the presence of realistic ecological contexts of the new environment into which the species is being released; it is not possible that we can be completely certain if our predictions of risk before release are valid. Post-release monitoring of newly colonized parasitoids in the new environment is, therefore, a necessary step for development of sound methods for nontarget impact analysis (Howarth 1991, Simberloff and Stiling 1996, Louda et al. 1997). However, post-release evaluations of introduced natural enemies for nontarget impacts have been rarely undertaken in previous classical biological control programs (Simberloff and Stiling 1996, Hawkins and Marino 1997). Follow-up studies on the ecological impacts of new entomophagous species are rarely conducted once a pest problem is solved or an introduced natural enemy is found ineffective.

Classical biological control programs against alien fruit fly pests during the first half of this century led to introductions of many hymenopteran parasitoids into Hawaii. Immediately following the parasitoid releases, extensive follow-up field studies were conducted in Hawaii to evaluate the effects of the introduced parasitoids on targeted tephritid pests (e.g., Willard and Manson 1937, Bess et al. 1961, Haramoto and Bess 1970, Wong and Ramadan 1987, Vargas et al. 1993). However, monitoring of the impacts on nontarget tephritids was largely neglected when the parasitoids were released. The first discovery of attack of the pamakani gall fly, *P. utilis*, by *D. tryoni* and *D. longicaudata* was the result of field surveys designed exclusively for studying field mortality factors affecting the weed control agents, rather than monitoring the nontarget impacts of the introduced parasitoids (Bess 1950, Bess and Haramoto 1959, 1972). The casual collecting and rearing of lantana stem galls by Clancy (1950) also led to the discovery of attack by *D. tryoni* on the lantana gall fly, *E. xanthochaeta*. Recently, retrospective studies specifically investigating potential interactions of previously introduced fruit fly parasitoids with nontarget tephritids have been conducted in Hawaii. Although these studies indicate that none of the established fruit fly parasitoids parasitize any endemic tephritid species, one parasitoid (*D. tryoni*) does frequently attack lantana gall flies, causing about 10% indispensable mortality of the fly populations (Duan et al. 1998). Another introduced opiine, *D. longicaudata*, was also found to attack lantana gall flies on the island of Kauai (Duan and Messing 1996; Duan et al. 1996).

D. tryoni was introduced into Hawaii from Australia in 1913 for control of the Mediterranean fruit fly (Silvestri 1914), and *D. longicaudata* was introduced from Southeast Asia in 1947 for control of the Oriental fruit fly (Clancy 1952). Currently, we do not know if *D. tryoni* and *D. longicaudata* co-evolved with or were pre-adapted to attack any gall-forming tephritids before their introduction. Associations with gall flies could raise the possibility of risk to Hawaiian endemic tephritids, because interactions with these flies could increase the likelihood of the parasitoid utilizing ecologically-similar types of hosts. Recent observations indicate that a small proportion (17%, n=12) of gravid *D. tryoni* females originating from lantana gall flies was capable of locating and probing the stem galls formed by the endemic fly *P. lortnocoibons* on the native plant, *Bidens cosmoides* when the parasitoid was placed directly on the galled stem (JJD, unpublished data).

Although Hawaiian endemic tephritids are associated with native composites distributed in forests at high elevations, the continued invasion of native forests by weeds such as lantana may bring exotic gall-forming tephritids and their associated parasitoids (including opiine braconids) closer in proximity to native tephritid habitats. Therefore, the potential interactions between these parasitoids and endemic gall-forming tephritids should be further studied in the context of long-term ecological impacts of biological control programs. Although such research is "post hoc," information gained from these studies will give us insights into the ecology of colonizing parasitoid populations, and thus promote development of sound protocols for analyzing nontarget risks in future fruit fly biological control programs.

ACKNOWLEDGMENTS

We thank John Herr (Department of Environmental Sciences and Pest Management, University of California, Berkeley) for allowing us to share his unpublished data, and two anonymous reviewers for helpful comments. The research mentioned in this article was supported in part by USDA-ARS Cooperative Agreement #58-5320-4549 and USDA-NRI Competitive Grant #9802864.

REFERENCES

Asquith, A. 1995. Alien species and the extinction crisis of Hawaii's invertebrates. Endangered Species Update. 12: 6–7.

Barratt, B. I. P., A. A. Evans, C. M. Ferguson, G. Barker, M. R. McNeill, and C. C. Philips. 1997. Laboratory nontarget host range of the introduced parasitoids *Microctonus aethiopoides* and *M. hyperodae* (Hymenoptera: Braconidae) compared with field parasitism in New Zealand. Environ. Entomol. 26: 694–720.

Baustista R. C., and E. J. Harris 1996. Effect of fruit substrates on parasitization of tephritid fruit flies (Diptera) by the parasitoid *Fopius arisanus* (Hymenoptera: Braconidae). Environ. Entomol. 25: 470–475.

Bess, H. A. 1950. Notes on parasites of tephritid flies. Proc. Hawaiian Entomol. Soc. 14: 26.

Bess, H. A., and F. H. Haramoto. 1959. Biological control of pamakani, *Eupatorium adenophorum*, in Hawaii by a tephritid gall fly, *Procecidochares utilis*. 2. Population studies of the weed, fly, and the parasites of the fly. Ecology 40: 244–249.

Bess, H. A., and F. H. Haramoto. 1972. Biological control of pamakani, *Eupatorium adenophorum*, in Hawaii by a tephritid gall fly, *Procecidochares utilis*. 3. Status of the weed, fly and parasites of the fly in 1966–1971 versus 1950–1957. Proc. Hawaiian Entomological Society 21: 165–178.

Bess, H. A., R. van den Bosch, and F. H. Haramoto. 1961. Fruit fly parasites and their activities in Hawaii. Proc. Hawaiian Entomol. Soc. 17: 367–578.

Clancy, D. W. 1950. Notes on parasites of tephritid flies. Proc. Hawaiian Entomol. Soc. 14:25–26.

Clancy, D. W. 1952. Notes on parasites of tephritid flies. Proc. Hawaiian Entomol. Soc. 14: 374–375.

Clausen, C. P. 1956. Biological control of fruit flies. J. Econ. Entomol. 49: 176–178.

Clausen, C. P. 1978. Introduced parasites and predators of arthropod pests and weeds: a world reviews. U.S. Dept. Agric., Agric. Handbk. 480. p. 320–335.

Duan, J. J., and R. H. Messing. 1996. Response of 2 opiine fruit fly parasitoids (Hymenoptera: Braconidae) to the lantana gall fly (Diptera: Tephritidae). Environ. Entomol. 25: 1428–1437.

Duan, J. J., and R. H. Messing. 1997a. Biological control of fruit flies in Hawaii: Factors affecting nontarget risk analysis. Agric. Human Values 14: 227–236.

Duan, J. J., and R. H. Messing. 1997b. Effects of two opiine parasitoids (Hymenoptera: Braconidae) introduced for fruit fly control on a nontarget native Hawaiian tephritid, *Trupanea dubautiae* (Diptera: Tephritidae). Biol. Control 8: 177–184.

Duan, J. J., and R. H. Mesing. 1998. Effect of *Tetrastichus giffardianus* (Hymenoptera: Eulophidae) on nontarget flowerhead-feeding tephritids (Diptera: Tephritidae). Environ. Entomol. (in press).

Duan, J. J., M. F. Purcell, and R. H. Messing. 1996. Parasitoids of nontarget tephritid flies in Hawaii: Implications for biological control of fruit fly pests. Entomophaga 41: 245–256.

Duan, J. J., A. Mukhtar, J. Kailash, and R. H. Messing. 1997a. Evaluation of the impact of the fruit fly parasitoid *Diachasmimorpha longicaudata* (Hymenoptera: Braconidae) on a nontarget tephritid, *Eutreta xanthochaeta* (Diptera: Tephritidae). Biol. Control 8:58–64.

Duan, J.J., M. F. Purcell, and R. H. Messing. 1997b. Ovipositional responses of 3 opiine fruit fly parasitoids (Hymenoptera: Braconidae) to gall-forming tephritids (Diptera: Tephritidae). Biol. Control 9: 81–88.

Duan, J. J., M. F. Purcell, and R. H. Messing. 1998. Association of the opiine parasitoid *Diachasmimorpha tryoni* (Hymenoptera: Braconidae) with the lantana gall fly (Diptera: Tephritidae) on Kauai. Environ.

Entomol. 27: 419–426.

Ehler, L. E., and R. W. Hall. 1982. Evidence for competitive exclusion of introduced natural enemies in biological control. Environ. Entomol. 11: 1–14.

Elliott, N. C., Kieckhefer, R. W., and W. C. Kauffman. 1996. Effects of an invading coccinellid on native coccinellids in an agricultural landscape. Oecologia 105: 537–544.

Fisher, M. 1971. Index of world Opiinae. Paris, Le Francois.

Fisher, M. 1973. Hymenoptera Braconidae (Opiinae I). Das Tierreich 91:1–620.

Fisher, M. 1977. Hymenoptera Braconidae (Opiinae II). Ibid. 96: 1–1001.

Field, R. P., and S. M. Darby. 1991. Host specificity of the parasitoid, *Sphecophaga vesparum* (Curtis) (Hymenoptera: Ichneumonidae), a potential biological control agent of the social wasps, *Vespula germanica* (Fabricius) and *V. vulgaris* (Linnaeus (Hymenoptera: Vespidae) in Australia. N.Z. J. Zool. 18: 193–197.

Funasaki, G. Y., P.-Y. Lai, L. M. Nakahara, J. W. Beardsley, and A. K. Ota. 1988. A review of biological control introductions in Hawaii: 1890 to 1985. Proc. Hawaiian Entomol. Soc. 28: 105–160.

Greany, P. D., J. H. Tumlinson, D. L. Chambers, and G. M. Boush. 1977. Chemically mediated host finding by *Fopius* (*Opius*) *longicaudatus*, a parasitoid of tephritid fruit fly larvae. J. Chem. Ecol. 3: 189–195.

Hardy, D. E., and M. D. Delfinado. 1980. Insects of Hawaii. vol. 13–Diptera: Cyclorrhapha III. Univ. Hawaii Press, Honolulu.

Harris, O. 1990. Environmental impact of introduced biological control agents. *In* M. Mackauer, L. E. Ehler, and J. Roland (eds.). Critical issues in biological control. Intercept Ltm., Andover, UK. p. 289–299.

Haramoto, R. H., and H. A. Bess. 1970. Recent studies on the abundance of the Oriental and Mediterranean fruit flies and the status of their parasites. Proc., Hawaiian Entomol. Soc. 20: 551–566.

Hawkins, B. A., and P. C. Marino. 1997. The colonization of native phytophagous insects in North America by exotic parasitoids. Oecologia 112: 566–571.

Heinz, K. M., and J. M. Nelson. 1996. Interspecific interactions among natural enemies of Bemisia in an inundative biological control program. Biol. Control 6: 384–393.

Howarth, F. G. 1991. Environmental impacts of classical biological control. Annu. Rev. of Entomol. 36: 485–509.

Louda, S. M., D. Kendall, J. Connor, and D. Simberloff. Ecological effects of an insect introduced for the biological control of weeds. Science 277: 1088–1090.

Leyva, J. L., H. W. Browning, and F. E. Gilstrap. 1991. Effect of host fruit species, size, and color on parasitization of *Anastrepha ludens* (Diptera: Tephritidae) by *Diachasmimorpha longicaudata* (Hymenoptera: Braconidae). Environ. Entomol. 20: 1469–1474.

Lockwood, J. A. 1996. The ethics of biological control: understanding the moral implications of our most powerful ecological technology. Agric. Human Values 13: 2–19.

Loockwood, J. A. 1997. Competing values and moral imperatives: an overview of ethical issues in biological control. Agric. Human Values 14: 205–210.

Marohasy, J. 1998. The design and interpretation of host-specificity tests for weed biological control with particular reference to insect behavior. Biol. Control News Info. 19: 13–20.

Messing, R H., and E. B. Jang. 1992. Response of the fruit fly parasitoid *Diachasmimorpha longicaudata* (Hymenoptera: Braconidae) to host fruit stimuli. Environ. Entomol. *21*: 1189–1195.

Messing, R. H., L. M. Klungness, E, B. Jang, and K. A. Nishijima. 1996. Response of the melon fly parasitoid *Psyttalia fletcheri* (Hym.: Braconidae) to host-habitat stimuli. J. Insect Behav. 9: 933–945

Newell, I. M., and F. H. Haramoto. 1968. Biotic factors influencing populations of *Dacus dorsalis* in Hawaii. Proc. Hawaiian Entomol. Soc. 20: 81–139.

Nishida, T. 1956. An experimental study of the ovipositional behavior of *Opius fletcheri*. Proc. Hawaiian Entomol. Soc. 16: 126–134.

Simberloff, D. 1992. Conservation of pristine habitats and unintended effects of biological control. *In* W. C. Kauffman and J. E. Nechols (eds.), Selection criteria and ecological consequences of importing natural enemies. Lanham, Maryland: Thomas Say Publications in Entomol., Proc. Entomol. Soc. Am. p. 103–117.

Simberloff, D., and P. Stiling. 1996. How risky is biological control? Ecology 77: 1965–1974.

Silvestri, F. 1914. Report of an expedition to Africa in search of the natural enemies of fruit flies (Trypetidae) with descriptions, observations, and biological notes. Territory of Hawaii, Board of Agric. and Forestry, Div. of Entomol., Bull. 3.

U.S. Fish and Wildlife Service. 1996. Endangered and threatened wildlife and plants: reviews of plant and animal taxa that are candidates for listing as endangered or threatened species. Federal Register 61 (40): 7596–7602.

Van Dreische, R. G., and T. S. Bellows, Jr. 1998. Use of host and parasitoid recruitment in quantifying losses from parasitism in insect populations. Ecol. Entomol. 13: 215–222.

Van Dreische, R. G., and T. S. Bellows, Jr. 1996. Biological control. Chapman and Hall, New York.

Van Driesche, R. G., and M. Hoddle. 1997. Should arthropod parasitoids and predators be subject to host range testing? Agric. Human Values 14: 211–226.

Van Driesche, R. G., and J. Benson. 1999. Invading nature: releasing exotic psecies to protect natural systems (in press).

Vargas, R. I., J. D. Stark, G. K. Uchida, and M. Purcell. 1993. Opiine parasitoids (Hymenoptera: Braconidae) of oriental fruit fly (Diptera: Tephritidae) on Kauai Island, Hawaii: islandwide relative abundance and parasitism rates in wild and orchard guava habitats. Environ. Entomol. 21: 246–253.

Vargas, R. I., W. A. Walsh, and T. Nishida. 1995. Colonization of newly planted coffee fields: dominance of Mediterranean fruit fly over oriental fruit fly (Diptera: Tephritidae). Journal of Economic Entomology 88: 620–627.

Vargas, R. I., J. D. Stark, R. J. Prokopy, and T. A. Green. 1991. Response of oriental fruit fly (Diptera: Tephritidae) and associated parasitoids (Hymenoptera: Braconidae) to different-color spheres. J. Econ. Entomol. 84: 1503–1507.

Wapshere, A. J. 1974. A strategy for evaluating the safety of organisms for biological weed control. Ann. Appl. Biol. 77: 201–211.

Wharton, R. A. 1989. Classical biological control of fruit infesting Tephritidae. *In* A. S. Robinson and G. Hooper (eds.), World crop pests: fruit flies, their biology, natural enemies, and control. vol. 3b. Elsevier Science, Amsterdam, Netherlands. p. 303–313.

Wharton, R. A. 1997. Generic relationships of opiine braconidae (Hymenoptera) parasitic on fruit-infesting tephritidae (Diptera). Contrib. Am. Entomol. Inst. 30: 3–53.

Wharton R. A. and F. E. Gilstrap. 1983. Key to and Status of opiine braconid (Hymenoptera) parasitoids used in biological control of *Ceratitis* and *Dacus s. l.* (Diptera: Tephritidae). Ann. Entomol. Soc. Am. 76: 721–742.

Willard, H. F., and A. C. Mason. 1937. Parasitization of the mediterranean fruit fly in Hawaii, 1914–33. U.S. Dept. Agric. Circ. 439.

Wong, T. T.Y., and M. M. Ramadan. 1987. Parasitization of the Mediterranean and oriental fruit flies (Diptera: Tephritidae) in the Kula area of Maui, Hawaii. J. Econ. Entomol. 80: 77–80.

Wong, T. T. Y., N. Mochizuki, and J. I. Nishimoto. 1984. Seasonal abundance of parasitoids of the Mediterranean and oriental fruit flies (Diptera: Tephritidae) in the Kula area of Maui, Hawaii. J. Econ. Entomol. 13: 140–145.

Wong, T. T. Y., M. M. Ramadan, D.O. McInnis, N. Mochizuki, J. A. Nishimoto, and J. C. Herr. 1991. Augmentative releases of *Diachasmimorpha tryoni* (Hymenoptera: Braconidae) to suppress a Mediterranean fruit fly population in Kula, Maui, Hawaii. Biol. Control 1: 2–7.

Zwölfer, H, and P. Harris. 1971. Host specificity determination of insects for biological control of weeds. Annu. Rev. Entomol. 16: 159–178.

8

TRICHOGRAMMA NONTARGET IMPACTS: A METHOD FOR BIOLOGICAL CONTROL RISK ASSESSMENT

David B. Orr
Department of Entomology, North Carolina State University, Raleigh, North Carolina

Carlos Garcia-Salazar
Department of Entomology, Pennsylvania State University, University Park, Pennsylvania

Douglas A. Landis
Department of Entomology and Pesticide Research Center, Michigan State University, E. Lansing, Michigan

INTRODUCTION

For much of its history, insect biological control has enjoyed a reputation as an environmentally safe and effective means of managing insect pests (Debach 1974, Doutt 1972, Caltagirone 1981). However, over the past 15 years, questions have been raised regarding the potential impact of biological control on nontarget organisms (Howarth 1983, 1991; Simberloff 1992; Lockwood 1993 a, b; Simberloff and Stiling 1996; Lockwood 1997). These concerns have prompted a much-needed discussion regarding the potential for nontarget effects and approaches for addressing this issue (Carruthers and Onsager 1993, Duan and Messing 1997, Van Driesche and Hoddle 1997, Frank 1998, Simberloff and Stiling 1998). Most of the discussion has focused on the risks posed by the importation and release of exotic natural enemies against exotic pests, i.e., classical biological control. Far less attention has been paid to the potential for nontarget effects from augmentative releases of natural enemies, even though this form of biological control may become more widespread in the future (Hoy et al. 1991, Moffat 1991, Parrella et al. 1992, Nordlund and Greenberg 1994, Nordlund et al. 1998).

Currently, *Trichogramma* species (Hymenoptera: Trichogrammatidae) are by far the most widely produced and released natural enemies for augmentative biological control. These tiny parasitoids have several advantages, including relative ease of rearing and the fact that they kill their host in the egg stage before it causes feeding injury (Wajnberg and Hassan 1994). There are approximately 20 species of *Trichogramma* regularly used in augmentative biological control programs to control insects, primarily Lepidoptera in at least 22 crops and trees (Li 1994). Most of these species are released in large numbers, i.e., inundative releases, in order to rapidly suppress the target pest. On a worldwide basis, an estimated 32 million ha of agriculture and forestry are estimated to be treated annually with *Trichogramma* for insect pest control (Li 1994). Because of the widespread use of *Trichogramma* and the large numbers released, it is important to understand the potential nontarget impacts of this form of biological control. We start by reviewing the biology and behavior of *Trichogramma* with respect to potential nontarget impacts and provide an overview of past nontarget studies. Secondly, we outline a framework for the study of nontarget impacts of *Trichogramma*. Finally, we describe a case study utilizing this framework to elucidate the potential nontarget impacts of *Trichogramma* releases in the midwestern United States.

TRICHOGRAMMA BIOLOGY AND POTENTIAL FOR NONTARGET IMPACTS

The *Trichogramma* are idiobiont parasitoids with the physiological ability to successfully utilize a variety of hosts. As a result, most of the better known genera of the Trichogrammatidae (including *Trichogramma*) have broad potential host ranges (Pinto and Stouthamer 1994). For example, *T. evanescens* has been reported to attack over 150 hosts in seven insect orders (Clausen 1940). This potential polyphagy is tempered, however, by the tendency of most *Trichogramma* to show strong preferences for certain host(s), crop(s) and climatic conditions (Hassan and Gou. 1991). Because of these preferences, *Trichogramma* are likely to have a limited host range in any given environment, decreasing the potential for nontarget impacts.

Habitat Selection and Host Acceptance

Host habitat selection and host acceptance behavior by *Trichogramma* play a major role in determining their potential to cause nontarget effects. *Trichogramma* host habitat finding may be influenced by cues received from host pheromones (Noldus and van Lenteren 1985, Zaki 1985), kairomones (Need and Burbutis 1979, Renou et al. 1992) and plant volatiles (Altieri et al. 1981, Nordlund et al. 1985, Kaiser et al. 1989a). For

Trichogramma, searching within a habitat is described as random (Bigler et al. 1988) with several studies demonstrating an inverse relationship between amount of surface area and rate of host discovery (Need and Burburtis 1979, Kanour and Burburtis 1984) or decreasing discovery with increasing distance between host eggs (Laing 1937). Increasing structural complexity reduced searching time and rate of discovery in *T. nubilale* (Andow and Prokrym 1990), and temperature, relative humidity and host species all affected rate of search by *T. evanescens* (Boldt 1974). Once an egg is found, the decision to oviposit depends on egg size (Salt 1934), surface chemistry, i.e., kairomones (De Jong and Pak 1984), and the host from which the female was reared (Kaiser et al. 1989a).

Learning and Pre-Imaginal Development

While females have innate preferences for certain hosts, their affinity for an alternative host is always higher after an oviposition on that species. Thus, if denied the preferred host and given an oviposition experience with an alternative host, they can develop equivalent levels of preference for the alternative host (Kaiser et al. 1989a), and learn to associate the presence of this host with olfactory cues (Kaiser et al. 1989b). Moth scales act in a relatively nonspecific manner (Thompson and Stinner 1990) to stimulate orientation and retention behaviors of *Trichogramma* towards host eggs (Shu and Jones 1989, Shu et al. 1990). The host species in which larval development takes place can have a dramatic impact on biological features such as size, longevity, fecundity, and host acceptance behavior of the adult wasp (Brenière 1965, Bigler et al. 1982, Kaiser et al. 1989a, Marston and Ertle 1973, Mayer and Quednau 1959, Taylor and Stern 1971, Salt 1935, 1940, Southard et al. 1982). The relative quality of these features determines the overall quality (parasitization capacity) of the parasitoids (Huettel 1976, Cerutti and Bigler 1991).

Dispersal Ability

A second factor mediating potential nontarget effects is the dispersal capabilities of *Trichogramma*. If wasps never left the release site, host selectivity would tend to minimize potential nontarget effects. While most studies show that *Trichogramma* dispersal is quite limited, (Bigler et al. 1989, McDougall and Mills 1997) others have found that at least some individuals can travel up to 1100 m (Stern et al. 1965). Bigler et al. (1990) found that 60–70% of *T. brassicae* (*evanescens*) Bezdenko migrated out of a release field within 4 d and demonstrated a close relationship between direction of movement and wind direction. They found that bait egg masses of the European corn borer, *Ostrinia nubilalis* Hübner, placed up to 400 m downwind from the release field were parasitized at the same level as those within the release field. From this they calculated that the effective area over which *T. brassicae* were distributed was at least two times as large as the initial release area, resulted in 36% parasitism of the *O. nubilalis* egg masses in nearby untreated fields (Bigler et al. 1990).

TRICHOGRAMMA NONTARGET IMPACTS: CURRENT KNOWLEDGE

Specific research on the nontarget impact of inundative releases of *Trichogramma* is limited. Curl and Burburtis (1978) worked with *T. nubilale* Ertle and Davis, a native North American species that they considered to be host-specific to *O. nubilalis*. They found however, that in no-choice tests with other Lepidoptera, 17 of 21 species in 5 of 7 families tested were successfully parasitized. In choice tests, *T. nubilale* preferred *O. nubilalis* in most cases but always parasitized at least 25% of the nonpreferred hosts. Additionally, two strains of *T. nubilale* did not discriminate at all between *O. nubilalis* and the alternate host. In the field, they found only one case where *T. nubilale* parasitized a host other than *O. nubilalis*. Given these data they termed *T. nubilale* host-selective rather than host-specific.

Andow et al. (1995) examined the nontarget impacts of inundative releases of *T. nubilale,* which has been proposed as an alternative to chemical to control for *O. nubilalis* infesting sweet and field corn in Minnesota. Since proposed release areas were in proximity to oak-savanna habitats utilized by the endangered Karner blue butterfly, the potential risk to nontargets was evaluated. They found that even following direct release of *T. nubilale* in an oak-savanna habitat, none of the sentinel eggs were parasitized by *T. nubilale* and he concluded that the nontarget impacts of *T. nubilale* on threatened and endangered Lepidoptera were minimal.

Recently, Suverkropp and Bigler (1997, personal communication) studied the nontarget impact of inundative releases of *T. brassicae* Bezdenko in cornfields of Switzerland. They found that parasitization of nontarget Lepidoptera by *T. brassicae* dispersing outside of the target area was approximately 4%, similar or less than that produced by native *Trichogramma*. Their preliminary conclusion was that *T. brassicae* does cause measurable mortality on nontargets, but that the impact was minimal and did not pose a high degree of risk for Lepidoptera or native egg parasitoids.

Trichogramma brassicae in North America

In the early 1990s, scientists in the Midwestern US were faced with the opportunity to examine the potential nontarget impacts of *T. brassicae*. This species is indigenous to western Asia and parts of Europe (Pintureau 1990) and is used extensively in Western Europe for control of *O. nubilalis,* in corn (Bigler 1986, 1989; Bigler et al. 1989, 1990; Hassan and Guo 1991). Interest was building in the USA regarding the use of *Trichogramma* for *O. nubilalis* control (Andow and Olson 1992, Prokrym et al. 1992) and evaluations of *T. brassicae* effectiveness against *O. nubilalis* in the Midwest had been very promising (Orr 1993, Orr et al. 1992). Due to interest in the technique by a major seed corn producer, and rising costs of chemical control, it appeared possible that *T. brassicae* use on corn in the Midwest, particularly seed corn, could be rapidly adopted.

Using *T. brassicae* to control *O. nubilalis* is a combination of inundative and inoculative release techniques. Within each host generation, large releases of adult *T. brassicae* suppress the initial cohort of *O. nubilalis* eggs. Adult wasps emerging from these early hosts then go on to parasitize host eggs laid later in the generation. This recycling of parasitoids is critical to the economics of the method as a whole, since it allows a limited number of released *T. brassicae* to control an entire generation of *O. nubilalis* (Frandon and Kabiri 1990, Kabiri et al. 1990). The downside is that large numbers of wasps emerging from the peak egg population are faced with limited host numbers and may perhaps be prone to leave the release field. Because many potential nontargets occur in habitats within the dispersal capabilities of *T. brassicae*, a method was needed to rapidly assess the probability for this to occur.

A GENERIC FRAMEWORK FOR *TRICHOGRAMMA* NONTARGET TESTING

There are several potential approaches to evaluating the risks posed by a natural enemy to nontarget organisms. These may range from expert opinion to simple host-range testing, population studies, or full-scale environmental impact assessments. While more in-depth testing seems intuitively better, a central problem is that not all nontarget organisms can be tested exhaustively. Therefore, it is desirable to develop a framework for rapidly eliminating those nontarget species for which biologically important impacts are less likely, and focusing efforts on those where the risk of biologically important impacts are greatest. Such a process itself involves tradeoffs, however, failure to prioritize may carry the greatest danger of all, i.e., that risk assessments are not conducted at all, or that they rely solely on expert opinion. Below, we present a framework to assess the potential risks of *Trichogramma* to nontarget species. The process assumes a basic understanding of the biological system into which the wasps will be released, and seeks to rapidly eliminate nontargets that are at little risk for impact.

The first step in the process is evaluation of the temporal overlap of potential nontargets with *Trichogramma* releases. This requires *a priori* knowledge of the range of hosts that could be attacked. Nontarget species that do not produce eggs during the time that released *Trichogramma* (or subsequent generations) are present in the field can be eliminated from further consideration. Temporal occurrence data may be available in the literature, or generated by field studies. The second step is for eggs of temporally available nontarget species to be evaluated in no-choice laboratory tests to determine their acceptability and suitability as hosts for *T. brassicae*. Further evaluations may focus on only suitable hosts (i.e., those that support complete development of the wasp) or on both acceptable and suitable hosts depending on the situation.

A nontarget that is attacked in the lab may or may not be attacked in nature. In part this will depend on where the nontarget occurs in relation to where the wasps are released and subsequently move to. Determining if females will accept nontargets under field conditions is critical. For those nontargets that occur in the target habitat (i.e., release area), direct testing of attack rates in the target habitat is the appropriate next step. For nontargets that do not occur in the target habitat, a two-step process is suggested. First, the ability of wasps to locate and attack nontarget hosts when released directly into a nontarget habitat is evaluated. If this occurs, then determining the ability of wasp to disperse from the target to the nontarget habitat is required to fully evaluate the potential for biologically significant risk. Finally, evaluating the potential for host acceptability to increase after development in a nontarget host and the ability of *Trichogramma* to overwinter in a region is also necessary. These steps are illustrated below for the case of *T. brassicae*.

CASE STUDY OF *TRICHOGRAMMA BRASSICAE* NONTARGET EFFECTS

The case study presented below summarizes investigations of *T. brassicae* nontarget effects conducted between 1993 and 1996 in corn and several adjacent habitats in southern Michigan (Orr et al., in prep.; Garcia-Salazar et al., in prep.). While the results only directly apply to these systems, we believe the approach has the potential to be generalized to other natural enemies and systems.

Temporal Overlap

The presence of nontarget Lepidoptera during periods of *T. brassicae* release against *O. nubilalis* was determined in 1993–94. The flight period of *T. brassicae* was estimated based on the range of dates on which experimental augmentative releases were conducted by the authors during 1993–1995 in southern Michigan inbred seed corn and hybrid field corn plots. Adult *T. brassicae* life span in the field was assumed to be 4 d, based on work with *T. nubilale* by Andow and Prokrym (1990). Adults produced by recycling were assumed to be present for 16 d following release of *O. nubilalis* (i.e., 4 d life span of released *T. brassicae*, 8 d [110 Celsius degree days] for pre-imaginal development, and 4 d life span of recycled *T. brassicae*).

Potential nontarget moth species were collected in a walk-in style black light, and nontarget butterfly species were collected using aerial and sweep nets. Habitats within range of the light trapping site included a cornfield, vegetable crops, oak-hickory forest, wetlands, old-field, and native prairie/bur oak savanna. Once identified to species, the potential flight period of each nontarget was estimated from collection data on museum holdings of specimens from southern Michigan.

In 1993–94, we collected 22 species of nontarget Lepidoptera with flight periods overlapping potential *T. brassicae* release periods. Because of its importance as a pest of seed corn, the corn earworm, *Helicoverpa zea* Boddie, was also included in our host range tests, even though its flight period in southern Michigan falls outside of the *T. brassicae* release period. There were undoubtedly additional species which were not attracted to light traps or not abundant enough to be sampled using our methods. This presents an inherent difficulty, in that these may be rare species and thus, potentially sensitive to population-level impacts.

Laboratory No-Choice Tests

For temporally available nontarget hosts, we determined which hosts were acceptable and suitable to *T. brassicae* under no-choice laboratory conditions. Adult moths and butterflies were held in paper cartons under ambient conditions in an outdoor insectary. A variety of oviposition substrates were provided (paper, cheesecloth, or plastic) in order to obtain eggs. The suitability of *O. nubilalis*, the target host, and of *Ephestia kuehniella* Zeller, the factitious host on which *T. brassicae* is commercially reared, were also included for comparison. Eggs or egg masses were presented to individual 1 d old mated *T. brassicae* females in plastic vials containing minute streaks of honey and water. Parasitoids had not been exposed to any host eggs prior to this study. Eggs were presented to wasps on the same substrates they were oviposited on, and which included host scales. Vials were held until eggs produced either larvae or parasitoids. For each test a separate control vial containing nontarget eggs not exposed to *T. brassicae* was used. To develop a host ranking, several quality variables (e.g. percent parasitism, percent emergence, sex ratio, and size) were measured for *T. brassicae* exposed to or emerged from nontarget eggs in laboratory no-choice tests. A relative value from 1 to 10 was assigned to each measured variable for each host species, then averaged across variables. Host species with rankings from 1 to 3 were considered poor quality hosts, those from 4 to 7 were considered moderate quality hosts, and those from 8 to 10 were considered high quality hosts.

Eighteen of the 23 (78.3%) potential nontarget host species tested in our study were accepted by *T. brassicae* and parasitized in the no-choice test. Percent parasitism of accepted hosts ranged from 10 to 100% with 12 species experiencing greater parasitism than the target host, *O. nubilalis* (72%). Fifteen of the accepted hosts were suitable for development of *T. brassicae*. Emergence from suitable nontarget hosts ranged from 12.5 to 100% with eight species having emergence rates which exceeded that of *O. nubilalis* (i.e., 85% emergence). Wasp production in nontarget species varied from 1.2 to 7.7 wasps per egg with half of the species producing more wasps per host egg than *O. nubilalis* and half less. Nontarget species produced from 46 to 69% female progeny, considerably lower than the 87% females produced from *O. nubilalis* eggs. Several of the nontarget species produced *Trichogramma* with elevated levels of brachyptery (2.2–58%), although only one exceeded the level of the factitious host *E. kuehniella* which had nearly 17% brachypterous adults.

Ten of the nontargets had overall host rankings that equaled or exceeded *O. nubilalis* (Table 1). Among these, *H. zea*, *Lacinipolia renigera* (Stephens), *Pseudaletia unipuncta* (Haworth), and *Xestia adela* Francelemont are widely distributed pest species. The remaining species with high potential host ranking include *Pyrrarctia isabella* (Smith and Abbott), *Haemetopis grataria* (F.), *Anagrapha falcifera* Kirby, *Caenurgina crassiuscula* (Haworth), *Leuconycta deptheroides*, and *Cercyonis pegala* (F.).

Developmental Conditioning

Both Kaiser et al. (1989a) and van Bergeijk et al. (1989) previously demonstrated increased levels of preference by *T. brassicae* for eggs of a non-preferred host in which pre-imaginal development had been completed. Both studies utilized *E. kuehniella*, the standard factitious laboratory host, compared with *O. nubilalis*, the target of biological control efforts with *T. brassicae*. We expanded this work to determine if altered host preference can take place following pre-imaginal development of *T. brassicae* through one generation in nontarget host species. This is critical for assessing whether *T. brassicae* that attack nontargets will produce progeny that are more likely to attack the same nontarget. Eggs of *O. nubilalis*, and two nontarget species, *P. unipuncta* and *Agrotis ipsilon* (Hufnagel) were used. *Agrotis ipsilon* was included in this and subsequent studies because eggs were commercially available and, while not collected in our light trap survey it is well known as a common cornfield Noctuid.

Trichogramma brassicae exhibited a ca. 5-fold greater preference for eggs of *P. unipuncta* than for eggs of either *O. nubilalis* or *A. ipsilon*. Preference for *P. unipuncta* was not significantly changed following development of *T. brassicae* in *P. unipuncta*

Table 1. Relative quality of southwest Michigan Lepidoptera species as hosts for *T. brassicae* in no-choice laboratory tests.

Host family	Host species	Ranking[a]
Arctiidae	*Holomelina aurantiaca* Hübner	6
	Pyrrharctia isabella (Smith & Abbot)	9
	Spilosoma virginica (F.)	3
Geometridae	*Haemetopis grataria* (F.)	8
	Xanthotype urticaria Swett	1
Hesperiidae	*Epargyreus clarus* (Cramer)	1
Lycaenidae	*Everes comyntas* Godart	2
Noctuidae	*Anagrapha falcifera* Kirby	8
	Caenurgina crassiuscula (Haworth)	8
	Heliocoverpa zea (Boddie)	8
	Lacinipolia renigera (Stephens)	8
	Leuconycta diphtheroides Guenée	8
	Melanchra picta (Harris)	1
	Pseudaletia unipuncta (Haworth)	8
	Xestia adela Francelemont	8
	Galgula partita Guenée	1
Pieridae	*Colias eurytheme* Boisduval	1
	Pieris rapae L.	3
Pyralidae	*Herpetogramma pertextalis* (Lederer)	4
	Ephestia kuehniella Zwolfer [b]	6
	Ostrinia nubilalis Hübner [c]	8
Satyridae	*Cercyonis pegala* (F.)	9
Sphingidae	*Hyles lineata* (F.)	1
Tortricidae	*Choristoneura rosaceana* (Harris)	1
Yponometridae	*Atteva punctella* (Cram.)	1

[a] Rankings 1–3 = poor quality hosts, 4–7 = moderate quality hosts, 8–10 = high quality hosts based on mean rankings of percent parasitism, percent emergence, sex ratio, and size.
[b] Factitious host utilized in commercial *Trichogramma* rearing.
[c] Target host.

eggs. However, rearing in *O. nubilalis* reduced *T. brassicae* preference for *P. unipuncta* eggs. Preference for either *O. nubilalis* or *A. ipsilon* eggs did not change, regardless of the rearing history of *T. brassicae* females. These data suggest that the host preference of *T. brassicae* which leave fields directly following a release would be unchanged by development in at least two nontarget hosts. In addition, *T. brassicae* that leave a field after developing in *O. nubilalis* (i.e., "recycled"), would actually have a reduced preference for *P. unipuncta*.

Table 2. Parasitism of nontarget lepidopteran eggs by *T. brassicae* under no-choice[a] field conditions in field corn.

Nontarget species [b]	Mean percentage ± SEM Egg swatches parasitized	Eggs parasitized within each swatch	Eggs parasitized in lab no-choice experiment
L. renigera	66.7 ± 32.1	57.4 ± 45.7	100 ± 0.0
P. unipuncta	55.6 ± 47.2	80.0 ± 29.9	65.6 ± 15.0
X. adela	56.7 ± 33.0	48.4 ± 47.8	98.9 ± 4.7

[a] Late-planted corn to avoid first generation *O. nubilalis* oviposition.
[b] Three replications, five 4-cm^2 swatches of cheesecloth with eggs for each species and each replication.

No-Choice Field Tests

For three of the most abundant and highly ranked nontarget hosts (*L. renigera*, *P. unipuncta*, and *X. adela*) we conducted field no-choice tests. No-choice conditions were created by late-planting a 0.405 ha corn plot so that attractiveness to first-generation *O. nubilalis* and numbers of naturally oviposited *O. nubilalis* egg masses would be low. These conditions provided the most conservative test of whether *T. brassicae* would attack nontarget species in the target habitat. Eggs of the three nontarget species oviposited on cheesecloth swatches were placed in the cornfield 24 h following release of ca. 86,000 female *T. brassicae* per hectare in a 5-m grid pattern. Egg swatches were recovered after 72 h and held to determine parasitism. These hosts were readily attacked by *T. brassicae* in the corn environment, although at lower levels than in lab no-choice tests (Table 2).

Nontarget Habitat Exploration

To determine if *T. brassicae* would search for hosts in nontarget habitats we released wasps directly into these habitats and measured the resulting parasitism on sentinel eggs. Three types of habitats: wetland, old field and forest were chosen as representative of a range of nontarget host habitats in the Midwestern United States. Diapausing *T. brassicae* (Trichocaps. BIOTOP, Valbonne, France) were reactivated in 30 ml clear plastic cups. A single Trichocap, containing approximately 500 *T. brassicae,* was placed in each cup and glued to the paper lid for handling purposes. Cups were supplied with honey and water to provide food to emerging adults. Upon emergence, insects were allowed to feed for 4–8 h before release in the selected habitat. In each plot, two releases totaling ca. 100,000 female *T. brassicae* per ha were conducted on two consecutive days as described above. *O. nubilalis, P. unipuncta,* and *A. ipsilon* eggs were used as the sentinels. The oviposition substrate on which eggs were laid was utilized to create sentinel egg swatches. One sentinel egg swatch per species was placed at each of three different heights (0.20, 1.00, and 2.00 m), in five equidistant locations within the sampling plots. Egg cards were placed side by side and pinned to existing vegetation. After 72 h, all cards were recovered and taken to the laboratory for incubation to determine the presence of parasitized eggs, predation, and for parasitoid identification. For these (and following) tests we did not analyze dates where we felt that weather was an overriding factor.

In corn, 16.7% of the *O. nubilalis* egg swatches were attacked by *T. brassicae*. Parasitism of *P. unipuncta* was 6% and *A. ipsilon* 25.9% in corn habitats. Sentinel egg swatch parasitism by *T. brassicae* released in nontarget habitats was significantly lower than that recorded in cornfields. Percent parasitism ranged from 2.08% in old fields to 5.49% in forest habitats.

Dispersal into Nontarget Habitats

In a preliminary examination of dispersal into nontarget habitats, sentinel egg swatches of three nontarget hosts were placed in a corn plot and adjacent habitats to assess whether *T. brassicae* released in corn would disperse into surrounding habitats. The corn plot was within 10 m of an apple orchard and poplar plantation so that prevailing winds could potentially carry dispersing *T. brassicae* into these nontarget habitats. Releases of ca. 86,000 *T. brassicae* females per ha were made 1 d prior to placing sentinel egg swatchs in the cornfield and adjacent habitats. Swatches were left in the field for 72 h, and each test was repeated three times. Despite high parasitism (70–90% of nontarget sentinel eggs in corn following release of *T. brassicae*, wasps did not locate and parasitize nontarget sentinel eggs in either the floor or canopy of the apple orchard or poplar plantation (Table 3).

In additional trials, cornfields adjacent to wetland, forest, and old field habitats were used to study post-release dispersal of *T. brassicae*. Fields were again selected such that adjacent nontarget habitats were located downwind of the prevailing wind. On two release dates in each habitat, parasitism of sentinel egg cards by *T. brassicae* was measured in nontarget habitats downwind from treated corn fields. Parasitism of egg swatches by *T. brassicae* was zero in the old field and wetland habitats. In the forest, parasitism occurred in only one of the three replicate habitats. In this site, two sentinel egg swatches of *A. ipsilon* were parasitized by *T. brassicae*, one ca. 20 m and the second ca. 32 m away from the release plot.

Overwintering

One factor that could reduce the potential for nontarget effects of a particular natural enemy is the failure of the enemy to overwinter in a given environment. An overwintering model was developed for diapausing *T. brassicae* based on a laboratory factorial experiment of temperature by exposure time, for temperatures ranging from 3 to –30°C and times ranging from 0 to 120 d. The model indicated that diapausing *T. brassicae* have the potential to overwinter successfully throughout most of the agricultural areas of the continental United States. However, no field data has been collected to substantiate this possibility. In addition, this model and any field data col-

Table 3. Parasitism of nontarget Lepidopteran eggs within corn and in adjacent down-wind habitats following release of *T. brassicae* into hybrid field corn.

	Habitat (mean ± SEM)[a]				
Nontarget species	Corn	Apple orchard		Poplar plantation	
	canopy	floor	canopy	floor	canopy
L. renigera	88.9 ± 26.5	0	0	0	0
P. unipuncta	90.0 ± 37.9	0	0	0	0
X. C. nigrum	70.0 ± 28.3	0	0	0	0

[a] Three replications, ten 4 cm^2 swatches of cheesecloth with eggs for each species and each replication.

lected using diapausing *T. brassicae* would include the assumption that the parasitoids are able to find a suitable host egg in a suitable habitat in which to enter diapause and successfully overwinter.

CONCLUSIONS

As a group, the *Trichogramma* clearly have biological characteristics that could allow them to impact nontarget organisms when used in augmentative releases. However, whether these characteristics actually result in nontarget impacts and the relative level and risk posed by those impacts is a question that can only be answered by empirical trials. We have attempted to develop a framework for characterizing the potential for attack of nontarget insects by *T. brassicae* and to address the relative likelihood of that occurring under field conditions. This framework uses biological criteria, i.e., temporal availability, coupled with simple laboratory tests and augmented by more realistic (and costly) field trials to characterize these risks. Ideally, one would like to minimize the number of large-scale trials by the development of simple and efficient tests that are predictive of what occurs in the field. This has proven to be an elusive goal for those involved in nontarget impact assessment.

Researchers have noted that in progressing from simple no-choice tests to more biologically realistic cage and field trials, the relative impact of parasitoids on nontarget organisms typically declines (Duan and Messing 1997 among others). We observed this to be the case with *T. brassicae*. For example, in no-choice lab tests *L. reginera* sustained 100% parasitism by *T. brassicae*, while field no-choice tests in the target habitat (corn), indicated 58–89% parasitism, and parasitism in nontarget habitats just 10 m away was zero. Similarly, for *P. unipuncta*, parasitism in lab no-choice tests was 100% and 80–90% in no-choice tests in corn habitats. However, when *P. unipuncta* was placed in nontarget habitats with *T. brassicae*, parasitism was reduced to 0–6.9%. Finally, when wasps were released in corn, no parasitism of *P. unipuncta* occurred in adjacent nontarget habitats.

Overall, we found little evidence that augmentative releases of *T. brassicae* are likely to be a persistent or important source of mortality to nontarget Lepidoptera outside of release habitats. However, the indication that overwintering is possible is a critical one, and must be explored in empirical trials. Our results indicate that for *T. brassicae*, development of predictive tests to determine potential for nontarget impacts should first focus on whether females will effectively search in nontarget habitats and secondly, if they will leave release sites to enter such nontarget habitats. Development of greenhouse or small cage studies to measure this effect is needed to make screening of larger numbers of nontargets cost efficient. If successful, these tests could be integrated into a general model of nontarget testing such as the one proposed by Van Driesche and Hoddle (1997).

ACKNOWLEDGMENTS

We thank M. Bell, D. Borchert, D. Carmona, A. Carvalho Silva, C. Fink, J. Forester, M. Haas, G. Hellmann, H. Martin, A. Preish and J. Zaslov for their assistance in the laboratory and field. Special thanks to J. Pinto for training on *Trichogramma* identification and F. Kabiri of BIOTOP, Valbonne, France for providing us with *T. brassicae*. Comments from J. Lee, F. Menalled, M. O'Neal and two anonymous reviewers have helped improve the manuscript. This work was part of the North Central Regional Project NC-205, Ecology and Management of European Corn Borer and other Stalk Boring Lepidoptera and funded in part by USDA NRI, Grant No. 9501531 to D.A.L. and D.B.O. and by the Michigan Agricultural Experiment Station.

REFERENCES

Altieri, M. A., W. J. Lewis, D. A. Nordlund, R. C. Gueldner, and J. W. Todd. 1981. Chemical interactions between plants and *Trichogramma* wasps in Georgia soybean fields. Prot. Ecol. 3: 259–263.

Andow, D., and D. Olson. 1992. Biological control of European corn borers with *Trichogramma*. Proc., Illinois Agric. Pesticides Conf. p. 52–65.

Andow, D. A., and D. R. Prokrym. 1990. Plant structural complexity and host–finding by a parasitoid. Oecologia. 82: 162–165.

Andow, D. A., C. P. Lane, and D. M. Olson. 1995. The use of *Trichogramma* in maize–estimating environmental risks. *In* H. M. T. Hokkanen and J. M. Lynch (eds.) Biological control: benefits and risks. Cambridge Univ. Press. p. 101–118.

Bigler, F. 1986. Mass production of *Trichogramma* maidis Pint. et Voeg. and its field application against *Ostrinia nubilalis* Hbn. in Switzerland. J. Appl. Entomol. 101: 23–29.

Bigler, F. 1989. Quality assessment and control in entomophagus insects used for biological control. J. Appl. Entomol. 108: 390–400.

Bigler, F., J. Baldinger, and L. Luisoni. 1982. L'impact de la methode d'elevage it de l'hote sur la qualite intrinseque de *Trichogramma evanescens* Westw. Les Trichogrammes, Colloques INRA 9: 167–180.

Bigler, F., M. Bieri, A. Fritschy, and K. Seidel. 1988. Variation in locomotion between laboratory strains of *Trichogramma maidis* and its impact on parasitism of eggs of *Ostrinia nubilalis* in the field. Entomol. Exp. Appl. 49:283–290.

Bigler, F., S. Bosshart, and M. Waldburger. 1989. Bisherige und neue Entwicklungen bei der biologischen Bekampfung des Maiszunslers, *Ostrinia nubilalis* Hbn., mit *Trichogramma maidis* Pint. et Voeg. in der Schweiz. Landwirtschaft Schweiz Band 2: 37–43.

Bigler, F., S. Bosshart, M. Waldburger, and M. Ingold. 1990. Dispersal of *Trichogramma evanescens* Westw. and its impact on parasitism of eggs of *Ostrinia nubilalis* Hbn. Bulletin de la Societe Entomologique Suisse 63: 381–388.

Boldt, P. E. 1974. Temperature, humidity, and host: effect on rate of search of *Trichogramma evanescens* and *T. minutum auctt.* (not Riley, 1871). *In* Biological control of insects research, U.S. Dept. Agric., Agric. Res. Svc., Colombia, Missouri. p. 706–708.

Brenière, J. 1965. Les Trichogrammes parasites de Proceras sacchariphages Boj. borer de la canne a sucre a Madagascar. Quatrieme partie: etude comparee de plusieurs especes de Trichogrammes. Entomophaga 10: 273–294.

Caltagirone, L. E. 1981. Landmark examples in classical biological control. Annu. Rev. Entomol. 26: 213–232.

Carruthers, R. I., and J. A. Onsager. 1993. Perspective on the use of exotic natural enemies for biological control of pest grasshoppers (Orthoptera: Acrididae). Environ. Entomol. 22: 885–903.

Cerutti, F., and F. Bigler. 1991. Methods for the quality evaluation of *Trichogramma evanescens* Westw. used against the European corn borer. Proc. 5th workshop of the global IOBC working group "Quality control of mass reared organisms." Wageningen, Netherlands, March 1991. p 119–126.

Clausen, C. P. 1940. Entomophagous Insects. Hafner Publishing Co., New York.

Curl, G., and P. P. Burbutis. 1978. Host-preference studies with *Trichogramma nubilale*. Environ. Entomol. 7: 541–543.

De Jong, E. J., and G. A. Pak. 1984. Factors determining host-egg recognition of two host-species by different *Trichogramma* spp. Mededelingen van de Faculteit Landbouwweten Shappen Rijksuniversiteit Gent 49: 815–825.

DeBach, P. 1974. Biological control by natural enemies. Cambridge Univ. Press, London, UK.

Doutt, R. L. 1972. Biological control: parasites and predators. *In* Pest control strategies for the future. Natl. Acad. Sci., Washington, D.C., USA.

Duan, J. J., and R. H. Messing. 1997. Biological control of fruit flies in Hawaii: factors affecting nontarget risk analysis. Agric. Human Values. 14: 227–236.

Frandon, J., and F. Kabiri. 1990. [Biological control of second generation European corn borer with *Trichogramma.*] Proc., ANPP Second International Conference on Agricultural Pests, Versailles, December 4–6, 1990. (in French, translated by P. Matteson).

Frank, J. H. 1998. How risky is biological control?; comment. Ecology 79: 1829–1834.

Garcia–Salazar, C., D. B. Orr, and D. A. Landis. Dispersal and nontarget habitat exploration by *Trichogramma brassicae* Bezdenko (Hymenoptera: Trichogrammatidae). In prep.

Hassan, S. A., and M. F. Guo. 1991. Selection of effective strains of egg parasites of the genus *Trichogramma* (Hym., Trichogrammatidae) to control the European corn borer *Ostrinia nubilalis* Hb. (Lep., Pyralidae). J. Appl. Entomol. 111: 335–341.

Howarth, F. G. 1983. Classical biological control: panacea or Pandora's box? Proc. Hawaiian Entomol. Soc. 24: 239–244.

Howarth, F. G. 1991. Environmental impacts of classical biological control. Annu. Rev. Entomol. 36: 485–509.

Hoy, M. A., R. M. Nowierski, M. W. Johnson, and J. L. Flexner. 1991. Issues and ethics in commercial releases of arthropod natural enemies. Am. Entomologist 37: 74–75.

Huettel, M. D. 1976. Monitoring the quality of laboratory–reared insects. A biological and behavioral perspective. Environ. Entomol. 5: 807–814.

Kabiri, F., J. Frandon, J. Voegele, N. Hawlitzky, and M. Stengel. 1990 [Evolution of a strategy for inundative releases of *Trichogramma brassicae* Bezd. (Hym.: Trichogrammatidae) against the European corn borer, *Ostrinia nubilalis* Hbn. (Lep.: Pyralidae)]. *In* Proc., ANPP—2nd Intl. Conf. on Agricultural Pests, Versailles, Dec. 4–6, 1990. (in French, translated by P. Matteson).

Kaiser, L., H. Pham–Delegue, and C. Masson. 1989a. Behavioral study of plasticity in host preferences of *Trichogramma maidis* (Hym.: Trichogrammatidae). Physiol. Entomol. 14: 53–60.

Kaiser, L., H. Pham–Delegue, E. Bakchine, and C. Masson. 1989b. Olfactory responses of *Trichogramma maidis* Pint. et Voeg.: Effects of chemical cues and behavioral plasticity. J. Insect Behav. 2: 701–712.

Kanour, W. W., and P. P. Burbutis. 1984. *Trichogramma nubilale* (Hymenoptera: Trichogrammatidae) field releases in corn and a hypothetical model for control of European corn borer (Lepidoptera: Pyralidae). J. Econ. Entomol. 77: 103–107.

Laing. J. 1937. Host finding by insect parasites; chances of *Trichogramma evanescens* finding its host. J. Biol. 15: 281–302.

Li, Li–Yang. 1994. Worldwide use of *Trichogramma* for biological control on different crops: a survey. p. 37–54. *In* E. Wajnberg, and S.A. Hassan (eds.) Biological control with egg parasitoids. CAB International, Oxon, UK.

Lockwood, J. A. 1993a. Environmental issues involved in biological control of rangeland grasshoppers

(Orthoptera: Acrididae) with exotic agents. Environ. Entomol. 22: 503–518.
Lockwood, J. A. 1993b. Benefits and costs of controlling rangeland grasshoppers (Orthoptera: Acrididae) with exotic organisms: Search for a null hypothesis and regulatory compromise. Environ. Entomol. 22: 904–914.
Lockwood. J. A. 1997. Competing values and moral imperatives: an overview of ethical issues in biological control. Agric. Human Values 14: 205–210.
Marston, N., and L. R. Ertle. 1973. Host influence on the bionomics of *Trichogramma minutum*. Ann. Entomol. Soc. Am. 66: 1155–1162.
Mayer, K., and W. Quednau. 1959. Verhaltensanderungen bei Eiparasiten der Gattung *Trichogramma* unter dem Einjluss des Wirtes. Z. fur Parasitenkunde 19:35–41.
McDougall, S. J., and N. J. Mills. 1997. Dispersal of *Trichogramma platneri* Nagarkatti (Hym., Trichogramatidae) from point–source releases in an apple orchard in California. J. Appl. Entomol. 121: 205–209.
Moffat, A. S. 1991. Research on biological pest control moves ahead. Science 252: 211–212.
Need, J. T., and P. P. Burburtis. 1979. Searching efficiency of *Trichogramma nubilale*. Environ. Entomol. 8: 224–227.
Noldus, L. P. J., and J. C. van Lenteren. 1985. Kairomones for the egg parasite
Trichogramma evanescens Westwood. II. Effect of contact chemicals produced by two of its hosts, *Pieris brassicae* L. and *Mamestra brassicae* L. J. Chem. Ecol. 11: 781–791.
Nordlund, D. A., and S. M. Greenberg. 1994. Facilities and automation for the mass production of arthropod predators and parasitoids. Biol. Control News Info. 4: 45–50.
Nordlund, D. A., R. B. Chalfant, and W. J. Lewis. 1985. Response of *Trichogramma pretiosum* females to volatile synomones from tomato plants. J. Entomol. Sci. 20: 372–376.
Nordlund, D. A., Z. X. Wu, A. C. Cohen, and S. M. Greenberg. 1998. Recent Advances in the In Vitro Rearing of *Trichogramma* spp. Proc. 5th Intl. Symp. "*Trichogramma* and other egg parasitoids." In press.
Orr, D. B. 1993. Biological control tactics for European corn borer. Proc. Illinois Crop Prot. Workshop, March 3–5, 1993, Champaign, Illinois.
Orr, D. B., L. C. Lewis, J. J. Obrycki, and M. E. Rice. 1992. Economic comparison of biological and chemical control of the European corn borer. Proc. 1992 Iowa Crop Prod. and Prot. Conf., December 3–4, 1992, Ames, Iowa., p.173–183.
Orr, D. B., D. A. Landis, and G. V. Manley. Range, quality and acceptance of Midwestern Lepidoptera species as hosts for *Trichogramma brassicae* (Hymenoptera: Trichogrammatidae). In prep.
Parrella, M. P., K. M. Heinz, and L. Nunney. 1992. Biological control through augmentative release of natural enemies: a strategy whose time has come. Am. Entomologist 38: 172–179.
Pinto, J. D., and R. Stouthamer. 1994. Systematics of the Trichogrammatidae with emphasis on *Trichogramma In* E. Wajnberg and S. A. Hassan (eds.) Biological control with egg parasitoids. CAB International, Oxon, UK. p. 1–36
Pintureau, B. 1990. Polymorphisme, biogeographie et specificite parasitaire des Trichogrammes europeens (Hym. Trichogrammatidae). Bull. Soc. Ent. Fr., 95: 17–38.
Prokrym, D. R., D. A. Andow, J. A. Ciborowski, and D. D. Sreenivasam. 1992. Suppression of *Ostrinia nubilalis* by *Trichogramma nubilale* in sweet corn. Entomol. Exp. Appl. 64: 74–85.
Renou, M., P. Nagnan, A. Berthier, and C. Durier. 1992. Identification of compounds form the eggs of *Ostrinia nubilalis* and *Mamestra brassicae* having kairomone activity on *Trichogramma brassicae*. 63: 291–303.
Salt, G. 1934. Experimental studies in insect parasitism. II. Superparasitism. Proc. Royal Soc. London, Series B, 114: 455–476.
Salt, G. 1935. Experimental studies in insect parasitism III: Host selection. Proc. Royal Soc. London, Series B, 117: 413–435.
Salt, G. 1940. Experimental studies in insect parasitism VII: The effects of different hosts on the parasite *Trichogramma evanescens* West. (Hym.: Chalcidoidae). Proc. Royal Soc. London, Series A, 15: 81–95.
Shu, S., and R.L. Jones. 1989. Kinetic effects of a kairomone in moth scales of the European corn borer on *Trichogramma nubilale* Ertle and Davis (Hymenoptera: Trichogrammatidae). J. Insect Behav. 2: 123–131.
Shu, S., P. D. Swedenborg, and R. L. Jones. 1990. A kairomone for *Trichogramma nubilale* (Hymenoptera: Trichogrammatidae) isolation, identification, and synthesis. J. Chem. Ecol. 16: 521–529.
Simberloff, D. 1992. Conservation of pristine habitats and unintended effects of biological control. *In* W.C.

Kauffman and J. E. Nechols (eds.), Thomas Say Proc.: Selection criteria and ecological consequences of importing natural enemies. Entomol. Soc. Amer. Lanham, Maryland. p. 103–17.

Simberloff, D., and P. Stiling. 1996. How risky is biological control? Ecology 77: 1865–1974.

Simberloff, D., and P. Stiling. 1998. How risky is biological control?; Reply. Ecology 79: 1834–36.

Southard, S. G., M. W. Houseweart, D. T. Jennings, and W.A. Halteman. 1982. Size differences of laboratory reared and wild populations of *Trichogramma minutum* (Hymenoptera: Trichogrammatidae). Can. Entomol. 114: 693–698.

Stern, V. M., E. I. Schlinger, and W. R. Bowen. 1965. Dispersal studies of *Trichogramma semifumatum* (Hymenoptera: Trichogrammatidae) tagged with radioactive phosphorous. Ann. Entomol. Soc. Am. 58: 234–240.

Taylor, T. A., and V. M. Stern. 1971. Host-preference studies with the egg parasite *Trichogramma semifumatum* (Hymenoptera: Trichogrammatidae). Ann. Entomol. Soc. Am. 64: 1381–1390.

Thompson, M. S. and R. E. Stinner. 1990. The scale response of *Trichogramma* (Hymenoptera: Trichogrammatidae) variation among species in host specificity and the effect of conditioning. Entomophaga, 35: 7–21.

van Bergeijk, K. E., F. Bigler, N. K. Kaashoek and G. A. Pak. 1989. Changes in host acceptance and host suitability as an effect of rearing *Trichogramma maidis* on a factitious host. Entomol. Exp. Appl. 52: 229–238.

Van Driesche, R.G. and M. Hoddle. 1997. Should arthropod parasitoids and predators be subject to host range testing when used as biological control agents? Agric. Human Values 14: 211–226.

Wajnberg, E. and S. A. Hassan. 1994. Biological control with egg parasitoids. CAB International, Oxon, UK. 286 p.

Zaki, P. N. 1985. Reactions of the egg parasitoid *Trichogramma evanescens* Westw. to certain insect sex pheromones. J. Appl. Entomol. 99: 448–453.

9

COCCINELLID INTRODUCTIONS: POTENTIAL FOR AND EVALUATION OF NONTARGET EFFECTS

John J. Obrycki
Department of Entomology, Iowa State University, Ames, Iowa

Norman C. Elliott
USDA-ARS Plant Sciences Research Laboratory, Stillwater, Oklahoma

Kristopher L. Giles
Department of Entomology and Plant Pathology, Oklahoma State University, Stillwater, Oklahoma

INTRODUCTION

Aphid-feeding Coccinellidae, subfamily Coccinellinae, are found in many terrestrial habitats and have been used extensively for biological control of aphid pest species (Hagen 1962, 1974; Frazer 1988; Hodek and Honek 1996; Obrycki and Kring 1998). Most if not all aphidophagous Coccinellidae feed on several prey items (Hodek 1993, Hodek and Honek 1996); the question is what level of prey specificity is required for the safe introduction of these predators? The selection of coccinellid species for importation is typically based on field observations of predation on the target pest and/or consistent associations of the coccinellid and prey species in their native range. While these observations are a logical first step, observed predation does not necessarily mean that the prey is preferred or suitable for development and reproduction (Strand and Obrycki 1996, Obrycki et al. 1997a).

In this article we discuss data on the effects of selected exotic Coccinellidae species that have established and spread in North America. We focus our discussion primarily on the potential of these exotics to displace native coccinellids because competitive exclusion is an acknowledged and well-documented consequence of successful biological control programs (Huffaker et al. 1976 and references therein). Furthermore, from the standpoint of biodiversity protection, there are often more species at risk on the same trophic level as the introduced species, due to competition, than are at risk on the next lower trophic level, due to the direct effects of predation (Simberloff and Stiling 1996, Samways 1997). Most of the experimental work has been conducted in the mid-western USA. We consider field data and laboratory studies that have examined the patterns and mechanisms underlying the interactions between introduced and native coccinellids. We also examine the factors that are known to influence coccinellid population dynamics. Additionally, we discuss the biology and ecology of selected native and introduced species to ascertain if the outcomes of their interactions can be predicted. Finally, we consider the role of aphidophagous Coccinellidae in importation biological control—Do the benefits outweigh the risks?

EFFECTS OF INTRODUCED COCCINELLIDS

To evaluate the benefits of introducing aphidophagous coccinellid species it is necessary to quantify the levels of existing biological control from native Coccinellidae. Because most aphid predators are not prey specific, an introduced aphid pest species is likely to be attacked to some degree by native predatory species (e.g., Goodarzy and Davis 1958, Archer and Bynum 1993, Wraight et al. 1993, Messina et al. 1995). Introductions of exotic predators are planned when existing levels of biological control do not meet economic expectations. Thus, data are needed to determine if the introduction of an exotic aphid predator has a positive effect on levels of biological control.

Because many aphid predators are not habitat or prey specific, introductions of these predators raise concerns about their effects on invertebrate biodiversity, possible disruption of existing biological control programs (Evans and England 1996), and alteration of existing trophic level interactions (see Obrycki and Kring 1998). The only example that we are aware of that illustrates a possible non-target effect on an endangered or threatened species is a study done by Horn (1991). He observed the introduced coccinellid, *Coccinella septempunctata*, in the wetland habitat of three endangered Lepidopteran species and in a laboratory study noted that larvae and adults consumed eggs of congeneric species of the endangered Lepidoptera.

Current diversity of coccinellid species in sections of the midwestern USA appears to be reduced compared to pre-1950 levels (e.g., Wingo 1952). These declines may be due to loss of habitat, or changes in agricultural practices, i.e., reduced diversity of crops and fewer rotations to multiyear pastures or alfalfa fields. However, in South Dakota, a steady reduction in alfalfa acreage over an 18-year period was unrelated to change in coccinellid community structure (Elliott and Kieckhefer 1990a, b). It is possible that introduced coccinellid species that compete with native species for resources, combined with loss of habitat, may cause declines of native species.

The above target and non-target effects need to be examined to assess the environmental impact of releases of predatory coccinellids. In addition, the examination of interactions between invading and indigenous predatory coccinellid species provides an understanding of population and community dynamics in ephemeral habitats.

RECENT COCCINELLID INVASIONS INTO NORTH AMERICA

During the twentieth century, at least eight aphidophagous coccinellid species have established in North America (Gordon 1985, Obrycki and Kring 1998, Day et al. 1994): *Coccinella undecimpunctata*, *Coccinella septempunctata*, *Harmonia axyridis*, *Harmonia dimidiata*, *Harmonia quadripunctata*, *Hippodamia variegata*, *Propylea quatuordecimpunctata*, and *Scymnus* (*Pullus*) *suturalis*. Although most were involved in some importation effort, it is possible that none of these species was established as a result of intentional releases.

In the late 1970s and 1980s, *C. septempunctata* rapidly expanded its range and is now widely distributed in North America (Schaefer et al. 1987, Elliott and Kieckhefer 1990a, b; Giles et al. 1994, Evans and England 1996). Several studies have examined the effects of *C. septempunctata* on native species and will be discussed in sections III and IV.

Starting in 1916, repeated releases of Asian *H. axyridis* were made in North America for aphid and scale pests (Gordon 1985, McClure 1987, Tedders and Schaefer 1994). In 1988 and 1991, the first collections of *H. axyridis* were made in Louisiana and Georgia, hundreds of kilometers from release sites (Chapin and Brou 1991, Tedders and Schaefer 1994). This species is now widely distributed in North America (LaMana and Miller 1996, Nalepa et al. 1996, Krafsur et al. 1997).

The first records of *P. quatuordecimpunctata* in Quebec, Canada in 1968 (Chantal 1972) suggest that adults were transported to North America on ships using the St. Lawrence Seaway (Wheeler 1993, Day et al. 1994); however, another introduction of *P. quatuordecimpunctata* likely occurred via the port of New York (Yanega 1996).

In contrast to *C. septempunctata* and *H. axyridis*, *P. quatuordecimpunctata* and *Hippodamia variegata* have spread relatively slowly in northeastern North America (Day et al. 1994, Hoebeke and Wheeler 1996, Wheeler and Stoops 1996, Yanega 1996). The remaining introduced species have not been extensively studied, and several have relatively restricted distributions in North America. As a result, very little is known about their biology and ecology in North America and their interactions with native aphids and coccinellids (see Gordon 1985).

OVERVIEW OF COCCINELLID POPULATION AND COMMUNITY ECOLOGY

An understanding of the factors involved in coccinellid population and community dynamics is needed to predict the outcome of introductions of exotic coccinellids. These factors are best considered from a multi-scale perspective, because coccinellid population and community dynamics are determined by factors operating at several spatial and temporal scales. Factors that appear to be important at a particular scale may not be relevant to the population dynamics of a species. For example, if agricultural crops are not preferred habitats, population fluctuations in agricultural fields would provide little insight into landscape scale population fluctuations, especially if the species did not reproduce in the fields.

Coccinellid populations fluctuate widely from year to year, and fluctuations of different species are often unrelated (Elliott and Kieckhefer 1990a, b; Kieckhefer and Elliott 1990). Thus, univariate explanations of population fluctuations, e.g. variation in prey density among years, are inadequate. Several abiotic and biotic factors involved in coccinellid population and community ecology are summarized below. Ranking the importance of these factors in particular circumstances is often not possible due to our limited understanding of the intricacies of coccinellid ecology.

General Patterns and Processes

Population fluctuations of some coccinellid species are correlated over great distances. For example, from 1988 to 1992, coccinellid populations in several alfalfa fields in eastern South Dakota and central Iowa were sampled during the growing season by sweep net (JJO and KLG unpublished, Elliott et al. 1996). These two locations are separated by over 300 km. Spearman rank correlations of abundance at the two locations were significant for three of the six species (Table 1), indicating that populations of these species interact via dispersal over long distances or respond similarly to environmental factors that are correlated over large geographic areas. Similarly, studies of gene flow among populations of several predaceous coccinellid species indicate long distance movement of individuals (Krafsur et al. 1992, 1995, 1996, 1997).

At the landscape scale, factors such as topography and spatial heterogeneity determine the size and diversity of the pool of potential colonists of a particular habitat (Honek 1985, Colunga-Garcia et al. 1997, Elliott et al. 1998a). Landscape structure influences the availability and spatial distribution of food and basic requirements such as overwintering habitat, but its effects are intertwined with the life history of a species. For example, dispersal behaviors differ within the seasonal life cycle of a species and among species (Hagen 1962, Ewert and Chiang 1966, Hodek et al. 1993, Elliott et al. 1998b). The successful use of resources by a species in a mosaic of landuse types is largely dependent on characteristics of movement in relation to landscape structure (Dunning et al. 1992). At a local, ecosystem scale, plant community composition and physical structure (due to its influence on microclimate), prey abundance, and interspecies interactions are key determinants of coccinellid abundance and species diversity (Honek 1982a, 1983, 1985).

Interactions among Coccinellidae

Interspecific competition for food or intraguild predation can occur among coccinellids during both the immature and adult stages. The degree to which such interactions occur depends on the extent to which species occupy the same habitats and prey on similar species. Information on the extent of niche overlap in aphidophagous Coccinellidae is limited. The range of habitats occupied by larvae often is more re-

Table 1. Spearman rank correlations of coccinellid abundances in eastern South Dakota and central Iowa from 1988 to 1992; (n = 5 years).

Species	Spearman rank correlation	P-value
Coleomegilla maculata	0.60	0.285
Cycloneda munda	0.97	0.005
Hippodamia convergens	0.90	0.037
Hippodamia tredecimpunctata	0.90	0.037
Hippodamia parenthesis	0.60	0.285
Coccinella septempunctata	0.60	0.285

stricted than that of adults because adults may occupy habitats in which they rarely, if ever, reproduce (Kieckhefer et al. 1992, Hodek and Honek 1996).

Interspecific interactions among coccinellids can occur among most life stages, but it is during the immature stages when competition and intra-guild predation may be most intense (Takahashi 1989, Ormord 1994, Triltsch 1997). For this reason, habitats in which a particular native and exotic species coexist and reproduce have the highest potential for adverse effects. If adults of two species forage in similar habitats, but one or both rarely reproduce in these habitats, negative effects of one species on another are less likely to occur. Additionally, coccinellid species often exhibit different spatial and temporal patterns of habitat use within a field (Ewert and Chiang 1966, Smith 1971, Coderre et al. 1987) which should promote coexistence.

In unpredictable habitats, where aphids exhibit patchy distributions in time and space, the invasion of an exotic coccinellid into a primarily aphidophagous predator guild may increase intra- and inter-specific interactions (Agarwala and Bhattacharya 1995). Consequently, competition for limited prey among coccinellids, which can manifest itself as intraguild predation, may reduce the survival, growth, and/or fecundity of coccinellid species on a local scale (Ives 1981, Connell 1983, Schoener 1983, Keddy 1989, Osawa 1989, Polis and Holt 1992, Rosenheim et al. 1995, Obrycki et al. 1998a).

Intraspecific Competition

The limited number of studies on coccinellid intraspecific competition suggests that levels and outcomes of competition are species dependent. The primarily arboreal coccinellid *Adalia bipunctata* shows ovipositional restraint in the presence of conspecific females or larvae (Hemptinne et al. 1992), reducing the potential for competition among offspring. Cannibalism of eggs and younger larvae by fourth instar *A. bipunctata* increases as prey levels decrease (Agarwala and Dixon 1992). Evans (1991) demonstrated intraspecific competition for *Hippodamia convergens* and *Hippodamia tredecimpunctata*; larvae gained less weight when paired with conspecifics. In a laboratory study conducted with limited prey, Obrycki et al. (1998a) observed no intraspecific competitive effects on survival, development time, and adult weight for *Coleomegilla maculata* and *Coccinella septempunctata*.

Evans (1991) suggests that intraspecific competition is more intense than interspecific competition because of species differences in seasonal abundance and foraging habits. The evidence for intraspecific competition among Nearctic coccinellid species suggests that the introduction of an exotic competitor may indirectly affect indigenous species. Reductions in prey abundance by an invading coccinellid may increase intraspecific competition within the predator guild and result in smaller populations of native species.

Interspecific Competition

Interspecific competition among coccinellids can take the form of competition for a limited resource and/or intraguild predation (Ives 1981, Polis and Holt 1992, Rosenheim et al. 1995, Obrycki et al. 1998a). As previously discussed, ovipositional restraint and sibling recognition to avoid cannibalism may result in reduced levels of intraspecific competition (Agarwala and Bhattacharya 1995). However, these behaviors are likely less important if an introduced competitor reduces prey abundance, out-competes indigenous species for limited prey, and/or exhibits significant intraguild predation.

The observed declines of *C. novemnotata*, *A. bipunctata*, *C. transversoguttata*, and other native coccinellid species in North America may be due to the activities of the introduced Palearctic coccinellid *C. septempunctata* and the Asian coccinellid *H. axyridis* (Wheeler and Hoebeke 1995, Elliott et al. 1996, Colunga-Garcia and Gage 1998). However, relatively few studies have quantified effects that introduced species might have on native species. The effects of interspecific competition on weight gain among third instar *C. septempunctata* and *H. convergens* was similar to the effects of intraspecific competition (Evans 1991). However, the high initial prey densities (eight aphids per predator) and the short duration of the study (48 hours) do not indicate that prey were limiting. Obrycki et al. (1998a) demonstrated that survival of *C. maculata* was significantly reduced, whereas *C. septempunctata* survival significantly increased when reared with a heterospecific larva under low prey densities (one aphid per predator per day). This asymmetry indicates that *C. septempunctata* may be a better competitor for limited prey and/or exhibits higher levels of intraguild predation. Previously, Agarwala and Dixon (1992) observed that starved fourth instar *C. septempunctata* are more likely to prey on heterospecific larvae.

However, in a two-year field cage study, no negative interspecific interactions were observed between *C. septempunctata* and *C. maculata* larvae over a wide range of prey densities (Obrycki et al. 1998b). Additionally, interspecific interactions between *C. septempunctata* and *C. maculata* larvae did not affect *Myzus persicae* density on potatoes (Obrycki et al. 1998b).

Quantitative studies of intra- and inter-specific interactions of coccinellids can provide insights into the mechanisms and outcomes of competitive interactions, the competitive advantage one species may have over another, and the dynamics of the predatory guild. Results from these types of studies would be most predictive when prey habitats are simulated experimentally by manipulating predator-prey densities (Agarwala and Bhattacharya 1995).

Interactions Among Aphidophaga

Coccinellidae occupy the same trophic level and often overlapping niches with many other aphidophagous organisms (Obrycki and Kring 1998, Phoofolo and Obrycki 1998, Rosenheim 1998). A broad taxonomic range of parasitoids, insect predators, and entomopathogens compete for aphid prey within the same spatial and temporal ecological boundaries. The interactions and competitive outcomes among these organisms are not well understood, but would likely be similar to those among Coccinellidae. Competitive exclusion, direct intraguild predation, or intensified intraspecific competition due to limited prey may result in lower populations of natural enemies that compete with Coccinellidae (Wheeler 1968, Samways 1997, Obrycki and Kring 1998, Rosenheim 1998).

The limited number of studies on competitive interaction between Coccinellidae and other aphidophagous organisms suggest species dependent outcomes for both competitive displacement and resulting biological control. Interactions between indigenous Coccinellidae and other aphidophagous organisms may reveal mechanisms that predict results of Coccinellidae introductions. In experimental exclusion cages, *Hippodamia convergens* consumed 73–100% of immature *Lysiphlebus testaceipes* within mummified *Aphis gossypii* on cotton (Colfer and Rosenheim 1995). This intraguild predation limited the effectiveness of *L. testaceipes* during the experiment, but did not disrupt regulation of *A. gossypii* populations; aphid numbers were lower when beetles, wasps, and aphids were present compared with cages with wasps and aphids only. In contrast, Ferguson and Stiling (1996) concluded that predation of parasitized saltmarsh aphids by *Cycloneda sanguinea* reduced the effectiveness of the parasitoid *Aphidius floridaensis* on march elder. Resulting saltmarsh aphid populations were lower in cages containing parasitoids and aphids compared to cages with parasitoids, beetles, and aphids.

The effects of introduced Coccinellidae on interspecific competition with other aphidophagous organisms and subsequent biological control has been little studied. Chang (1996) found no clear evidence of intraguild predation by *Coccinella septempunctata* on *Chysoperla plorabunda* or that the presence of both predators affected biological control of bean aphid populations. Evans and England (1996) concluded that *C. septempunctata* may indirectly influence existing biological control of the alfalfa weevil by reducing pea aphid populations and thus available honeydew for the weevil parasitoid *Bathyplectes curculionis*. Parasitism levels of weevils by *B. curculionis* were reduced when *C. septempunctata* were present, however, predation by beetles decreased the number of late maturing weevil larvae.

In a unique study examining the effect of *C. septempunctata* feeding on pea aphids that were infected with the fungal entomopathogen *Erynia neoaphidis*, Roy et al. (1998) concluded that beetle feeding affects sporulation. When aphids were damaged (fed upon) within 2 days of infection by *E. neoaphidis* resulting cadavers did not sporulate. Feeding on 3-day or older infected aphids reduced conidia density, but did not affect levels of transmission to healthy aphids. In addition, foraging adults increased transmission levels of *E. neoaphidis* among aphids, likely by mechanically spreading conidia.

COCCINELLID COMMUNITIES IN AGRICULTURAL CROPS IN THE MIDWESTERN USA

Coccinellid communities in agricultural crops are generally comprised of relatively few species (Hodek and Honek 1996). The coccinellid communities in alfalfa and small grain crops in the midwestern USA consist of fewer than 10 native species, most of which occur in both crops (Table 2). Community composition differs somewhat geographically within the region. For example, *Hippodamia tredecimpunctata* is common in the Northern Plains region but declines in abundance in the Central Plains and is entirely absent from the Southern Plains. Conversely, *Hippodamia sinuata* is abundant in the southern Great Plains but is absent from the Northern Plains. Several species occur throughout the region; for example, the native species *H. convergens* and *C. maculata* and the introduced *C. septempunctata.*

The coccinellid species found in agricultural crops in the midwestern USA are highly mobile and have relatively broad habitat ranges. As a result, their populations may exhibit complex interactions related to spatial patterns of habitat use and oviposition, and differences in seasonal phenology (Wright and Laing 1980; Coderre et al. 1987; Elliott and Kieckhefer 1990a, b; Kieckhefer et al. 1992; Colunga-Garcia et al. 1997; Colunga-Garcia and Gage 1998; Elliott et al. 1998a). The reasons why coccinellids prefer certain habitats and avoid others are poorly understood. Furthermore, plasticity in patterns of habitat use has been observed for geographic populations of some species.

Many species of Coccinellidae are influenced by spatial heterogeneity occurring at the landscape scale (Colunga-Garcia et al. 1997, Elliott et al. 1998a). For example, because of their life history characteristics, *Coleomegilla maculata* and *Hippodamia parenthesis* might be expected to respond to landscape variability at a relatively fine spatial scale (Ewert and Chiang 1966, Hemptinne 1988, Roach and Thomas 1991, Kieckhefer et al. 1992, Elliott et al. 1998a). The high mobility of *C. septempunctata* (Honek 1982b, 1990) and especially *H. convergens* (Hagen 1962), relative to *C. maculata* and *H. parenthesis*, suggests that populations of these species may be more dependent on heterogeneity at a broader spatial scale.

Potential for Displacement of Native Species

The exotic coccinellid *C. septempunctata* has been implicated in the decline in abundance of *A. bipunctata* and *C. transversoguttata* in agricultural crops in eastern South Dakota (Elliott et al. 1996). Populations of the latter two native species decreased from historical levels of abundance precisely when *C. septempunctata* invaded the region and remained low for five years. Several other species occur in agricultural crops in eastern South Dakota and other parts of the midwestern USA (Table 2), but five-year studies in South Dakota and Iowa failed to demonstrate population declines for these species (JJO and KLG unpublished, Elliott et al. 1996).

Based upon trends in sampling data in eastern North America, a decline in the indigenous species *Coccinella novemnotata* appears to be correlated with the introduction of *C. septempunctata* (Wheeler and Hoebeke 1995). The distribution of *C. novemnotata* includes the midwest (Gordon 1985), but this species has not been ob-

Table 2. Coccinellids of wheat and alfalfa at several locations in the midwestern United States.

Alfalfa

Location	S. Dakota	Iowa	Oklahoma	Texas
Years	1988–92	1988–92	1991–92	1991–92
Hippodamia convergens	Common[a]	Common	Common	Common
H. glacialis	Rare	Rare	Absent	Absent
H. tredecimpunctata	Common	Uncommon	Absent	Absent
H. parenthesis	Common	Common	Uncommon	Uncommon
H. sinuata	Absent	Absent	Absent	Common
Coleomegilla maculata	Common	Common	Common	Common
Coccinella tranversoguttata	Rare	Absent	Absent	Absent
C. septempunctata	Common	Common	Common	Common
Cycloneda munda	Rare	Uncommon	Uncommon	Uncommon
Adalia bipunctata	Absent	Rare	Rare	Rare
Olla v-nigrum	Absent	Absent	Rare	Uncommon
Scymnus spp.	Absent	Rare	Uncommon	Uncommon

Wheat

Location	S. Dakota	Iowa	Oklahoma	Texas
Years	1988–92	no data	1991–92	1991–92
H. convergens	Common		Common	Common
H. glacialis	Absent		Absent	Absent
H. tredecimpunctata	Common		Absent	Absent
H. parenthesis	Common		Uncommon	Uncommon
H. sinuata	Absent		Absent	Common
C. maculata	Common		Common	Uncommon
C. tranversoguttata	Rare		Absent	Absent
C. septempunctata	Common		Common	Common
C. munda	Rare		Rare	Rare
A. bipunctata	Absent		Absent	Absent
O. v-nigrum	Absent		Rare	Uncommon
Scymnus spp.	Absent		Uncommon	Uncommon

[a] Common > 5 % of total; Uncommon 1–5% of total; Rare < 1% of total; Absent 0% of total.

served in our study sites (Table 2). In Arkansas, densities of *C. novemnotata* have declined during the past four years (T. Kring, pers. communication).

In the following sections we review the ecology and habitat associations of *A. bipunctata* and *C. transveroguttata*, followed by species that have shown no apparent decline in abundance since the invasion of *C. septempunctata*. Considerable information is available on coccinellid communities in agricultural crops in the midwestern

USA, but very little is known about species associations in semi-natural and natural habitats. Therefore, we discuss information from other regions where necessary to help determine the potential for declines of native species resulting from interactions with *C. septempunctata.*

Adalia bipunctata

Coccinella septempunctata and *A. bipunctata* show considerable plasticity in habitat use in response to prey availability and macro-climatic factors that render the microclimate of particular habitats suitable (Honek 1985, Hodek and Honek 1996). In Europe, adults of *C. septempunctata* and *A. bipunctata* forage for prey in a broad range of habitats, from agricultural crops to semi-natural habitats dominated by herbaceous and graminaceous vegetation to arboreal habitats (Bode 1980, Stechmann 1982, Honek 1982a, 1982b, 1985, Radwan and Lovei 1983, Lovei et al. 1991, Zhou and Carter 1992, Hodek and Honek 1996). Thus, there appears to be potential for interspecific interaction between *C. septempunctata* and *A. bipunctata* in most habitats occupied by the two species. However, the number of habitats the species share may not predict the potential for interaction.

Adalia bipunctata larvae have a pronounced adhesive organ on the 10th abdominal segment that permits them to stick tightly to foliage, an adaptation important for survival of arboreal larvae. *Coccinella septempunctata* larvae do not have this morphological character, and tend to fall from plants when disturbed (Hodek 1973).

Adults of the two species differ somewhat in the temporal sequence of their flights between various habitats (Zhou et al. 1994), potentially reducing the level of interspecies interaction. Honek (1985) observed that *A. bipunctata* closely tracked habitats with dense aphid populations, whereas *C. septempunctata* was more tolerant of habitats with low-density aphid populations. *Adalia bipunctata* oviposits primarily in arboreal habitats, whereas *C. septempunctata* generally does not reproduce in arboreal habitats, instead preferring to reproduce in agricultural fields and herbaceous semi-natural habitats (Wratten 1973; Bode 1980; Honek 1982a, 1982b, 1983, 1989; Stechmann 1982; Hemptinne and Naisse 1988; Lovei et al. 1991; Zhou and Carter 1992; Hodek 1993; Hodek and Honek 1996). *Coccinella septempunctata* is sometimes the most abundant adult coccinellid in orchards (Hodek and Honek 1996), but Lovei et al. (1991) found that larvae of *C. septempunctata* had a lower survival rate in orchards than *A. bipunctata.* Thus, although *C. septempunctata* may be less fit in orchards than *A. bipunctata* the large numbers of adults observed in orchards are primarily immigrants from other habitats rather than the result of *in situ* recruitment (Lovei et al. 1991). The fitness of *C. septempunctata* may be greater than that of *A. bipunctata* in non-arboreal habitats, but this relationship has not been studied.

Based on data from Honek (1985) on the abundance of adult coccinellids in numerous habitats, Hodek and Honek (1996) calculated a niche overlap index (Pianka 1973) that indicated a low level of spatial and temporal overlap in the distributions of adult *C. septempunctata* and *A. bipunctata.* If differences in habitats used for reproduction were considered, niche overlap would probably be further reduced.

Because of the plasticity in life histories of many coccinellids (Hodek 1993), extrapolation of European data to the USA is not necessarily valid. *Coccinella septempunctata* has a broad habitat range in the USA. In pecan orchards in the southeastern USA, Bugg et al. (1991) and Tedders and Angalet (1981) found that *C. septempunctata* primarily inhabited leguminous ground cover, but was uncommon in pecan trees. Maredia et al. (1992) and Colunga-Garcia et al. (1997) observed *C. septempunctata* adults in arboreal habitats, but the species was more abundant in agricultural fields and old-fields dominated by herbaceous vegetation. *Coccinella septempunctata* was the most abundant coccinellid in apple orchards in West Virginia, from 1990 through 1993 (Brown and Miller 1998). Interestingly, *C. septempunctata* decreased in abundance in apple orchards, when the primarily arboreal exotic *H. axyridis* invaded West Virginia (Brown and Miller 1998).

Adult *A. bipunctata* forage in agricultural crops and semi-natural herbaceous and arboreal habitats (Kieckhefer et al. 1992, Campbell and Cone 1994, Giles et al. 1994, Colunga-Garcia et al. 1997), but apparently do not reproduce in agricultural crops or in herbaceous or graminaceous semi-natural habitats (Kieckhefer et al. 1992, Maredia et al. 1992). On the other hand, *A. bipunctata* larvae accounted for a greater percentage (69%) of total coccinellid larvae in an arboreal setting (peach orchards) in Ontario, Canada, than that represented by adults (46%). All fourth instar coccinellid larvae were *A. bipunctata* indicating that immature survival of this species in peach was greater than that of other species (Putman 1957, 1964).

Because of temporal and spatial differences in habitat use, the extent of interaction between *C. septempunctata* and *A. bipunctata* is probably limited. Therefore, the observation of Elliott et al. (1996) that *A. bipunctata* suffered population declines as *C. septempunctata* increased in abundance, was not anticipated. The limited availability of arboreal habitats in eastern South Dakota may be related to the observed decline in *A. bipunctata.* Wooded habitats comprise only 3% of the land area in eastern South Dakota, compared to 74% for agriculture and 18% for grasslands (Elliott et al. 1998a). The reduction in abundance of *A. bipunctata* in eastern South Dakota may be related to the paucity of preferred habitat for *A. bipunctata,* which often occurs in small isolated tracts surrounded by habitat preferred by *C. septempunctata.* Under these circumstances, a habitat generalist such as *C. septempunctata* might exert greater impact on *A. bipunctata* populations than would be realized if wooded habitats were abundant. Since many areas of the midwestern USA are more heavily wooded than eastern South Dakota, we predict that the apparent adverse affect on *A. bipunctata* is a local phenomenon.

Coccinella transversoguttata

Populations of the Nearctic species, *Coccinella transversoguttata,* declined in South Dakota, presumably as the result of interaction with *C. septempunctata* (Elliott et al. 1996). This species is a habitat generalist that occurs in several agricultural crops including corn, alfalfa, small grains, potatoes, hops, and fruit orchards (Smith 1971, Shands and Simpson 1972, Kieckhefer et al. 1992, Campbell and Cone 1994, Putman 1964, Brown and Miller 1998). It apparently prefers seminatural habitats such as old-fields, especially where woody vegetation is regenerating, and forest edges (Gagne and Martin 1968). Information on the reproductive ecology of *C. transversoguttata* is generally lacking. Larvae of *C. transversoguttata* have been reported from old-fields, recently established pine plantations, and forest edges (Gagne and Martin 1968). Larvae were not found in peach orchards even though adults were common (Putman 1957, 1964). *Coccinella transversoguttata* larvae were never collected in corn, alfalfa, or small grain fields during 18 years of relatively intensive sampling of these crops, even though adults were collected from the three crops in most years (Kieckhefer et al. 1992). In contrast, larvae of *C. septempunctata* are routinely observed in corn, alfalfa, and small grains (Elliott et al. 1996).

Within agricultural crops, *C. septempunctata* adults are more abundant and lay more eggs in open vegetation than in dense vegetation (Honek 1982a, 1983). *Coccinella transversoguttata* shows a similar aversion for dense plant canopies, at least in corn fields where adults are most abundant in fields with low plant density (Smith 1971). The effects of photoperiod on diapause induction and temperature on preimaginal development are similar in *C. transversoguttata* and *C. septempunctata* (Hodek and Cerkasov 1961, Storch 1973, Obrycki and Tauber 1981). Larvae of the two species consume about the same number of aphids during immature development (Shands and Simpson 1972). Consumption of aphids by adults has not been compared, nor have oviposition rates, total fecundity, or survival rates.

The autecologies of *C. transversoguttata*, and to a lesser degree *C. septempunctata*, are poorly known, resulting in speculation on niche overlap between the species. Furthermore, the nature and intensity of interspecific interactions between the two species are completely unknown. However, aspects of the biology and pattern of habitat use by *C. transversoguttata* appear to be similar enough to that of *C. septempunctata* that population declines such as those observed by Elliott et. al. (1996) may have been predicted following the invasion by *C. septempunctata* in the midwestern USA. We hypothesize that *C. transversoguttata* suffered from direct interaction with *C. septempunctata* in many of its primary habitats. The observation that coccinellid population fluctuations may be correlated over long distances (Table 1) lends further cause for concern that *C. transversoguttata* could be displaced on a regional scale.

Elliott et al. (1996) did not observe a numerical increase in coccinellid abundance in agricultural crops in eastern South Dakota as a result of invasion of *C. septempunctata*. We conclude that there is a good possibility that invasion by *C. septempunctata* has caused or at least contributed to the decline in abundance of *C. transversoguttata*. There is no evidence that *C. septempunctata* differs in its ability to reduce aphid pests compared to *C. transversoguttata*.

Native Species Co-existing with *Coccinella septempunctata*

Coleomegilla maculata feeds and reproduces in virtually the entire spectrum of plant communities in most landscapes, from old-fields to crops, to areas dominated by trees and shrubs (Warren and Tadic 1967, Gagne and Martin 1968, Weber and Holman 1976, Kieckhefer et al. 1992, Giles et al. 1994). *Coleomegilla maculata* is highly polyphagous and makes extensive use of pollen as a food source for adults and larvae (Atallah and Newsom 1966, Weber and Holman 1976, Warren and Tadic 1967, Shepard et al. 1974, Andow and Risch 1985, Hazzard et al. 1991, Pilcher et al. 1997). *Coccinella septempunctata* does not appear to demonstrate as wide a range of acceptable foods (Triltsch 1997). Differences in prey specificity may partially explain why populations of *C. maculata* have been unaffected by *C. septempunctata*, in spite of the fact that larvae suffer disproportionally in interspecific laboratory studies (Obrycki et al. 1998a). Results from a field cage study suggest that when *Coleomegilla maculata* and *Coccinella septempunctata* feed on the same aphid species in the same crop, they may coexist because aphid densities fluctuate within and between years (Obrycki et al. 1998b). Larvae and adults of *C. septempunctata* and *C. maculata* overlap spatially and temporally in spring alfalfa in Iowa. In 1993, first to third instar mortality was high for both species, 93% for *C. maculata*, and 91% for *C. septempunctata* (Obrycki et al. 1997). In habitats of low aphid densities, *Coleomegilla maculata* may have a competitive advantage because of its smaller size and lower food requirements. At higher aphid densities, the larger *Coccinella septempunctata* would have a competitive advantage, via interference competition and intraguild predation (Obrycki et al. 1998a).

Cycloneda munda is also a habitat generalist that occurs in several crops (Kieckhefer et al. 1992, Giles et al. 1994), but prefers arboreal habitats (Gagne and Martin 1968, Maredia et al. 1992, Colunga-Garcia et al. 1997). Even though *C. munda* occurs in many habitats, it apparently does not reproduce in most agricultural crops (Kieckhefer et al. 1992, Maredia et al. 1992). Information is limited allowing no more than speculation on why *C. munda* appears unaffected by *C. septempunctata.*

Hippodamia tredecimpunctata also occurs in a wide variety of habitats including field crops, old-fields, orchards, and forest. But records of occurrence of its larval stage come mostly from grassy and herbaceous habitats, including agricultural crops (Putman 1964, Wright and Laing 1980, Kieckhefer et al. 1992, Maredia et al. 1992, Giles et al. 1994). *Hippodamia parenthesis* is found in a variety of habitats including old-fields and field crops (Gagne and Martin 1968, Kieckhefer et al. 1992, Maredia et al. 1992, Colunga-Garcia et al. 1997), but there are no records of its occurrence in arboreal habitats. This species and *H. tredecimpunctata* may have been most strongly associated with grasses and sedges prior to the spread of agriculture through the region (Palmer 1914, Cutright 1924).

Abundance of *Hippodamia convergens* has apparently not been affected by the invasion of *C. septempunctata* (Nalepa et al. 1997). *Hippodamia convergens* has a very broad habitat range (Hodek and Honek 1996) and might be predicted to suffer from interaction with *C. septempunctata*. For *H. convergens*, and the other species discussed in this section, the inability to explain why reductions in abundance have not occurred in spite of significant niche overlap with *C. septempunctata* illustrates the lack of knowledge of coccinellid ecology.

CONCLUSIONS

There has never been a recorded occurrence of a global extinction of a coccinellid species (Hodek and Honek 1996). However, the extirpation of *H. tredecimpunctata* from Britain, where it has not been recorded since 1950 (Majerus and Kearns 1989), indicates that coccinellids are susceptible to forces that cause species extinction in spite of their high mobility and adaptability.

During the past two decades, several exotic aphidophagous coccinellid species have increased in abundance and distribution in North America. The primary focus of this chapter has been on the effects of one of these species, *C. septempunctata*, in the midwestern USA. In South Dakota, where long-term ecological studies have been conducted (Elliott et al. 1996), there is no evidence that the addition of *C. septempunctata* increased overall coccinellid densities, thus it appears unlikely that levels of biological control have been increased. We are not aware of any studies in North America that have demonstrated a benefit from the presence of *C. septempunctata* on aphid suppression and there is evidence that it may interfere with biological control of the alfalfa weevil in Utah (Evans and England 1996). Recently, *H. axyridis* rapidly expanded its range across the USA and appears to be adversely affecting some native species (Colunga-Garcia and Gage 1998). *Harmonia axyridis* is reducing aphids in pecan orchards in the southeastern USA (Tedders and Schaefer 1994), but this benefit needs to be weighed against its effects on native species and its status as an overwintering invader of homes.

Our present level of knowledge of coccinellid ecology does not allow for predictions of the interactions and effects of an introduced coccinellid species. This was demonstrated by the apparent effect of *C. septempunctata* on *A. bipunctata* in South Dakota and the lack of measurable effects on several other species of native coccinellids. We conclude that the benefits of introducing aphidophagous coccinellid species for suppression of exotic aphid pests may not outweigh the potential risks. Presently, we have limited data showing benefits of the introduction of *H. axyridis* but no data demonstrating a positive effect of the introduction of *C. septempunctata*. The ecological effects of the remaining introduced aphid feeding species have not been studied. Additional data are needed before we can predict the outcome of introductions of aphid-feeding coccinellid species.

ACKNOWLEDGMENTS

We thank T.J. Kring, Department of Entomology, University of Arkansas, for his critical review of this manuscript and C.A. Nalepa, North Carolina Department of Agriculture for providing information from a poster presentation. The support of USDA NRI Grants 90-37250-5357 and 94-37312-0673, and the Leopold Center for Sustainable Agriculture is acknowledged. Journal Paper No. J-18042 of the Iowa Agricultural and Home Economics Experiment Station, Ames, Iowa. Project No. 3437, and supported by Hatch Act and State of Iowa funds.

REFERENCES

Agarwala, B. K., and S. Bhattacharya. 1995. Adaptive features of the predatory ladybird beetles in uncertain and competitive habitats. Current Sci. 69: 587–591.

Agarwala, B. K., and A. F. G. Dixon. 1992. Laboratory study of cannibalism and interspecific predation in ladybirds. Ecol. Entomol. 17: 303–309.

Andow, D. A., and S. J. Risch. 1985. Predation in diversified agroecosystems: relations between a coccinellid predator *Coleomegilla maculata* and its food. J. Appl. Ecol. 22: 357–372.

Atallah, Y. H., and L. D. Newsom. 1966. Ecological and nutritional studies on *Coleomegilla maculata* De Geer (Coleoptera: Coccinellidae). I. The development of an artificial diet and a laboratory rearing technique. J. Econ. Entomol. 39: 1173–1179.

Archer, T. L., and E. D. Bynum, Jr. 1993. Ecology of the Russian wheat aphid (Homoptera: Aphididae) on dryland winter wheat in the southern United States. J. Kansas Entomol. Soc. 66: 60–68.

Bode, H. 1980. Untersuchungen zum auftreten der haferblattlaus *Rhopalosiphum padi* (L.) (Homoptera: Aphididae) an ihrem winterwirt *Prunus padus* L. I. Biologie der haferblattaus *Rhopalosiphum padi* (L.) am winterwirt. Z. Angew. Entomol. 89: 363–377.

Brown, M. W., and S. S. Miller. 1998. Coccinellidae (Coleoptera) in apple orchards of eastern West Virginia and the impact of invasion by *Harmonia axyridis*. Entomol. News 109: 143–151.

Bugg, R. L., J. D. Dutcher, and P. J. McNeill. 1991. Cool-season cover crops in the pecan orchard understory: effects on Coccinellidae (Coleoptera) and pecan aphids (Homoptera: Aphididae). Biol. Control 1: 8–15.

Campbell, C. A., and W. W. Cone. 1994. Influence of predators on population development of *Phorodon humuli* (Homoptera: Aphididae) on hops. Environ. Entomol. 23: 1391–1396.

Chang, G. C. 1996. Comparison of single versus multiple species of generalist predators for biological control. Environ. Entomol. 25: 207–212.

Chantal, C. 1972. Additions a la faune coleopterique du Quebec. Nat. Can. 99:243–244.

Chapin J. B., and V. A. Brou. 1991. *Harmonia axyridis* (Pallas), the third species of the genus to be found in the United States (Coleoptera: Coccinellidae). Proc. Entomol. Soc. Washington 93: 630–635.

Coderre, D., L. Provencher, and J. C. Tourneur. 1987. Oviposition and niche partitioning in aphidophagous insects on maize. Can. Entomol. 119:195–203.

Colfer, R. G., and J. A. Rosenheim. 1995. Intraguild predation by coccinellid beetles on an aphid parasitoid, *Lysiphlebus testaceipes*. Proc. Beltwide Cotton Conf. 2: 1033–1036.

Colunga–Garcia, M., and S. H. Gage. 1998. Arrival, establishment, and habitat use of the multicolored Asian lady beetle (Coleoptera: Coccinellidae) in a Michigan landscape. Environ. Entomol. 27: 1574–1580.

Colunga–Garcia, M., S. H. Gage, and D. A. Landis. 1997. Response of an assemblage of Coccinellidae (Coleoptera) to a diverse agricultural landscape. Environ. Entomol. 26: 797–804.

Connell, J. H. 1983. On the prevalence and relative importance of interspecific competition: evidence from field experiments. Amer. Nat. 122: 661–696.

Cutright, C. R. 1924. Bionomics of *Hippodamia tredecimpunctata*. Ann. Entomol. Soc. Am. 17: 188–192.

Day, W. H., D. R. Prokrym, D. R. Ellis, and R. J. Chianese. 1994. The known distribution of the predator *Propylea quatuordecimpunctata* (Coleoptera: Coccinellidae) in the United States, and thoughts on the origin of this species and five other exotic lady beetles in eastern North America. Entomol. News 105: 244–56.

Dunning, J. B., B. J. Danielson, and H. R. Pulliam. 1992. Ecological processes that affect populations in complex landscapes. Oikos 65: 169–175.

Elliott N. C., and R. W. Kieckhefer. 1990a. Dynamics of aphidophagous coccinellid assemblages in small grain fields in eastern South Dakota. Environ. Entomol. 19:1320–1329.

Elliott, N. C., and R. W. Kieckhefer. 1990b. A thirteen-year survey of the aphidophagous insects of alfalfa. Prairie Natural. 22: 87–96.

Elliott, N. C., R. W. Kieckhefer, and W. C. Kauffman. 1996. Effects of an invading coccinellid on native coccinellids in an agricultural landscape. Oecologia 105: 537–544.

Elliott, N. C., R. W. Kieckhefer, J. H. Lee, and B. W. French. 1998a. Influence of habitat and landscape related factors on aphid predator populations in wheat. Landscape Ecol. In press.

Elliott, N. C., R. W. Kieckhefer, B. W. French, and J. H. Lee. 1998b. Variation in size, sex ratio, and dispersal in three coccinellid (Coleoptera) species. Southwest. Entomol. 23: 137–145.

Evans, E. W. 1991. Intra versus interspecific interactions of ladybeetles (Coleoptera: Coccinellidae) attacking aphids. Oecologia. 87: 401–408.

Evans E. W., and S. England. 1996. Indirect interactions in biological control of insects: Pests and natural enemies in alfalfa. Ecol. Appl. 6: 920–930.

Ewert, M. A., and H. C. Chiang. 1966. Dispersal of three species of coccinellids in corn fields. Can. Entomol. 98: 999–1003.

Ferguson, K. I., and P. Stiling. 1996. Non-additive effects of multiple natural enemies on aphid populations. Oecologia 108: 375–379.

Frazer, B. D. 1988. Coccinellidae., *In* A. K. Minks and P. Harrewijn (eds.), Aphids—their biology, natural enemies and control, vol. B. Elsevier Press, New York. p. 231–247.

Gagne, W. C., and J. L. Martin. 1968. The insect ecology of red pine plantations in central Ontario. Can. Entomol. 100: 835–846.

Giles, K. L., J. J. Obrycki, and T. A. Degooyer. 1994. Prevalence of predators associated with *Acyrthosiphon pisum* (Homoptera: Aphididae) and *Hypera postica* Gyllenhal (Coleoptera: Curculionidae) during growth

of the first crop of alfalfa. Biol. Control 4: 170–177.
Goodarzy, K., and D.W. Davis. 1958. Natural enemies of the spotted alfalfa aphid in Utah. J. Econ. Entomol. 51: 612–616.
Gordon, R. D. 1985. The Coccinellidae (Coleoptera) of America north of Mexico. J. N.Y. Entomol. Soc. 93: 1–912
Hagen, K. S. 1962. Biology and ecology of predaceous Coccinellidae. Annu. Rev. Entomol. 7: 289–326
Hagen, K. S. 1974. The significance of predaceous Coccinellidae in biological and integrated control of insects. Entomophaga Mem. H.S. 7: 25–44.
Hazzard, R. V., D. N. Ferro, R. G. Van Driesche, and A. F. Tuttle. 1991. Mortality of eggs of Colorado potato beetle (Coleoptera: Chrysomelidae) from predation by *Coleomegilla maculata* (Coleoptera: Coccinellidae). Environ. Entomol. 20: 841–848.
Hemptinne, J. L. 1988. Ecological requirements for hibernating *Propylea quatuordecimpunctata* [L.] and *Coccinella septempunctata* [Col.: Coccinellidae]. Entomophaga 33: 505–515.
Hemptinne, J. L., and J. Naisse. 1988. Life cycle strategy of *Adalia bipunctata* (L.) (Col., Coccinellidae) in a temperate country. *In* E. Niemczyk and A. F. G. Dixon (eds.), Ecology and effectiveness of Aphidophaga. SPB Academic Publishers, The Hague, Netherlands. p. 71–77.
Hemptinne, J. L., A. F. G. Dixon and J. Coffin. 1992. Attack strategy of ladybird beetles (Coccinellidae): factors shaping their numerical response. Oecologia 90: 238–245.
Hodek, I. 1973. Biology of Coccinellidae. Dr. W. Junk N. V. Publishers, The Hague, Netherlands.
Hodek, I. 1993. Habitat and food specificity in aphidophagous predators. Biocontrol Sci. Tech. 3: 91–100.
Hodek, I., and J. Cerkasov. 1961. Prevention and artificial induction of diapause in *Coccinella septempunctata* L. (Col., Coccinellidae). Entomol. Exp. Appl. 4: 179–190.
Hodek, I., G. Iperti, and M. Hodkova. 1993. Long-distance flights in Coccinellidae (Coleoptera). Eur. J. Entomol. 90: 403–414.
Hodek, I., and A. Honek. 1996. Ecology of Coccinellidae. Kluwer Academic Publishers, Dordrecht, Netherlands.
Hoebeke E. R., and A. G. Wheeler. 1996. Adventive lady beetles (Coleoptera: Coccinellidae) in the Canadian maritime provinces, with new eastern U.S. records of *Harmonia quadripunctata*. Entomol. News 107: 281–290.
Honek, A. 1982a Factors which determine the composition of field communities of adult aphidophagous Coccinellidae (Coleoptera). Z. Ang. Entomol. 94:157–168.
Honek, A. 1982b. The distribution of overwintered *Coccinella septempunctata* L. (Col., Coccinellidae) adults in agricultural crops. Z. Ang. Entomol. 94: 311–319.
Honek, A. 1983. Factors affecting the distribution of larvae of aphid predators (Col., Coccinellid and Dipt., Syrphidae) in cereal stands. Z. Ang. Entomol. 95: 336–345.
Honek, A. 1985. Habitat preferences of aphidophagous coccinellids [Coleoptera]. Entomophaga 30: 253–264.
Honek, A. 1989. Overwintering and annual changes in abundance of *Coccinella septempunctata* in Czechoslovakia (Coleoptera: Coccinellidae). Acta Entomol. Bohemoslov. 86: 179–193.
Honek, A. 1990. Seasonal changes in flight activity of *Coccinella septempunctata* L. (Coleoptera, Coccinellidae). Acta Entomol. Bohemoslov. 87: 336–341.
Horn, D. 1991. Potential impact of *Coccinella septempunctata* on endangered Lycaenidae (Lepidoptera) in northwestern Ohio, USA. *In* L. Polgar, R. J. Chambers, A. F. G. Dixon, and I. Hodek (eds.), Behavior and impact of Aphidophaga. SPB Academic Publishing bv, The Hague, Netherlands. p. 159–162.
Huffaker, C. B., F. J. Simmonds, and J. E. Laing. 1976. The theoretical and empirical basis of biological control. *In* C. B. Huffaker and P. S. Messenger (eds.), Theory and practice of biological control. Academic Press, New York. p. 41–78.
Ives, P. M. 1981. Estimation of coccinellid numbers and movement in the field. Can. Entomol. 113: 981–987.
Keddy, P. A. 1989. Competition. Chapman and Hall, London, UK.
Kieckhefer, R. W., and N. C. Elliott. 1990. A 13-year survey of the aphidophagous Coccinellidae in maize fields in eastern South Dakota. Can. Entomol. 122: 579–581.
Kieckhefer, R. W., N. C. Elliott, and D. A. Beck. 1992. Aphidophagous coccinellids in alfalfa, small grains, and maize in eastern South Dakota. Great Lakes Entomol. 25: 15–23.
Krafsur, E. S., J. J. Obrycki, and R. V. Flanders. 1992. Gene flow in populations of the seven–spotted lady beetle, *Coccinella septempunctata*. J. Heredity. 83: 440–444.
Krafsur, E. S., J. J. Obrycki, and P. W. Schaefer. 1995. Genetic heterozygosity and gene flow in *Coleomegilla*

maculata Degeer (Coleoptera: Coccinellidae). Biol. Control 5: 104–111.

Krafsur, E. S., P. Nariboli, and J.J. Obrycki. 1996. Gene flow and diversity at allozyme loci in the twospotted lady beetle (Coleoptera: Coccinellidae). Ann. Entomol. Soc. Am. 89: 410–419.

Krafsur E. S., T. J. Kring, J. C. Miller, P. Nariboli, J. J. Obrycki, J. R. Ruberson, and P. W. Schaefer. 1997. Gene flow in the exotic colonizing ladybird *Harmonia axyridis* in North America. Biol. Control 8: 207–214.

LaMana, M. L., and J. C. Miller. 1996. Field observations on *Harmonia axyridis* Pallas (Coleoptera: Coccinellidae) in Oregon. Biol. Control 6: 232–237.

Lovei, G. L., M. Sarospataki, and Z. A. Radwan. 1991. Structure of Ladybird (Coleoptera: Coccinellidae) Assemblages in apple: changes through developmental stages. Environ. Entomol. 20: 1301–1308.

Majerus, M., and P. Kearns. 1989. Ladybirds. Richmond Publishing Co., London.

Maredia, K. M., S. H. Gage, D. A. Landis, and J. M. Scriber. 1992. Habitat use patterns by the seven-spotted lady beetle (Coleoptera: Coccinellidae) in a diverse agricultural landscape. Biol. Control 2: 159–165.

McClure M. S. 1987. Potential of the Asian predator, *Harmonia axyridis* Pallas (Coleoptera: Coccinellidae), to control *Matsucoccus resinosae* Bean and Godwin (Homoptera: Margarodidae) in the United States. Environ. Entomol. 16: 224–230.

Messina, F. J., T. A. Jones, and D. C. Nelson. 1995. Host plant affects the interaction between the Russian wheat aphid and a generalist predator, Chrysoperla carnea. J. Kansas Entomol. Soc. 68: 313–319.

Nalepa, C.A., K.R. Ahlstrom, J.L. Williams, and R. Gardner. 1997. Coccinellidae in pitcher plants: identification, seasonal activity, and comparison to historic data. Poster D557. Entomol. Soc. Am. Nat. Mtg. Nashville, Tennessee.

Nalepa, C. A., K. A. Kidd, and K. R. Ahlstrom. 1996. Biology of *Harmonia axyridis* (Coleoptera: Coccinellidae) in winter aggregations. Ann. Entomol. Soc. Am. 89: 681–685.

Obrycki, J. J., K. L. Giles, and A. M. Ormord. 1998a. Interactions between an introduced and indigenous coccinellid species at different prey densities. Oecologia 117:279–285.

Obrycki, J. J., K. L. Giles, and A. M. Ormord. 1998b. Experimental assessment of interactions between larval *Coleomegilla maculata* and *Coccinella septempunctata* (Coleoptera: Coccinellidae) in field cages. Environ. Entomol. 27:1280–1288.

Obrycki, J. J., and T. J. Kring. 1998. Predaceous Coccinellidae in biological control. Annu. Rev. Entomol. 43: 295–321.

Obrycki, J. J., A. M. Ormord, and K. L. Giles. 1997. Partial life table analysis for larval *Coleomegilla maculata* (Degeer) and *Coccinella septempunctata* L. (Coleoptera: Coccinellidae) in alfalfa. J. Kansas Entomol. Soc. 70: 339–346.

Obrycki, J. J., and M. J. Tauber. 1981. Phenology of three Coccinellid species: thermal requirements for development. Ann. Entomol. Soc. Am. 74: 31–36.

Obrycki, J. J., M.J. Tauber, C.A. Tauber, and J.R. Ruberson. 1997a. Prey specialization in insect predators. http://ipmworld.umn.edu/chapters/obrycki.htm.

Ormord, A. M. 1994. Interactions between larval *Coccinella septempunctata* L., and *Coleomegilla maculata* (Degeer) Coleoptera: Coccinellidae). M.S. Thesis, Iowa State Univ., Ames, Iowa. 122 p.

Osawa N. 1989. Sibling and non–sibling cannibalism by larvae of a lady beetle *Harmonia axyridis* Pallas (Coleoptera: Coccinellidae) in the field. Res. Popul. Ecol. 31: 153–160.

Palmer, M. A. 1914. Some notes on the life history of ladybeetles. Ann. Entomol. Soc. Am. 7: 213–238.

Pianka, E. R. 1973. The structure of lizard communities. Annu Rev. Ecol. Syst. 4: 53–74.

Phoofolo, M.W., and J.J. Obrycki. 1998. Potential for intraguild predation and competition among predatory Coccinellidae and Chrysopidae. Entomol. Exp. Appl. 89: 47–55.

Pilcher, C. D., J. J. Obrycki, M. E. Rice, and L. C. Lewis. 1997. Preimaginal development, survival, and field abundance of insect predators on transgenic *Bacillus thuringiensis* corn. Environ. Entomol. 26: 446–454.

Polis, G. A., and R. D. Holt. 1992. Intraguild predation: the dynamics of complex trophic interactions. Tree 7: 151–154.

Putman, W. L. 1957. Laboratory studies on the food of some coccinellids (Coleoptera) found in Ontario peach orchards. Can. Entomol. 89: 527–579.

Putman, W. L. 1964. Occurrence and food of some coccinellids (Coleoptera) in Ontario peach orchards. Can. Entomol. 96: 1149–1155.

Radwan, Z., and G. L. Lovei. 1983. Structure and seasonal dynamics of larval, pupal, and adult coccinellid (Col., Coccinellidae) assemblages in two types of maize fields in Hungary. Z. Ang. Entomol. 96: 396–408.

Roach, S. H., and W. M. Thomas. 1991. Overwintering and spring emergency of three coccinellid species on the coastal plain of South Carolina. Environ. Entomol. 20: 540–544.
Rosenheim, J. A., H. K. Kaya, L. E. Ehler, J. J. Marois, and B. A. Jaffee. 1995. Intraguild predation among biological control agents: theory and evidence. Biol. Control 5: 303–335.
Rosenheim, J. A. 1998. Higher–order predators and the regulation of insect herbivore populations. Ann. Rev. Entomol. 43: 421–447.
Roy, H. E., J. K. Pell, S. J. Clark, and P. G. Alderson. 1998. Implications of predator foraging on aphid pathogen dynamics. J. Invert. Pathol. 71: 236–247.
Samways, M. J. 1986. Combined effect of natural enemies (Hymenoptera: Aphelinidae and Coleoptera: Coccinellidae) with different niche breadths in reducing high populations of red scale, *Aonidiella aurantii* (Maskell) (Hemiptera: Diaspididae). Bull. Entomol. Res. 76: 671–683.
Samways, M. J. 1997. Classical biological control and biodiversity conservation: what risks are we prepared to accept? Biodiversity and Conservation 6: 1309–1316.
Schaefer, P. W., R. J. Dysart, and H. B. Specht. 1987. North American distribution of *Coccinella septempunctata* (Coleoptera: Coccinellidae) and its mass appearance in coastal Delaware. Environ. Entomol. 16: 368–73.
Schoener, T. W. 1983. Field experiments on interspecific competition. Amer. Nat. 122: 240–285.
Shands, W. A., and G. W. Simpson. 1972. Insect predators for controlling aphids on potatoes: green peach aphid consumption by *Coccinella septempunctata* and *C. transversoguttata*. J. Econ. Entomol. 65: 1388–1392.
Shepard, M., G. R. Carner, and S. G. Turnipseed. 1974. Seasonal abundance of predaceous arthropods in soybeans. Environ. Entomol. 3: 985–988.
Simberloff, D., and P. Stiling. 1996. How risky is biological control? Ecology 77: 1965–1974.
Smith, B. C. 1971. Effects of various factors on the local distribution and density of coccinellid adults on corn (Coleoptera: Coccinellidae). Can. Entomol. 103: 1115–1120.
Stechmann, D. H. 1982. Zur kologie aphidophager insekten in hecken und feldern oberfrankens: beobachtungen an coccinelliden in den jahren 1978/79. Jber. Naturwiss. Verein Wuppertal 35: 38–42.
Storch, R. H. 1973. The effect of photoperiod on *Coccinella transversoguttata* (Coleoptera: Coccinellidae). Entomol. Exp. et Appl. 16: 77–82.
Strand, M. R., and J. J. Obrycki. 1996. Host specificity of insect parasitoids and predators. Bioscience 46: 422–429.
Takahashi, K. 1989. Intra- and inter-specific predations of lady beetles in spring alfalfa fields. Japan J. Entomol. 57: 199–203.
Tedders, W. L., and G. W. Angalet. 1981. Colonization of *Coccinella septempunctata* (L.) in Georgia. J. Georgia Entomol. Soc. 16: 41–47.
Tedders, W. L., and P. W. Schaefer. 1994. Release and establishment of *Harmonia axyridis* (Coleoptera: Coccinellidae) in the southeastern United States. Entomol. News 105: 228–43.
Triltsch, H. 1997. Contents in field sampled adults of *Coccinella septempunctata* (Col.: Coccinellidae). Entomophaga. 42: 125–131.
Warren, L. O., and M. Tadic. 1967. Biological observations on *Coleomegilla maculata* and its role as a predator of the fall webworm. J. Econ. Entomol. 60: 1492–1496.
Weber, R. G., and M. V. Holman. 1976. *Coleomegilla maculata* (Degeer), a new coleopteran predator of the elm leaf beetle, *Pyrrhalta luteola* (Mueller). J. Kansas Entomol. Soc. 49: 160.
Wheeler, A. G. 1993. Establishment of *Hippodamia variegata* and new records of *Propylea quatuordecimpunctata* (Coleoptera, Coccinellidae) in the eastern United States. Entomol. News 104: 102–110.
Wheeler, A. G., Jr., J. T. Hayes, and J. L. Stephens. 1968. Insect predators of mummified pea aphids. Can. Entomol. 100: 221–222.
Wheeler, A. G., and E. R. Hoebeke. 1995. *Coccinella novemnotata* in northeastern North America: Historical occurrence and current status (Coleoptera: Coccinellidae). Proc. Entomol. Soc. Washington. 97: 701–716.
Wheeler, A. G., and C. A. Stoops. 1996. Status and spread of the Palearctic lady beetles *Hippodamia variegata* and *Propylea quatuordecimpunctata* (Coleoptera: Coccinellidae) in Pennsylvania, 1993–1995. Entomol. News 107: 291–298.
Wingo, C.W. 1952. Monograph of the Coccinellidae (Coleoptera) of the upper Mississippi basin. PhD Thesis. Iowa State College. 269 p.
Wraight, S. P., T. J. Poprawski, W. L. Meyer, and F. B. Peairs. 1993. Natural enemies of Russian wheat

aphid (Homoptera: Aphididae) and associated cereal aphid species in spring–planted wheat and barley in Colorado. Environ. Entomol. 22: 1383–1391.

Wratten, S. D. 1973. The effectiveness of the coccinellid beetle, *Adalia bipunctata* (L.), as a predator of the lime aphid, *Eucallipterus tiliae* L. J. Anim. Ecol. 42: 785–802.

Wright, E. J., and J. E. Laing. 1980. Numerical response to coccinellids of aphids in corn in southern Ontario. Can. Entomol. 112: 977–988.

Yanega, D. 1996. Records of *Propylea quatuordecimpunctata* (Coleoptera: Coccinellidae) from Long Island, New York: Evidence for a naturalized population before 1991. Entomol. News 107: 36–38.

Zhou, X., and N. Carter. 1992. The ecology of coccinellids on farmland. Asp. Appl. Biol. 31: 133–138.

Zhou, X., N. Carter, and W. Powell. 1994. Seasonal distribution and aerial movement of adult coccinellids on farmland. Biocontrol Sci. Tech. 4: 167–175.

10

FOOD WEBS AS A TOOL FOR STUDYING NONTARGET EFFECTS IN BIOLOGICAL CONTROL

Jane Memmott
School of Biological Sciences, University of Bristol, United Kingdom

INTRODUCTION

Biological control has been described as the most powerful ecological technology for pest management (Lockwood 1996). However, biological control practitioners are under growing pressure from both science and society to improve upon their predictions of safety and effectiveness of control-organism introductions (McEvoy 1996). This is because over the last 10 to 15 years, conservation biologists have become increasingly concerned that biological control can lead to the extinction of native species (Bennett 1993, Howarth 1985, Howarth 1991, Gagne and Howarth 1982) and potentially damage whole ecosystems (Simberloff and Stiling 1996a, Simberloff and Stiling 1996b, Miller and Aplet 1993).

Responding to the concerns of conservationists, biological control practitioners state that there are few documented examples of biological control agents attacking indigenous organisms, relative to the number of releases. While this assertion is correct, it does not necessarily mean that biological control is safe, because the extent that biological control agents infiltrate natural communities is rarely investigated. Moreover, the impact of biological control agents upon the population dynamics of their nontarget hosts has not been addressed. Studies investigating whether biological control agents have an impact upon community and ecosystem functions are rather rare in the literature. The paucity of information on the extent and impact of nontarget effects means that the debate about whether biological control is a threat to native habitats remains unresolved.

If the impact of alien arthropods on natural communities is to become predictable, biologists must ask the appropriate questions and gather the relevant data. For example, while it is widely known that some biological control agents attack nontarget species, this information does not assess their impact on the nontarget populations. This impact potentially ranges from the incidental (the occasional case of nontarget "use") to the catastrophic (the extinction of the native organisms). Recently, it has been proposed that the appropriate model for research on nontarget interactions is a reticulate web of species interactions rather than a simple unbranched food chain (Strong 1997). Consequently, nontarget effects will be embedded in a complex web of trophic interactions. While there are a number of recorded cases of interactions between biological control agents and their nontarget hosts, nobody has, to date, used a food web to quantify the extent to which biological control agents have infiltrated natural communities. If food webs were available for communities containing biological control agents attacking nontarget flora or fauna, the following questions could be answered: How many nontarget interactions are there in comparison to the total number of interactions in a community? What proportion of the nontarget species does the biological control agent consume, relative to native natural enemies? Do the biological control agents feed on one or several nontarget organisms? How do biological control agents behave in comparison to other alien arthropods? Furthermore once a web is available, experiments can be designed to determine whether nontarget interactions are dynamically important.

Understanding the impact of biological control agents on nontarget hosts requires information about the distribution and the strength of resource-consumer interactions. Such information is readily available from food webs. For example, Memmott, Godfray and Gauld (1994) constructed food webs that described the trophic interactions for a Costa Rican tropical dry forest community. These webs were used to test a variety of hypotheses about the diversity and structure of parasitoid communities. The techniques developed for the Costa Rican project could be used to determine the extent and the importance of nontarget effects in biological control. The aim of such a food web would not be to incriminate practitioners of biological control, rather the aim would be to quantify the impact of biological control agents on natural communities. This will set the scene for an informed debate on the subject. At present, the literature on the environmental impact of biological control encompasses a wide variety of viewpoints. This variety is illustrated by the following four quotes:

- Lai (1988): "There is absolutely no data indicating that biological control introductions have caused any extinctions of native biota."
- Funasaki et al. (1988): "Biological control introductions have been recorded as having attacked relatively few species of the native biota."
- Gagne and Howarth (1982): "Biological control agents were responsible for the demise of 15 species of native moths."

- Zimmerman (1958): "Biological control has resulted in the wholesale slaughter and near, or complete extermination, of countless species."

This level of confusion is not uncommon in countries that practice biological control.

Food webs describe the feeding relationships between different trophic levels within a community as a picture and, more formally, by mathematical matrices. Summary statistics can be calculated from the web and used to compare different webs. These statistics include species number: the total number of species in the web, connectance: the proportion of all possible interactions that are realized (Gardner and Ashby 1970), predator/prey ratios: the total of the top and intermediate species, divided by the total of the intermediate and basal species (Martinez 1991), interaction strength: the intensity of interactions (Paine 1980) and average chain length: the average number of links in every chain connecting all consumers to basal species (Martinez 1991). Further statistics could be calculated from webs specifically used to study nontarget effects, for example the number of herbivore species per plant, number of natural enemies per herbivore, mortality rates and parasitoid/predator diet breadth. Originally, food webs simply showed which species were resources and which species were consumers. In pictures of food webs, consumer and resource species were shown connected by a line if the former feeds on the latter. These types of webs are known as connectance webs. Food webs can, however, be quantitative and these show not only who eats whom, but the size of that interaction relative to other feeding relationships. Quantitative webs are superior to connectance webs for addressing intra-system population dynamics because they provide an insight into the relative influence of shared consumers and the degree to which one or a few species affect population dynamics within the system (Memmott et al. 1994).

While food webs are extremely useful for retrospective analysis, they are not directly suited to pre-release testing of agents. However, they are likely to improve our ability to predict the outcomes of future releases, through a better understanding of the ecology of biological control. In this chapter I propose that quantitative food webs serve two specific purposes in studies of nontarget effects. First, they provide a means of quantifying the extent that weed and pest biological control agents have infiltrated native communities. Second, they can be used to generate testable hypotheses concerning the impact of biological control agents upon the population dynamics of nontarget hosts. This combined approach, of observational data from food webs and experimental data from web manipulations, offers a unique opportunity to advance research on nontarget interactions and to improve the science of biological control.

THE USE OF FOOD WEBS IN NONTARGET RESEARCH

In this section, I suggest that food webs can be used to answer four questions critically important to the field of biological control. First, do the recorded examples of nontarget effects represent the "tip of the iceberg" as suggested by Simberloff and Stiling (1996)? Second, can biological control agents lead to the extinction of native species? Third, can biological control agents become keystone species? And finally, do biological control agents change community structure?

Are the Recorded Examples of Nontarget Effects the Tip of the Iceberg?

Simberloff and Stiling (1996a, b) noted that because monitoring of nontarget interactions on indigenous species, communities and ecosystems is relatively rare, the reported incidents of such interactions probably under-estimate the size of the problem. Nontarget interactions have been published for parasitoid biological control agents in both Hawaii and mainland North America. In Hawaii 32% of introduced parasitoid species (Funasaki et al. 1988) and in North America 16% of introduced parasitoid species (Hawkins and Marino 1997) have been recorded on nontarget hosts. Whether any of these parasitoids affect the population dynamics of their nontarget hosts is unknown. For weed agents, information on nontarget effects is more limited: a fact that may reflect the more extensive host range testing of weed biological control agents and the subsequent rejection of non-specific agents (Simberloff and Stiling 1996b, Thomas and Willis 1998). There are however, a number of cases of weed agents affecting nontarget native species, for example the weevil, *Rhinocyllus conicus* Froeh., which was introduced to control weedy thistles in North America. Subsequently it has spread to native thistles, and significantly reduces their seed production (Louda et al. 1997). In contrast, Willis et al. (1993, 1995, 1998) demonstrated that although the mite, *Aculus hyperici* Liro, introduced to control St. John's Wort (*Hypericum perforatum* L.) in Australia, infests field populations of the nontarget native, *H. gramineum* J.Forst, its impact on the growth and reproduction of *H. gramineum* is minimal.

While data exists on nontarget effects, it is important to note that the available information is usually based on literature surveys, rather than on systematic sampling of communities. As such, the available data should be treated as a very rough estimate of the level of nontarget interactions in communities. At present, the actual level of nontarget interactions in native habitats is unknown. Because food webs describe all trophic relationships within a given community, they can provide a thorough method of evaluating the extent to which biological control agents have infiltrated natural communities. For example, the number of nontarget interactions could be expressed as a proportion of all interactions, or the number of biological control agents in the community expressed as a proportion of the total number of species in the community.

Can Biological Control Agents Lead to the Extinction of Native Species?

In no other aspect is classical biological control strategy more in conflict with environmental protection than as the cause of species extinctions (Howarth 1991). A biological control agent maintained at high densities on a common weed or pest can potentially drive a rare nontarget species to extinction (Simberloff and Stiling 1996). This effect is an example of apparent competition (Holt 1977, 1984). Here, if the target species increases in abundance, the shared consumer will also increase in abundance and the nontarget species will be subject to increased mortality. Consequently, the population growth rate of both target and nontarget suffers when the other becomes more common, just as happens in true competition. As in true competition though, environmental heterogeneity may ameliorate the intensity of the interaction that may allow the coexistence of the target and the nontarget species.

While the term apparent competition has been mostly used with respect to invertebrate species (e.g., two herbivores and a parasitoid), obviously similar dynamics could operate between target and nontarget plants and a shared weed biological control agent. In this case, if the weed increases in abundance, the weed biological control agent will increase in density and the nontarget plant may be subject to increased mortality.

Food webs can be extremely useful when studying apparent competition. This is because by quantitatively describing the extent to which species share common consumers, they reveal the potential for indirect interactions. Food webs, however, provide only indirect evidence for the presence and importance of indirect interactions such as apparent competition. Their value is that they point the way towards the experiments that can reveal the dynamic consequences of nontarget feeding. For example, if a food web reveals that a biological control agent commonly attacks its target species and rarely attacks a nontarget species, it is reasonable to assume that the abundance of the agent is determined by its interaction with the target species. However, even though only a small fraction of the agent population develops on the nontarget host, the agent may be an important source of mortality for this nontarget species. Here, the obvious experiment would be to manipulate the density of the target species and monitor the impact on the nontarget species. If apparent competition is important, the high density of the target is predicted to have a negative effect on populations of the nontarget, the effect being mediated by the shared consumer. Such interactions may be asymmetric (Bonsall and Hassell 1997) and ideally the densities of both target and nontarget species should be manipulated if the dynamics are to be fully understood. To run this type of experiment effectively, detailed information on shared parasitism is essential. Food webs are the best means of providing this data.

The experiment could be run under field conditions, as seen with the manipulation of aphid species in England (Muller and Godfray 1996). Here, the density of one aphid was manipulated and apparent competition, mediated by a shared predatory coccinelid, led to a decline in the second aphid species. Alternatively, the experiments could be carried out in the laboratory. For example, two species of moth and a shared parasitoid were laboratory-reared in a controlled multi-generational laboratory experiment (Bonsall and Hassell 1997). This experiment clearly demonstrated that while the two species system was stable, the three species system was unstable with one of the hosts being eliminated from the interaction.

If apparent competition is widespread in natural communities, then the environmental impact of biological control agents could, potentially, be significant. Gagne and Howarth (1982) suggested that biological control introductions were the major factor in the extinction of at least 15 species of the larger native moths of Hawaii. This may be correct, but their evidence is conjectural which makes their claim difficult to substantiate. If we had food webs which showed the extent of shared consumption in communities, then we could design the experiments that could calculate the probability of extinction of a nontarget species. Conjecture could then be replaced by scientific rigour.

Can Biological Control Agents Become Keystone Species?

While it is obvious that very abundant species may play a major part in regulating community and ecosystem processes, experiments have revealed that less abundant species may also have strong effects on their communities and ecosystems. These species are known as keystone species (Paine 1969). A keystone species differs from dominant species in that their effects are much larger than would be predicted on the basis of their abundance. The keystone-species concept may provide an entree into the morass of predicting major environmental impacts of introduced species (Simberloff 1991). If biological control agents are either very abundant or have become keystone species in native habitats, they could be more problematic than presently envisaged. Both diversity and trophic level considerations suggest that keystone species are likely to occur near the top of the food chain (Power et al. 1996), consequently arthropod biological control agents are perhaps more likely to become keystone species than weed control agents.

The term keystone has been criticized for being so vague as to be meaningless (Mills et al. 1993). Recently however, Power et al. (1996) developed a new method of testing for the presence of keystone species. Here, the "community importance" of each species within a community is calculated. This is the change in the community or ecosystem trait per unit change in the abundance of the species. Potential traits to use include nutrient cycling, productivity, species richness or the abundance of one or more functional groups of species or dominant species. If food web data were available for communities containing biological control agents, experiments could be designed to determine whether or not biological control agents have become keystone species.

Do Biological Control Agents Change the Structure of Natural Communities?

Food webs provide a comprehensive means of quantifying the impact of alien species on community structure as they record both species number and the trophic links between these species. In addition, quantitative webs also record species abundance and interaction strength. The analysis of food web data can be used to determine whether biological control agents change community structure. This could be done by comparing food web data such as the number of species, connectance, trophic ratios and links per species in communities with and without biological control agents. Given that the underlying rationale of biological control is that it disrupts established populations, biological control agents could potentially change food web structure. The question is however, do they have unexpected, undesirable effects or do they just change the structure by reducing the abundance of the pest species?

In addition to its application in agricultural, horticultural and silvicultural settings, biological control can be used in natural ecosystems as a conservation tool. For example, an introduced mealybug threatening the last remaining populations of gumwoods on St. Helena, was successfully controlled by the introduction of predatory coccinellid (IIBC 1996). In this situation, food webs could provide an ideal means of quantifying the damage caused by the alien pest to a natural habitat. Thus, the food web statistics listed above could be calculated for plots with and without the mealybug. The impact of the pest upon food web structure could then be compared to the impact of the biological control agent on food web structure. If, in general, the former is greater than the latter, biological control would appear to be a risk well worth taking.

HOW TO CONSTRUCT A QUANTITATIVE FOOD WEB

There is now considerable advice on how best to construct webs (for example, Pimm and Kitching 1988, Cohen et al. 1993, Lawton 1995, Memmott and Godfray 1994). This means that there is the opportunity to produce a collection of high quality food webs containing nontarget interactions in addition to other trophic interactions that may be important within the community. In this section, I will explain how to construct a quantitative food web. This is done by describing the construction of a food web that quantifies the interactions between plants, leaf miners, and their parasitoids in a Costa Rican tropical dry forest (Memmott et al. 1994). The Costa Rican web was used to test a variety of hypotheses about the diversity and structure of parasitoid communities, rather than investigating nontarget effects associated with a biological control program. In fact, no nontarget effects have been recorded from this area. However, the general approach and methodology used to construct this web is very similar to that needed to construct webs for studying nontarget effects.

Which Species to Sample?

Since terrestrial communities typically contain considerable numbers of plants, herbivores and natural enemies, a food web containing every species at a single locality is probably an unrealistic objective. Rather, most webs are likely to concentrate on a subset of organisms. When studying nontarget effects, this subset will obviously contain the biological control agent and any species it consumes at the chosen study site. Starting with these few species soon leads to a web of interactions.

The Sampling Protocol

The first stage in any food web project is to delimit the study area: in the Costa Rican project an area of forest c.100 m x 170 m was chosen. This size was chosen for practical reasons: it was deemed to be an area that could be sampled rigorously and one side of the study plot was bordered by a track that made access simple. The inferences that can be drawn about community structure from a web are influenced markedly by the form of data used in the web's assembly (Memmott and Godfray 1994). In particular, summing data collected at different times or from different places may hide patterns in the web. The use of a plot, or ideally replicated plots, reduces this problem.

The plants, herbivores and parasitoids were sampled by the use of transect lines. These were 100 m in length and were placed at 10 m intervals across the plot. The order in which they were used was randomised and the lines were sampled every ten days over two seven-month field seasons. Seven months is the length of the wet season at this field site, leaf miners were absent from the plot during the dry season. Each transect line took between half a day and three days to complete. How the transect lines sampled each trophic level is outlined below:

(1) Sampling the plants: The identity and number of the plants growing in 1m x 1m quadrats alongside two transect lines were recorded to provide data on plant abundance. Ideally, a measure of absolute abundance, such as biomass or leaf area per

cubic meter of forest, would have been used. This proved, however, to be too labor intensive, especially as the project was specifically interested in the role of parasitoids within the web. Consequently, in the web described here, the relative abundance of plant species, rather than the absolute abundance, is presented. This is a crude measure of plant abundance and is used only for purposes of illustration.

(2) Sampling the herbivores: All leaf miners found within 1m of the transect line and up to 3 m in height were collected. Additional miners were collected from within the plot in order to increase the number of insects reared. Leaf miners were also gathered from other areas of secondary forest in order to assess how representative the plot miner-parasitoid interactions were of the area. The miners were reared individually in plastic rearing pots or in small plastic bags with specific methods for individual species being developed throughout the field season. During the project, approximately 7000 leaf miners were reared.

(3) Sampling the parasitoids: Parasitoids were collected by rearing them from the leaf miners collected along the transects and from more general collections made in the plots. For rarer leaf miner species, or species on rare plants, these collections were not large enough to reveal the parasitoid complex and so we also collected material from similar secondary growth close to the study site.

Most plants were identified in the field, but grasses and composites were sent to specialists for identification. Many of the insects reared were species new to science and several of the parasitoids represented new genera. New taxa were not problematic as the Costa Rican project was a collaborative venture between a taxonomic institute (The Natural History Museum, UK) and an ecological institute (Imperial College at Silwood Park, UK). Because of the taxonomic participation, all species of insect could be identified to morphotype. Identification to morphotype is sufficient for the purposes of constructing quantitative food webs (Memmott and Godfray 1993).

The Results of the Sampling

Sampling the plot revealed 88 plant species, 92 species of leaf miner and 93 parasitoid species. The leaf miners were from insect orders: Coleoptera, Lepidoptera and Diptera. A detailed report on the food web has been published elsewhere (Memmott et al. 1996). Using the Lepidoptera data from the Costa Rica web, I will illustrate how a food web approach can be used to study nontarget effects.

This subset of the Costa Rican data is chosen as 32% of biological control programs are targeted at lepidopteran pests, while only 11% and 10% are targeted at coleopteran and dipteran pests, respectively (Greathead and Greathead 1992). Moreover, recorded cases of nontarget effects (for example, Gagne and Howarth 1982, Duan et al. 1996, Funasaki et al. 1988) indicate that nontarget effects mostly occur within the taxonomic order of the target species. Consequently, an implication is that about one third of nontarget effects are likely to occur within the Lepidoptera. Obviously, the available data may be biased; it is easy to imagine that more effort may have been directed into looking for nontarget effects within the same order as the target host.

Web Construction

To illustrate exactly how food webs are drawn, model data has been presented for a hypothetical eight-species community of plants, Lepidoptera, and parasitoids (Table 1 and Table 2). These insects were collected from a hypothetical plot 100 m x 100 m. Table 1 lists plant species, plant abundance, Lepidoptera species, Lepidoptera abundance and the number of each Lepidoptera species reared from each plant species. Table 2 lists parasitoid species, parasitoid abundance, host species and the number of parasitoids reared from each host species. To draw a food web, each species is represented by a box, the boxes are arranged in three trophic levels (plants, Lepidoptera, and parasitoids), and then boxes which interact as resource and consumer are connected. Two types of web can be constructed, connectance webs which simply show which species are linked as resource, and consumer and quantitative webs which quantify species abundance and interaction frequency. The model data in Table 1 and Table 2 has been used to construct a connectance and a quantitative webs to illustrate exactly how raw data is turned into a food web diagram (Figure 1a, 1b).

Space restraints preclude a detailed account of how to calculate food web statistics from food webs. Rather, attention is directed to the following publications that outline how such statistics were calculated for a variety of communities: Warren (1989), Winemiller (1990), Schoenly and Cohen (1991), and Martinez (1991).

The Costa Rican data on plants, lepidopteran leaf miners, and parasitoids were used to draw both a connectance and a quantitative web. The connectance web shows all of the interactions observed in the plot between plants, Lepidoptera and parasitoids during the field seasons (Figure 2). In contrast to the connectance web, the quantified web shows only those interactions observed during quantitative sampling (Figure 3). Rare interactions are thus excluded from the quantitative web. The quantified web is a more meaningful description of host-parasitoid interactions in the plot as it provides a snapshot of species abundance and interaction frequency at the time of sampling.

The above methodology was developed specifically for an indigenous leaf miner community. Exactly the same methods could be used to look at nontarget effects in either biological control programs for pest leaf miners or for programs using a leaf miner as a weed biological control agent. However, there may be some differences in the methods required for other arthropod pests. For example, line transects were used for the leaf miner community, but point transects, random quadrats, or some other method may be more suitable for different types of arthropods. Likewise, differences in the vegetation structure, density of insects, ease of rearing, length of the field season and location of the plot may affect the sampling regime. The salient point is that plants, herbivores and their natural enemies must be quantitatively sampled: thus the density of each can be expressed in common units. While serving as an ideal at which to aim, this level of information may prove to be too much of a burden to collect for all three trophic levels and a compromise will be needed. For example, in the case of the Costa Rican web, the plant data were not as detailed as the miner and the parasitoid data. Given that the project was investigating miner-parasitoid relationships, the compromise was not particularly limiting.

Table 1. Rearing data from a model community of 8 species: the plant and Lepidoptera data.

Lepidoptera species	Lepidoptera abundance	Plant species	Plant abundance	No. Lepidoptera reared from each plant species
Lep A	500	Species X	1120	500
Lep B	1500	Species Y	250	1500
Lep C	7800	Species Y	250	3000
Lep C	7800	Species Z	240	4800

Table 2. Rearing data from a model community of 8 species: the Lepidoptera and parasitoid data.

Parasitoid	Host species	Parasitoid	Number of parasitoids reared from each host species
Species 1	Lep A	74	2
Species 1	Lep B	74	15
Species 1	Lep C	74	57
Species 2	Lep A	25	4
Species 2	Lep C	25	21

Predators Versus Parasitoids as Biological Control Agents

Parasitoids and herbivores are not the only types of biological control agents to be accused of nontarget effects, predators have been implicated as well. For example, in some parts of the United States native ladybirds have been replaced by the European species, *Coccinella decipunctata* L. (Elliott et al. 1996). This species was released to biologically control Russian wheat aphid. Unfortunately, predator nontarget effects are harder to quantity than parasitoid nontarget effects. This is because it is relatively easy to determine a parasitoid's host range by rearing the hosts. In contrast, the hit-and-run style of insect predation makes it much more difficult to quantify predator host range. There are various ways of approaching this problem. One way is to enclose the predator and prey in the laboratory and observe which prey species are consumed. Cohen et al. (1994) used this approach to work out the host range of the predators of the herbivores in rice fields. This is also a common method used to determine trophic relationships in aquatic food webs (e.g., Warren and Lawton 1987). Another approach would be to use modern molecular techniques, such as enzyme-linked immunosorbent assay (ELISA) to determine predator gut contents. This technique was successfully used to determine the concentration and quantity of slug haemolymph in the crops of carabid beetles (Symondson et al. 1996).

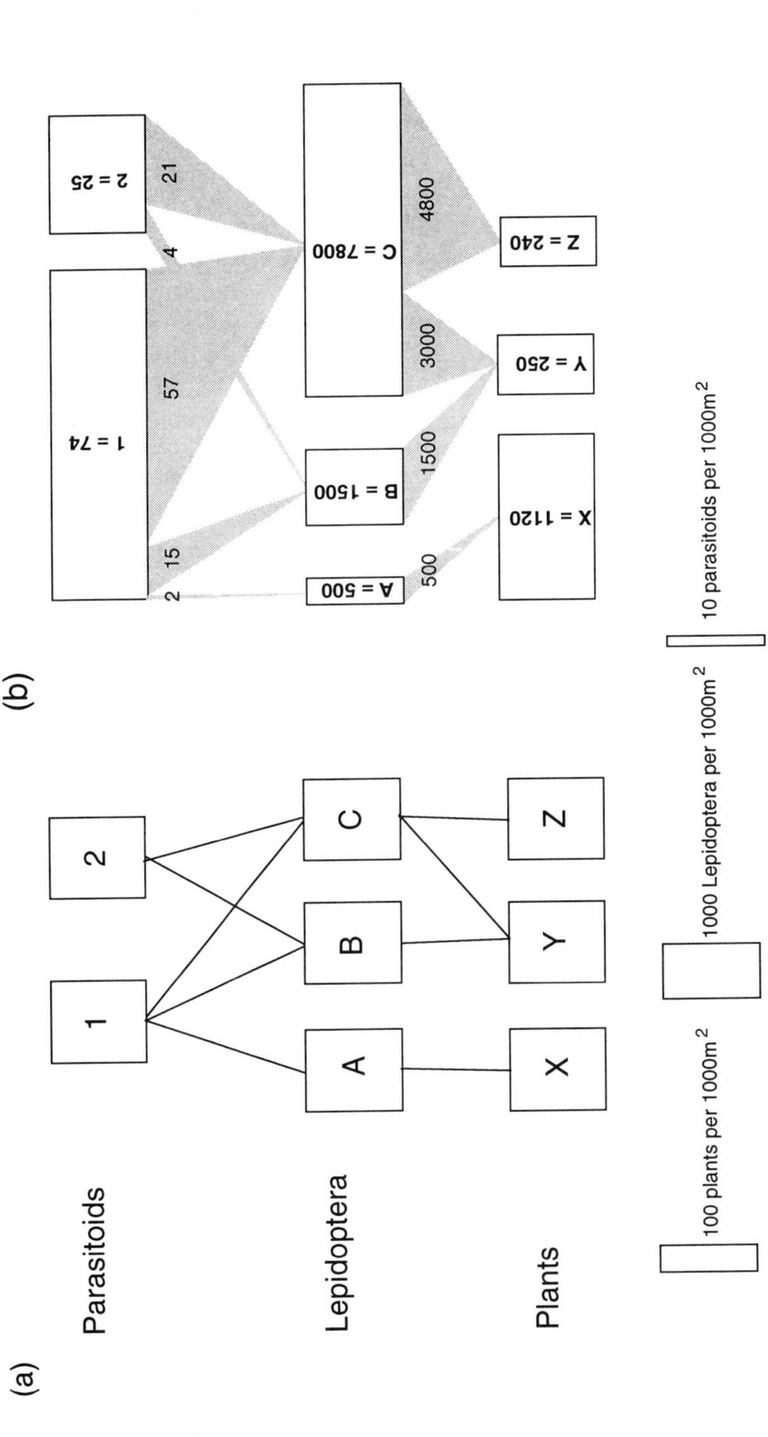

Figure 1. Food webs constructed using the model data in Table 1 and Table 2. (a) connectance web, (b) quantitative web. The raw data is shown on the quantitative web in order to make it clear how the raw data and the figure are related.

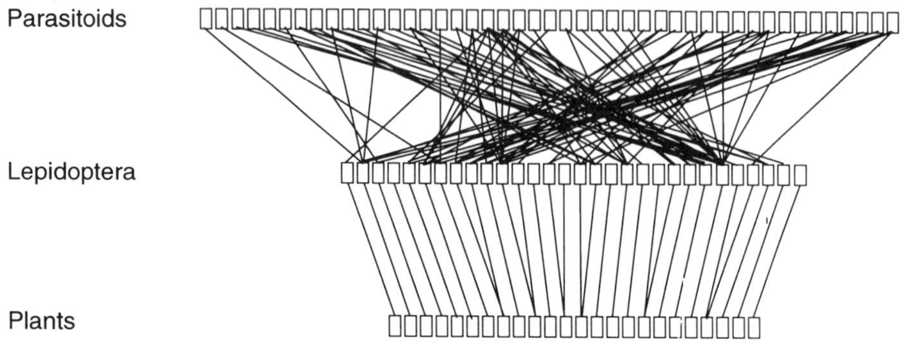

Figure 2. The connectance web summarizing the trophic links between the plants, Lepidoptera and parasitoids in the plot. Each species is represented by a rectangle, the plants at the bottom, the Lepidoptera in the middle and the parasitoids at the top. The names of the species are given in Memmott et al. (1996).

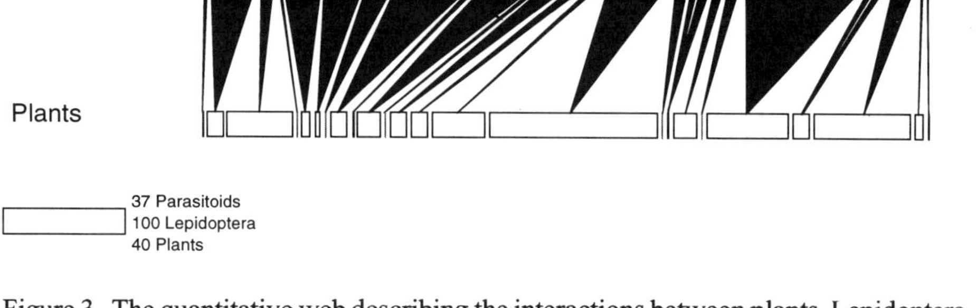

Figure 3. The quantitative web describing the interactions between plants, Lepidoptera and parasitoids. As before, each species is represented by a rectangle with the plants at the bottom, the Lepidoptera in the middle and the parasitoids at the top. The widths of rectangles depicting plants, Lepidoptera and parasitoids is proportional to their abundance at the study site, although for clarity the scales for the three trophic levels is different. The width of the plant rectangles depicts the number of plants rather than the number of leaves (the part of the plant used by the leaf miner).

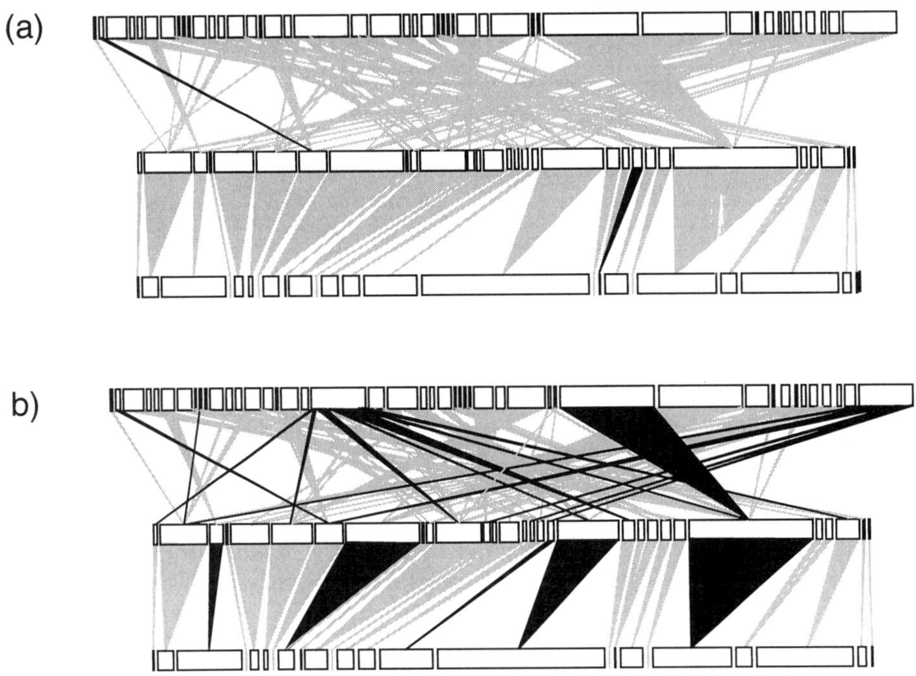

Figure 4. The quantitative web invaded by hypothetical biological control agents in a habitat far from the point of release. The interactions between native species are shown in grey, the interactions between biological control agents and native species are shown in black. (a) A single parasitoid biological control agent and a single weed biological control agent have infiltrated the web. (b) Five parasitoids biological control agents and five weed biological control agents have infiltrated the web.

How to Quantify the Level of Nontarget Interactions in Natural Communities

Food webs of the structure seen in Figure 3 are probably typical of natural communities. For the sake of example, imagine that this web describes the patterns of abundance in a community that has been invaded by biological control agents. Further, imagine that the nontarget species affected by the agent (and their respective nontarget interactions) are highlighted in a different colour. Figure 4 shows such a lepidopteran community, hypothetically invaded by biological control agents. Figure 4a shows a low level of infiltration - one weed and one pest biological control agent have attacked species in the community. Figure 4b shows a much higher level of infiltration, several weed and several pest biological control agents are attacking species in the community. At present we have no idea of which two of these scenarios is the more accurate description of communities host to biological control agents. Biological control practitioners err toward Figure 4a, conservationists towards Figure 4b. Neither group has the data to clearly substantiate their claim.

If food webs of the type shown in Figure 3 were available for real natural communities then we would know the level of nontarget interactions. For example, the number of nontarget species attacked would be known, along with the frequency of these interactions.

THE DRAWBACKS OF WORKING WITH FOOD WEBS

While quantitative food webs offer the most comprehensive means of advancing research on nontarget impacts of biological control agents, they are not without drawbacks. In this final section, I address these drawbacks and compare them to those of other methods of studying nontarget effects.

Probably the most obvious drawback of food webs, is that they are time consuming to construct. For example, the Costa Rican web took two years to complete. However, the time taken for construction depends, in part, on the questions being asked. In Costa Rica we used, in part, the approach of many food web projects: we summed more than one field seasons' data into a single food web in an effort to gain the maximum information about the community. While this was suited to this particular project, two years may not be the appropriate time scale for understanding the dynamics of interactions in food webs. Unfortunately, the appropriate time scale for studying the dynamics of agent- nontarget interactions remains unknown. It may, however, be less than two years. For example, if a biological control agent attacks a nontarget host for two months of the year only, then two months could be a suitable time-scale for the project; or at least a starting point. A food web based on two months of sampling would provide a "snap shot" description of the web at the time of the nontarget interaction. This web could be compared to webs from communities without biological control agents and which, therefore, have no such nontarget impacts on indigenous flora or fauna. Obviously, there is a chance that some effects may be missed over such a short time period. Only by constructing webs containing nontarget interactions, will the appropriate time scale for analysis become apparent, and this is likely to vary between both communities and biological control agents

There are three additional disadvantages to food webs. First, communities may contain a variety of taxonomically intractable organisms. Second, food webs are purely observational studies although, as I have stressed throughout this chapter, webs are an extremely useful tool for developing hypotheses that can be experimentally tested. Thirdly, there are a number of potential biases that need to be considered if the web contains parasitoids.

Taxonomic Problems

Taxonomic intractability is best overcomed by involving a taxonomic institute in the project: it is almost impossible for one person to be an expert in all fields of identification required in tri-trophic web construction. The Costa Rican web, for example, collated data from 35 plant families, 18 herbivore families, 10 parasitoid families and 3 predator families. Once a web has been constructed, however, further work on the community is much simpler as most species will be known.

The Observational Nature of Food Webs

Food webs are often criticized because they use purely observational data, which may limit the conclusions that can be drawn from them as the dynamically important interactions in a community are not always those that are the most apparent from the food web (Paine 1992). Webs constructed around biological control agents exhibiting nontarget interactions are no exception to this rule. However, experimental manipulations of food webs have proved an extremely successful means of elucidating the processes that underlie the observed patterns. For example, Paine (1992) used removal experiments to show which species were the strong interactors in a littoral community, and Memmott et al. (1993) used exclusion experiments to prove that predator and parasitoid guilds were not competing over leaf miners in a tropical dry forest. Experimental manipulations of webs are likely to be equally useful in determining the importance of nontarget effects in biological control programs.

Sources of Bias When Sampling Parasitoid Communities

The potential sources of bias encountered when sampling a parasitoid complex are discussed in Memmott et al. (1996). A primary source of bias is that the size of a host parasitoid complex is influenced by host abundance. Moreover, there may be differences in the probability of collecting parasitized and unparasitized hosts.

CONCLUSIONS

Food webs are a fundamental component of any attempt to describe how natural communities are structured or how complexes of species interact. They do not, however, provide a flawless method for studying nontarget effects of biological control initiatives. This is because the inferences that can be drawn from static "snapshots" of trophic relationships is limited (Memmott and Godfray 1994). Nonetheless, the construction of food webs has many advantages over the methods by which nontarget effects have been studied in the past, which is largely through the use of imperfect literature surveys.

Recently, some impressive field data on the nontarget effects of biological control agents has emerged. For example, in the study of Louda et al. (1997) which was mentioned earlier, a biological control weevil (*R. conicus*) reduced the seed set of a nontarget native plant. Further, the weevil affected the abundance of native seed feeding insects. The complex relationship among *R. conicus*, its target weed, and its nontarget native hosts illustrates the inadequacies of a simple food chain model for biological control. Such simplistic models ignore other species that are inevitably involved, both directly and indirectly, in food webs (Strong 1997).

While constructing quantitative food webs may be time consuming, it provides indisputable evidence on the extent to which natural habitats host non-indigenous biological control agents. At present virtually nothing is known about the extent of nontarget interactions in natural communities. To date, no webs have been constructed showing nontarget effects, although one is presently under construction in Hawaii (running from 1998 to 2001: for further information contact J. Memmott). Food webs will become particularly valuable to the science of biological control when a data base of webs from different biological control programs is available for comparison. It will then be easier to ask the fundamental questions that practitioners of biological control have, so far, been unable to pose, but which are central to refining the success of biological control: for example, are certain arthropod or plant families particularly susceptible to nontarget effects? Are certain biological control agents particularly prone to attacking nontarget species?

As pointed out by Strong (1997), biological control conducted without regard for environmental costs could result in such great resistance from society that even the sensible application of biological control could be prevented. Biological control practitioners are under enormous pressure from society to provide a scientific underpinning to their assertions that biological control agents pose no threat to the environment. Soc. needs information on the risks associated with biological control in a form that it can assimilate. The concept of a food web is widely understood. Consequently, food webs should become a more widely used vehicle for explaining the risks associated with biological control. If food webs show that on average nontarget effects are relatively rare and usually of little importance, then society will probably deem biological control an acceptable risk, especially when compared to the risks associated with other forms of pest or weed control.

REFERENCES

Bennett, F. D. 1993. Do introduced parasitoids displace native ones. Florida Entomologist 76: 54–63.

Bonsall, M. B., and M. P. Hassell. 1997. Apparent competition structures ecological assemblages. Nature 388: 371–373.

Cohen, J. E., R. A. Beaver, S. H. Cousins, D. L. Deangelis, L. Goldwasser, K. L. Heong, R. D. Holt, A. J. Kohn, J. H. Lawton, N. Martinez, R. Omalley, L. M. Page, B. C. Patten, S. L. Pimm, G. A. Polis, M. Rejmanek, T. W. Schoener, K. Schoenly, W. G. Sprules, J. M. Teal, R. E. Ulanowicz, P. H. Warren, H. M. Wilbur, and P. Yodzis. 1993. Improving food webs. Ecology 74: 252–258.

Cohen, J. E., K. Schoenly, K. L. Heong, H. Justo, G. Arida, A. T. Barrion, and J. A. Litsinger. 1994. A food-web approach to evaluating the effect of insecticide spraying on insect pest population-dynamics in a philippine irrigated rice ecosystem. J. Appl. Ecol. 31: 747–763.

Duan, J. J., M. F. Purcell, and H. Messing. 1996. Parasitoids of nontarget tephritid flies in Hawaii: implications for biological control of fruit fly pests. Entomophaga 41: 245–256.

Elliott, N., R. Kieckhefer, and W. Kauffman. 1996. Effects of an invading coccinellid on native coccinellids in an agricultural landscape. Oecologia 105: 537–544.

Funasaki, G., P. Y. Lai, L. M. Nakahara, J. W. Beardsley, and A. K. Ota, 1988. A review of biological control introductions in Hawaii: 1890–1985. Proc. Hawaiian Entomol. Soc. 28: 105–160.

Gagne, W. C., and F.G. Howarth. 1982. Conservation status of endemic Hawaiian Lepidoptera. In: Proc. 3rd Congress on European Lepidoptera, Cambridge. p. 74–84.

Gardner, M. R., and W. R. Ashby. 1970. Connectance of large dynamical (cybernetics) systems: critical values for stability. Nature 228: 784.

Greathead, D. J., and A. H. Greathead. 1992. Biological control of insect pests by insect parasitoids and predators: the BIOCAT database. Biol. Control News Info. 13: 61N–68N.

Hawkins, B. A., and P. C. Marino. 1997. The colonization of native phytophagous insects in North America by exotic parasitoids. Oecologia 112: 566–571.

Holt, R. D. 1977. Predation, apparent competition and the structure of prey communities. Theor. Pop. Biol. 12: 197–229.

Holt, R. D. 1984. Spatial heterogeneity, indirect interactions, and the coexistence of prey species. Am. Nat. 124: 377–406.

Howarth, R. G. 1985. Classical biological control: panacea or Pandora's Box? Proc. Hawaiian Entomol. Soc. 24: 239–244.

Howarth, R. G. 1991. Environmental impacts of classical biological control. Annu. Rev. Entomol. 36: 485–

509.
International Institute of Biological Control 1996. Annual report (1995) of the International Institute of Biological Control. CAB International.
Lai, P. Y. 1988. Biological control: a positive point of view. Proc. Hawaiian Entomol. Soc. 28: 179–191.
Lawton, J. H. 1995. Webbing and wiwacs. Oikos 72: 305–306.
Lockwood, J. A. 1996. The ethics of biological control: understanding the moral implications of our most powerful ecological technology. Agric. Human Values 13: 2–19.
Louda, S. M., D. Kendall, J. Connor, and D. Simberloff, 1997. Ecological effects of an insect introduced for the biological control of weeds. Science 277: 1088–1090.
Martinez, N. D. 1991. Artifacts or attributes—effects of resolution on the little-rock lake food web. Ecological Monographs 61: 367–392.
McEvoy, P. 1996. Evaluation and economics: synthesis of session 7. *In* Proc. IX Intl. Sym. Biol. Control of Weeds, January 19–26, 1996, Stellenbosch, South Africa.
Memmott, J., and H. C. J. Godfray. 1993. Parasitoid webs. *In* J. Lasalle and I. D. Gould (eds.), Hymenoptera and biodiversity. CAB International, Wallingford, UK. p. 217–234.
Memmott, J., and H. C. J. Godfray. 1994. The use of parasitoid webs in community ecology. *In* B. A. Hawkins and W. Sheehan (eds.), Parasitoid community ecology. Blackwell Univ. Press. p. 300–319.
Memmott, J., H. C. J. Godfray, and B. Bolton. 1993. Predation and parasitism in a tropical herbivore community. Ecol. Ent. 18: 348–352.
Memmott, J., H. C. J. Godfray, and I. D. Gauld. 1994. The structure of a tropical host parasitoid community. J. Anim. Ecol. 63: 521–540.
Miller, M., and G. Aplet. 1993. Biological control: a little knowledge is a dangerous thing. Rutgers Law Rev. 45: 285–335.
Mills, L. S., M. E. Soule, and D. F. Doak. 1993. The keystone-species concept in ecology and conservation. Bioscience 43: 219–224.
Muller, C. B., and H. C. J. Godfray 1996. Apparent competition between two aphid species. J. Anim. Ecol. 66: 57–64.
Paine, R. T. 1969. A note on trophic complexity and species stability. Am. Nat. 103: 91–93.
Paine, R. T. 1980. Food web: linkage, interaction strength and community infrastructure. J. Anim. Ecol. 49: 667–685.
Paine, R. T. 1992. Food–web analysis through field measurement of per-capita interaction strength. Nature 355: 73–75.
Pimm, S., and R. L. Kitching, 1988. Food web patterns—trivial flaws or the basis of an active research–program. Ecology 69: 1669–1672.
Power, M. E., D. Tilman, J. A. Estes, B. A. Menge, W. J. Bond, L. S. Mills, G. Daily, J. C. Castilla, J. Lubchenco, and R. T. Paine. 1996. Challenges in the quest for keystones. Bioscience 46: 609–620.
Schoenly, K., and J. E. Cohen. 1991. Temporal variation in food web structure—16 empirical cases. Ecological Monographs 61: 267–298.
Simberloff, D. 1991. Keystone species an community effects of biological introductions. *In* L. R. Ginzburg (ed.), Assessing ecological risks of biotechnology. Butterworth-Heinnemann, Boston. p. 1–19.
Simberloff, D., and P. Stiling. 1996a. How risky is biological control. Ecology 77: 1965–1974.
Simberloff, D., and P. Stiling. 1996b. Risks of species introduced for biological-control. Biol. Cons. 78: 185–192.
Strong, D.R. 1997. Ecology—fear no weevil? Science 277: 1058–1059.
Symondson, W. O. C., D. M. Glen, C. W. Wiltshire, C. J. Langdon, and J. E. Liddell. 1996. Effects of cultivation techniques and methods of straw disposal on predation by pterostichus-melanarius (Coleoptera, Carabidae) upon slugs (Gastropoda, Pulmonata) in an arable field. J. Appl. Ecol. 33: 741–753.
Thomas, M. B., and A. J. Willis. 1998. Biological control—risky but necessary? Tree 13: 325–329.
Warren, P. H. 1989. Spatial and temporal variation in the structure of a fresh-water food web. Oikos 55: 299–311.
Warren, P. H., and J. H. Lawton. 1987. Invertebrate predator-prey body size relationships—an explanation for upper-triangular food webs and patterns in food web structure. Oecologia 74: 231–235.
Willis, A. J., J. E. Ash, and R. H. Groves. 1993. Combined effects of 2 arthropod herbivores and water-stress on growth of hypericum species. Oecologia 96: 517–525.
Willis, A. J., J. E. Ash, and R. H. Groves. 1995. The effects of herbivory by a mite, aculus–hyperici nutrient deficiency on growth in hypericumspecies. Austral. J. Bot. 43: 305–316.
Zimmerman, E. C. 1958. Insects of Hawaii. vol. 7. (Macrolepidoptera). Univ. Hawaii Press, Honolulu.

PART 3

WEEDS

11

WHY THINGS BITE BACK: UNINTENDED CONSEQUENCES OF BIOLOGICAL WEED CONTROL

P. B. McEvoy
Department of Entomology, Oregon State University, Corvallis, Oregon

E. M. Coombs
Oregon Department of Agriculture, Salem, Oregon

INTRODUCTION

There is an upside and a downside to the current "boom market" in biological control, the use of living organisms to control pests. The upside is that biological control provides a potentially effective control technique, an alternative to chemical pesticides, and an ecological foundation for pest control strategies. As a method of weed control, biological control can claim a growing number of successes. As of the end of 1996, there have been at least 1150 planned releases of 365 species of invertebrates and fungi on 133 weed species in 75 countries (Julien and Griffiths 1998). Julien (1989) estimates that 25% of all releases made up until 1985 contributed to control; some recent estimates of success rates are higher (McFadyen 1998). As an alternative to chemical pesticides, biological control potentially reduces pesticide use and its undesirable effects on human health and the environment (OTA—U.S. Congress Office of Technology Assessment 1995). In contrast to chemicals, biological control organisms pose negligible human health risk, circumvent the problem of pest resistance (Holt and Hochberg 1997), and reduce the need for repeated and costly interventions to control invasive species, plants and animals that are not native to ecosystems they invade (OTA—U.S. Congress Office of Technology Assessment 1995). As an exercise in applied ecology, the scientific study of biological control systems has helped transform perspectives in pest control from an industrial to an ecological model (Levins 1986), fostering a sustainable pest control technology (Karban et al. 1997) that potentially (1) is effective, (2) requires low inputs of resources, (3) is self-perpetuating, (4) produces minimal pollution, (5) produces minimal effects on nontarget organisms, and (6) is compatible with other management practices.

Increasing concern that biological control is often ineffective or possibly unsafe marks the downside to biological control. Safety concerns center on classical biological control in which alien organisms are introduced to control alien pests (Howarth 1983, Ehler 1991, Howarth 1991, Simberloff 1992, Lockwood 1993, McEvoy 1996, Secord and Kareiva 1996, Simberloff and Stiling 1996a, Simberloff and Stiling 1996b, Louda et al. 1997, Louda 1998, Thomas and Willis 1998). Introducing control organisms in classical biological control programs is generally irreversible. Unlike pesticides, control organisms cannot be withdrawn and discontinued because they have proven to be ineffective or because society has become less tolerant of their harmful side effects on nontarget organisms. There is growing public and scientific awareness that importing potentially harmful alien organisms into new environments can lead to irretrievable loss of native biodiversity and disruption of ecosystem processes (Levin 1989, Levin 1990, OTA—U.S. Congress Office of Technology Assessment 1993). Given concerns about alien species, biological control introductions seem like solutions from one point of view, potential problems from another. Biological control scientists find themselves caught up in a debate about adverse environmental effects, protocols for avoiding them, and regulations to restrict the flow of potentially harmful alien organisms across the borders of biogeographic regions.

Classical weed biological control is subject to some of the strictest evaluations currently made in the United States for planned release of alien organisms (Ruesink et al. 1995), and it has been held to a higher safety standard than biological control of insects and other arthropods. Many alien species other than biological control organisms (including potential plant invaders) are allowed to enter the United States as "innocent until proven guilty," but weed biological control organisms must meet a stricter standard of "guilty until proven innocent" (Ruesink et al. 1995). To ensure beyond a reasonable doubt that biological controls are necessary, effective, and safe, there are protocols for (1) selecting target organisms (McClay 1989, Peschken and McClay 1995), (2) selecting effective control organisms (Harris 1973, Goeden 1983, Wapshere 1985), (3) testing control organisms for safety (Zwölfer and Harris 1971, Wapshere 1974, Klingman and Coulson 1983, Coulson and Soper 1989, Cullen 1990, Weidemann and Tebeest 1990, Coulson 1992, Harris and McEvoy 1995, McEvoy 1996, Marohasy 1998, USDA 1999), and (4) encouraging high standards of professional conduct (FAO 1994). Most scientists believe that environmental risks of weed biological control using host-specific control organisms are largely hypothetical and that current safeguards are adequate. However, even ardent supporters of biological control call for more research on environmental effects and for introduction of monitoring systems that would allow the early detection of any long-term problems.

It is traditionally assumed that adequate assurances of safety in biological control reduce the need for assurances of effectiveness. Yet each new biological control organism that fails to establish or achieve control wastes scarce resources and may decrease public support (Harris 1988, Howarth 1991). In the worst case scenario, biological control organisms may become pests themselves (Howarth 1991, Simberloff and Stiling 1996a, Louda et al. 1997), although a recent worldwide survey found only eight instances of weed-biological control organisms directly damaging nontarget plants (McFadyen 1998). Fortunately, host specificity still provides the best assurance that biological control will suppress the target organism without harmful side effects. Experience with past introductions has shown that problem species tend to have broad host ranges. To determine whether a planned introduction of a biological control organism is safe, new control organisms now require extensive lab and field trials to demonstrate host specificity before approval. Yet host specificity has its limitations as a safety criterion. The most important uncertainties that remain have to do with forecasting control organism movement, evolution, and indirect effects, plus refining estimates of the severity, probability, and consequences of nontarget effects (McEvoy 1996, Secord and Kareiva 1996).

In this chapter, we put safety issues into a general context of technology and its unforeseen consequences, and we suggest ways to resolve thorny scientific and policy dilemmas. The chapter has four parts:

- First, we address the belief that biological control resembles a lottery, in which more and more control organisms must be added to speed the discovery of the few that are effective.
- Second, we briefly review the scientific and sociological factors that make the lottery so seductive, leading scientists to speed up the screening of control organisms for safety while neglecting to screen species efficiently for effectiveness.
- Third, we ask whether over-prescription of biological control introductions under the lottery model leads to "revenge effects" (Tenner 1996). That is, in our rush to solve local and acute pest problems, are we creating diffuse and chronic problems that are even harder to solve?
- Fourth, we suggest ways to avoid revenge effects by applying the Precautionary Principle (O'Riordan and Cameron 1994) and following simple guidelines for decision making under uncertainty (Ludwig et al. 1993).

Biological Control Currently Resembles a Lottery

Discovery of biological control organisms is largely a matter of trial-and-error influenced by an individual scientist's storehouse of knowledge and experience. Designing biological control systems has been compared to a lottery (Myers 1985): the outcomes of control organism introductions are so dominated by chance that the best course is to import many control organisms (after they have passed the host specificity test), leaving the organisms to sort it out for themselves which one or combination will prove most effective. The Lottery Model continues to guide weed biological control decisions today, endorsed by influential investigators (Müller and Schroeder 1989). In a recent article (McEvoy and Coombs 1999), we identified two symptoms of biological control under the Lottery Model as "runaway importation-rates" and the "monitoring and evaluation gap."

"Runaway importation-rates" are indicated by the curves describing the cumulative number of imported control organisms pulling away from the curves describing the cumulative number of target organisms (Fig. 1A). World-wide trends show control organism species increasing faster than target organism species (Julien and Griffiths 1998) and biological control efforts in Oregon exemplify that trend. In the three decades before 1977, an average of four new control organisms was introduced per decade in Oregon. Since 1977, the average has been 24 per decade. The increase in biological control activity is an expected response to the relentless increase in plant invaders and the economic and environmental harm they cause. Controlling plant invaders can help protect ecosystems against the irretrievable loss of native biodiversity, a global phenomenon caused principally by habitat destruction and invasion by alien species (Wilcove et al. 1998). Yet it is perhaps surprising to find that the number of control-organism species is increasing faster than the number of target-organism species, skewed by the more profligate control programs for knapweeds *Centaurea* and leafy spurge *Euphorbia* (McEvoy and Coombs 1999). The number of control-organisms per target-organism is increasing, and some biological control systems are growing more, not less, complex in their structure. Either weed problems are becoming harder to solve, scientific effort is not responding to meet increasing demand for research on biological control safety and effectiveness, or our learning curve may be getting worse. If practice makes perfect, then we may be practicing the wrong things, speeding up the screening of new introductions for safety while neglecting to efficiently screen species for effectiveness.

The "monitoring and evaluation gap" describes the skewed priorities characteristic of some biological control programs: the activities of finding, screening, releasing, and redistributing control organisms command far more attention than the more mundane but essential activities of monitoring and evaluating effects on target and nontarget organisms (Fig. 1B). Perhaps the greatest advances in improving our understanding, prediction, and management of biological control could be made through more timely and informative post-release evaluation of its effects (Thomas and Willis 1998, McEvoy and Coombs 1999). Biological control programs develop by stages from finding, screening, releasing, increasing, redistributing control organisms; to damaging and suppressing target organisms; to fostering replacement of target weeds with desirable plants. It is useful to monitor progress and improve performance at each step. For example, systematic attempts to improve success in biological control have helped raise establishment rates in Oregon, where 81% of imported control organisms have become established (Fig. 1B). Rising establishment rates probably reflect continuity of program support, learning from experience, plus developing and exchanging new technology for releasing, establishing, and redistributing control organisms. Global average establishment rates continue to be lower, remaining around 55%, and an analysis by decade shows that there has been no overall increase in establishment rates (Sheppard 1992).

Global establishment rates in biological control of weeds and arthropods may be low, but if done with the right control organism, right host, and right method, establishment rates can be much higher. Following earlier recommendations (Beirne 1985),

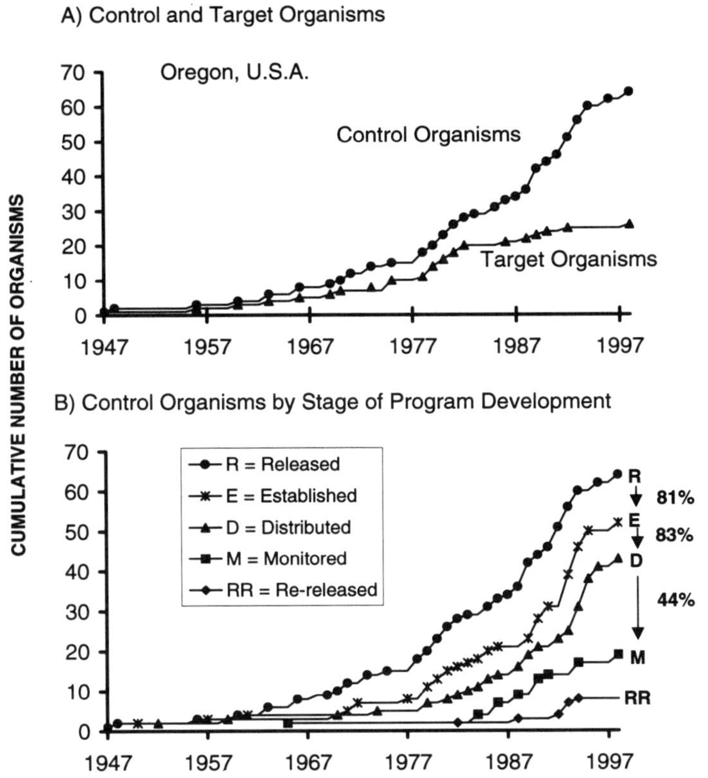

Figure 1. Cumulative number of imported control organisms for invasive plants in Oregon, USA, 1947–1998. (A) Cumulative numbers of target and control organism species. (B) Cumulative numbers of control organisms progressing through the stages of release, establishment, redistribution, and evaluation. Number re-released refers to the number of control organism species established after repeated attempts rather than a single attempt. The percentages to the right of the figure refer to probabilities of transition from stage to stage for each step in the sequence from release through monitoring and evaluation. Note that 81 percent of released organisms become established, 83 percent of established organisms are redistributed, and only 44 percent of redistributed organisms are monitored and evaluated. From (McEvoy and Coombs 1999) with permission.

mathematical models and field experiments addressing the colonization process are being used to devise optimal release and redistribution strategies (Grevstad 1996, Memmott et al. 1996, Grevstad 1999). How to balance the inherent tradeoff between the size of each release and the number of releases (given a finite initial release stock) depends on the susceptibility of the particular species to factors that put small populations at risk of extinction. These factors include demographic stochasticity, environmental variability, and Allee effects (reduced population growth at low density). In most cases, we are unlikely to have estimates of the required parameters for determining the optimum release size (fecundity, survivorship, variance in year to year survivorship, Allee effect intensity). There are logistical as well as ecological constraints operating on the number, size, timing, and spacing of releases. Nevertheless, Grevstad (1999) suggests that we can use simple rules of thumb that can tell us whether releases should be "on the large side" or "on the small side" — a significant improvement over the current use of arbitrary release sizes

While establishment rates are climbing in some areas of the world, rates of control by established organisms are hard to gauge. Field measurements of establishment rates tend to be more reliable than those of weed suppression, which tend to be subjective and impressionistic; and attributes that promote establishment have been easier to identify than attributes that promote successful control (Crawley 1989c). It invites trouble to continue to import alien control organisms under the Lottery Model without equal efforts to thoroughly monitor and evaluate their effects on target and nontarget organisms and update assessments accordingly.

WHY DOES THE LOTTERY MODEL REMAIN SO SEDUCTIVE?

The rush to introduce control organisms has as much to do with human behavior and psychology as with ecology and economics. Some scientists and organizations have strong vested interests in making introductions. Their emphasis is on advertising the release and redistribution of new organisms rather than dispassionately evaluating impacts on target and nontarget organisms and exposing all their acquired knowledge to criticism by others in scientific publications. This tendency has generated a debate about whether scientists can be objective about the effects of control organisms and promote them (through personal assurances of safety and effectiveness) at the same time. There is an important distinction between science and technology: between knowledge of the world and how it is used (Wolpert 1999). It is not for scientists to make moral or ethical decisions about the use of biological control, as they have no special rights or skills in this regard. The strength of a scientist is not in being an advocate, but in making scientific judgements based on evidence. Scientists should not be given the right or authority to make importation decisions on their own; once they get caught up in making value judgements, they become reluctant to accept the findings of their own research when findings contradict their beliefs. Given the public concern and potential for conflicts of interest, there seems to be no alternative but to rely on regulatory bodies plus rigorous public and peer review to assess whether controls are necessary, effective, and safe.

While importing new control organisms creates excitement and growth, revelations about effectiveness are often highly subjective. Crawley (1989c) reviewed 627 biological weed-control program and found few provided an objective measure of the depression in weed abundance because pre-release plant abundance was rarely measured and little follow-up work was carried out. Subjective evaluation reveals little of how a population is affected by abiotic factors, by age or other aspects of its structure, by population density feedbacks, or by biotic interaction with other populations. When experts were asked in an opinion survey in the Silwood International Project on Biological Control of Weeds (Moran 1985) what factors limit the effectiveness of control organisms in their weed biological control programs, they implicated a variety of possible causes. The percentage of all cases implicating each factor (some cases implicating more than one factor) was Climate 44%, Predators 22%, Parasitoids 11%, Disease 8%, Host incompatibility 33%, and Competition 12% (Crawley 1986). The percentage of all cases is low, but the percentage of known cases is higher. This is because the all-too-common response to questions in this survey was "unknown" — the percent of cases for which the influence of each factor was unknown was Climate 46%, Predators 56%, Parasitoids 49%, Host incompatibility 16%, and Competition 13%. The picture of what factors limit the effectiveness of control organisms when only the known cases are considered changes to Climate 81%, Predators 50%, Host incompatibility 39%, Parasitoids 22%, Competition 14% (data on Disease was of such poor quality that it is excluded from the analysis). We conclude that biological control scientists often have very limited knowledge of the factors that limit effectiveness of control organisms, and much of that knowledge is subjective. Subjective evaluation of control organisms in the absence of standard, quantitative measures becomes essentially a comparative exercise limited to the scope of one person's experience. Relying on expert opinion requires scrutiny of the scientific authority of the expert. Over-reliance on expert opinion and subjective evaluation currently hinders progress in understanding, predicting, and managing weed biological control systems. Quantitative evaluation is needed to (1) determine whether present control organisms are adequate or whether additional control organisms are needed, (2) create a more reliable basis for comparison, interpolation, and extrapolation, (3) develop and refine principles, concepts, and predictions regarding control organism specificity and effectiveness (McEvoy and Coombs 1999).

Possible reasons for a lack of effective monitoring and evaluation include time delays in the emergence of impacts, too little time, funding, will, and scientific know-how.

- ***Time delays.*** If we accept that it takes about 10–20 years for impact of biological control organisms to emerge (McFadyen 1998), and we observe that most control organisms in the western USA have been introduced in the last two decades (McEvoy and Coombs 1999), then some might conclude that most weed control organisms have not had enough time to reveal their impacts and evaluation can be put off until later. McFadyen (1998) pointed out that some evaluations have been undertaken too early, before control organisms have had a chance to achieve their full potential. Our view is that local evaluation can start early on after release and proceed step by step through the phases of establishment, increase, spread of control organisms, suppression of the target, and plant succession. Local evaluation can help create a more reliable basis for extrapolating these effects and processes to more global scales of space and time.
- ***Time and funding.*** Dollars that fuel the effort to find, screen, and release new control organisms dry up in the later stages of evaluation. Also, it is easier to fund studies of successes rather than failures, leading to bias in the overall assessment of outcomes. In our view, if funding continues to be a problem, importers and users of biological control organisms should cooperate to post the money up front to cover monitoring costs (McEvoy and Coombs 1999).

- ***Will power.*** A lack of will stems from scientists' dim view of monitoring — monitoring is unattractive to researchers because it takes too long, is too boring, and frequently produces a lot of negative results (e.g. no effect, no change) with little intellectual market-value related to publication, fund-raising, or career advancement (Harris 1997). Conversely, our cooperation in the study of ragwort biological control shows that it is possible to investigate applied problems in a fundamental way, to balance the concerns of practical management and basic ecology (McEvoy et al. 1991, Coombs et al. 1996, McEvoy and Coombs 1999).
- ***Scientific know-how.*** In these days of specialization, few scientists combine the zoological and botanical skills to work effectively on plant-herbivore interactions and effectively bridge the divide between botanical and zoological perspectives on herbivore-plant interactions (Harper 1977, Müller 1990, McClay 1995). A zoological perspective emphasizes the resources consumed and their effects on the consumer. A botanical perspective emphasizes the plant parts left over after herbivory and their capacity to maintain a plant population. Biological control of weeds is a specialized business run mostly by entomologists, who must enlist the help of plant population biologists to evaluate potential and actual effects (McClay 1995). While reason, argument, and evidence play crucial roles in directing biological control science, we suspect these social factors play important roles as well.

The steepness in the control organism curve in the last two decades (Fig. 1A) looks very much like the work of momentum investors: additional control organisms are being introduced before prior introductions have had a chance to work. An economic argument supporting the Lottery Model is that some economies of scale (lower production costs per control-organism species) can be achieved by screening many control-organism species simultaneously rather than sequentially, even though lower rates of introduction generally mean lower aggregate economic and ecological costs. An ecological argument supporting the Lottery Model is that unpredictable outcomes favor imaginative trial-and-error as the best approach. Predicting biological control has been the "holy grail" of ecological analysis, but researchers have had difficulty finding attributes of organisms or environments that reliably predict a species' behavior outside its native home. Some obvious factors to consider are the intrinsic rate of increase (r), abundance in the native habitat, taxonomic isolation of the target organism from potential nontarget organisms, climate and habitat matching, and vacant niches (Williamson 1996). However, with some notable exceptions among forecasts of plant invasions (Rejmánek and Richardson 1996, Reichard and Hamilton 1997), statistical generalizations about invasions have so far tended to be weak and unreliable (Williamson 1996). Also, a coherent mathematical framework for analyzing invasions and their ecological effects has only recently emerged (Williamson 1996, Shigesada and Kawasaki 1997). As a result, practitioners of biological control who actually make the decisions and place the bets in the biological control lottery find themselves whipsawed by conflicting and often unsubstantiated claims made by analysts: argument by example leads to rebuttal by counter-example, and the confused practitioner returns to simple trial-and-error as the best approach (McFadyen 1998).

Over-reliance on trial-and-error involving many control organisms delays the development of rational alternatives. Alternatives include experimental screening of each candidate control organism prior to release or making educated guesses based on expert intuition, data bases, and mathematical and experimental models. Two basic approaches have been used alone or in combination when assessing a weed control organism's potential for suppressing the target organism (Wapshere et al. 1989): extrapolating measurements from the area of origin to the area of introduction using carefully matched field situations (Wapshere 1985) or identifying important and researchable attributes of control organism and target organism biology (Harris 1973, Goeden 1983). Alternative ways to develop biological control programs using a minimum number of control organism species include critical attributes, targeted disruption of pest life cycles, and combinatorial ecology.

- ***Critical Attributes.*** Species with desirable biological properties can be selected based on expert intuition, data bases, and mathematical and experimental models. The more these approaches relate to each other, the better each becomes. Research and development of weed biological control organisms has remained rather empirical, or worse, anecdotal. The few attempts to develop model experimental systems to investigate attributes shared by an entire class of biological control systems include studies of Scot's broom (Fowler et al. 1996, Memmott et al. 1996, Rees and Paynter 1997), ragwort (McEvoy and Rudd 1993, McEvoy et al. 1993, McEvoy and Coombs 1999), and purple loosestrife (Blossey 1995, Blossey 1996, Grevstad 1996, Grevstad 1999). Data bases like the Julien Catalog (Julien and Griffiths 1998) and Silwood Project (Moran 1985, Crawley 1986, Bergelson and Crawley 1989, Crawley 1989c, Crawley 1989b, Crawley 1989a, Crawley 1990) are being used to catalog and analyze large amounts of data on biological weed control organisms. One of the best predictors of control effectiveness is whether or not a species was known to be effective elsewhere in the world (Crawley 1989c). This finding is enough to justify efforts to create and maintain global data bases of weed biological control organisms. Further investment in data bases may increase understanding of underlying causes and the practical development of new, unprecedented biological control systems in the future.
- ***Targeted disruption of pest life cycles.*** A second alternative is targeted disruption of weed life cycles, which means using field studies and mathematical models to identify which life-cycle transitions and pathways contribute most to population growth, which are most easily disrupted, and using this information to improve targeting of pest vulnerabilities (Müller 1990, Lonsdale et al. 1995, Rees and Paynter 1997, Shea and Kelly 1998, McEvoy and Coombs 1999). Analyzing the sensitivity of model outputs to changes in parameter values (Caswell 1997, Tuljapurkar and Caswell 1997), scientists ask "what if questions" such as this: if a control organism or other mechanism were to block this life cycle transition, what would be the effect on the growth rate of the pest population? This approach, is founded on the classical notions in biology of stimulus, to describe a change in the environment, and response, to describe the resulting change in the organism. It avoids the confusion about stimulus and response that creeps into writing about biological weed control (Harris 1981) and ecology (Harper 1982) with the use of anthropomorphic and emotive concepts like "stress." Targeted disruption of life cycles also refines the concept of a target site, from plant parts to plant life-cycle-transitions. Attacking a plant part is very different from attacking life cycle transitions, just as killing

an individual plant is very different from suppressing and regulating a weed population. Only structured population models make an explicit link between individual behavior or demography and population dynamics, and such models can be combined with factorial experiments to achieve more reliable inferences about cause and effect, as in studies of ragwort biological control (McEvoy and Coombs 1999).

- ***Combinatorial ecology.*** A third alternative exploits the combinatorial ecology of biological control through coordinated manipulation of disturbance, plant competition, and natural enemy regimes (McEvoy and Coombs 1999). Outbreaks of weeds result from an imbalance in activators and inhibitors of population growth. Top-down control using low numbers and low diversity of control organisms can be effective in suppressing target-organism populations provided care is taken (1) to avoid introducing propagules and disturbances that trigger weed recruitment, and (2) to conserve bottom-up control mechanisms that reduce the supply of resources fueling weed population growth either by lowering resource-renewal rates (by avoiding disturbances) or by increasing demand for resources (by enhancing competition with other plant species) (McEvoy et al. 1993, Rees and Paynter 1997, McEvoy and Coombs 1999). This approach promises to address the need for a community ecology perspective (identified in the discussion of critical attributes above) in researching and developing biological control organisms for weeds.

In summary, biological control of weeds continues to resemble a lottery, emphasizing rapid screening of candidate control organisms for safety while neglecting to screen them efficiently for effectiveness. Even those skeptical about predicting effectiveness of biological control organisms concede that advances in understanding, prediction, and management of biological control could be made through more timely and informative post-release evaluation of impacts on target and nontarget organisms. It is useful to recognize stages in the development of biological control (release, establishment, increase and spread of control organisms, suppression of the target organism, and plant succession replacing the weed with other plants), and then analyze progress step by step. Factors working against effective monitoring and evaluation of the impacts on target organisms and nontarget organisms in biological weed control programs include time delays in the control process plus too little time, funding, will, and scientific know-how. These factors explain but do not excuse the low level of quantitative monitoring and evaluation in biological weed control.

While some practitioners have been playing the lottery in weed biological control, alternative frameworks for predicting effects of introductions have been neglected. Between the extremes of introducing all control organism species at once, or introducing them one at a time, lies a worthy alternative: introducing a carefully chosen subset of candidates based on pre-release studies of their potential effectiveness. Such studies could be based on appropriate application of all the tools of the trade, including experimental studies of each species and educated guesses informed by expert intuition, databases, and mathematical and experimental models.

REVENGE EFFECTS AND THE LOTTERY MODEL

Over-prescription of biological control introductions encouraged under the Lottery Model can lead to "revenge effects" (Tenner 1996). In our rush to solve local and acute pest problems we may be creating diffuse and chronic problems that are even harder to solve. By adding more and more control organisms to make pest control more reliable, we may end up making it less effective and more risky (McEvoy and Coombs 1999). Revenge effects could arise in four possible ways: (1) scarce resources are diverted from more profitable alternatives for managing pests, (2) one control organism undermines another, more effective control organism, leading to increase in pest density, (3) one pest is replaced by another pest that can be even harder to control, (4) control organisms introduced to promote environmental and economic health end up undermining it by harming nontarget organisms. Each of these revenge effects is reviewed below.

Scarce Resources Are Diverted from More Profitable Alternatives for Managing Pests

The Precautionary Principle (O'Riordan and Cameron 1994) requires examination of the full range of alternatives, including no action. Revenge effects of the first kind arise if (1) by focussing on introducing new control organisms we neglect to conserve and augment established ones, (2) by relying on foreign control organisms we overlook the use of natives, (3) by emphasizing biological control we disregard integrated control combining biological, mechanical, cultural, chemical control in appropriate ways, or (4) by concentrating on mitigating harm caused by established invaders we neglect to predict and prevent new invaders.

Conserving and augmenting established control organisms

One way to reduce demand for new control organisms is to augment and conserve existing ones. However, biological control has become so complex and technologically fragmented that introducing, augmenting, and conserving control organisms are regarded not merely as separate approaches, but separate subcultures. Biological control workers frequently spend their professional careers in just one of the subcultures. For example, biological weed control emphasizes introducing new control organisms and the approaches of augmenting or conserving control organisms already resident in an area are rarely used (Wapshere et al. 1989, McFadyen 1998). Conservation techniques, which involve the identification and manipulation of factors that influence the abundance and effectiveness of control agents, have only recently received serious attention as a way to enhance weed control (Newman et al. 1998).

Using native control organisms

A related way to be more frugal with introductions is to make better use of native species. Native insects commonly cause considerable damage to weed species, but their use in biological control is often overlooked. Those who favor classical biological control are inclined to believe that natives are already doing the best they can, and it is not good enough. However, there is increasing experimental as well as observational evidence that native insect herbivores can control plant abundance and distribution in some cases (Nowierski et al. 1999), and use of native organisms for biological control is growing more common (Julien and Griffiths 1998). There are examples where native insects are artificially increased or otherwise manipulated for the control of exotic weeds (Sheldon 1997, McFadyen 1998). The native weevil *Euhrychiopsis lecontei* (Dietz) appears to be a promising agent for control of exotic Eurasian watermilfoil (*Myriophyllum spicatum* L.), but factors that limit success need to be identified and manipulated for control to become predictably effective (Newman et al. 1998). We can expect to see more examples of weed control using native control organisms in the future.

Integrated control strategies

A third way to enhance biological control using fewer natural enemies is to use biological control in combination with chemical, cultural, mechanical control as part of an Integrated Weed Management program (Wapshere et al. 1989). Few studies have attempted and reported the research needed to determine how these tactics work together in practice (Briese 1996), perhaps because users of biological control continue to hope for "complete control" by control organisms unaided by other tactics. Complete control is biological control's notion of a silver bullet, defined by Webster's as "a magical weapon...that instantly solves a long-standing problem."

Predicting and preventing invasions

A fourth way to reduce demand for new control organisms is to prevent plant invasions from occurring in the first place. There are a variety of ways to manage biological invasions, ranging from prediction, prevention, early detection, and treatment. Inadequate or misdirected research or regulatory oversight may leave potentially serious risks unaddressed (Pemberton 1995). With scientific attention focused on using introduced control organisms to mitigate harm by current plant invaders, we may invite trouble by neglecting adequate screening and quarantine procedures to predict and prevent the arrival of future plant invaders. Plant introductions, such as woody plants (Hughes 1994) and pastoral plant species (Lonsdale 1994), are largely unregulated and seldom screened for weed potential, yet they entail well-known risks that have not yet been effectively addressed in North America (OTA—U.S. Congress Office of Technology Assessment 1993). Awakened to the risks of a previous open-door policy toward alien species, scientists and policy makers have spawned a growing number of theoretical and practical frameworks for diagnosing invasions and prescribing remedial measures (Williamson 1996, Shigesada and Kawasaki 1997).

One Control Organism Undermines Another, More Effective Control Organism, Leading to Increase in Pest Density

Adding more and more control organism species increases the chance that some interactions will be antagonistic, as in cases where (1) one control organism preys on another, (2) one interferes with another's access to the shared resource, or (3) one reduces the resources available to another. An assessment of the implications for weed control posed by antagonistic interactions must determine whether direct, negative effects on the plant by the antagonist outweigh the indirect, positive effect on the plant created by the antagonist acting via intermediate species (Fig. 2).

One control organism preys on another

Most scientists agree that weed-biological control organisms work best if free of predators, parasites, or pathogens (Goeden and Louda 1976). Ridding the control organism of natural enemies is a good way to cut down on the number of control organism introductions required for biological control. In the case of spotted knapweed *Centaurea maculosa* Lamarck, Harris (1990) estimates that one seed-destroying insect species freed from natural enemies abroad in Canada can do the damage of seven back home in Europe.

Despite the advantages of releasing herbivores free from predators, parasites, and pathogens, these carnivorous organisms commonly reduce the impact of herbivores introduced for weed biological control. Three categories of interaction commonly occur in classical biological control programs in which one control organism species consumes another.

First, some species used to control arthropod pests have adverse effects on a species used to control plant pests. Mites in the family Phytoseiidae, including specialist (i.e., *Phytoseiulus persimilis* A.H.) and generalist (i.e., *Typhlodromus pyri* Scheuten) predators used for biological control of pest spider mites, suppress Oregon populations of a beneficial spider mite *Tetranychus lintearius* Dufour imported for control of gorse (*Ulex europaeus* L.) (Paul Pratt, personal communication). *Bacillus thuringiensis* var. *kurstaki* applied for control of forest-pest insects harms the cinnabar moth *Tyria jacobaeae* (Lepidoptera: Arctiidae) used to control ragwort *Senecio jacobaea* (Asteraceae) (James et al. 1993).

Second, some species used to control plant pests function at two trophic levels as carnivores and herbivores. Most scientists believe that facultative carnivores should not be used for biological weed control. Some practitioners are not so sure, holding out hope that the control organism's direct, negative effect will outweigh its indirect, positive effect on the target plant (Fig. 2A). Take the case of spotted knapweed *C. maculosa* and several of its control organisms, a moth *Metzneria paucipunctella* Zeller (Lepidoptera: Gelechiidae) (the superior interference competitor that functions as a predator and a seed-feeder) and two gall-flies *Urophora affinis* (Frauenfeld) and *U. quadrifasciata* (Meigen) (Diptera: Tephritidae) (the superior colonizers believed to be the more desirable biological control organisms). An observational study of the pattern of association among the three species within 10 field sites suggests that these species are distributed randomly and independently among heads (Story et al. 1991).

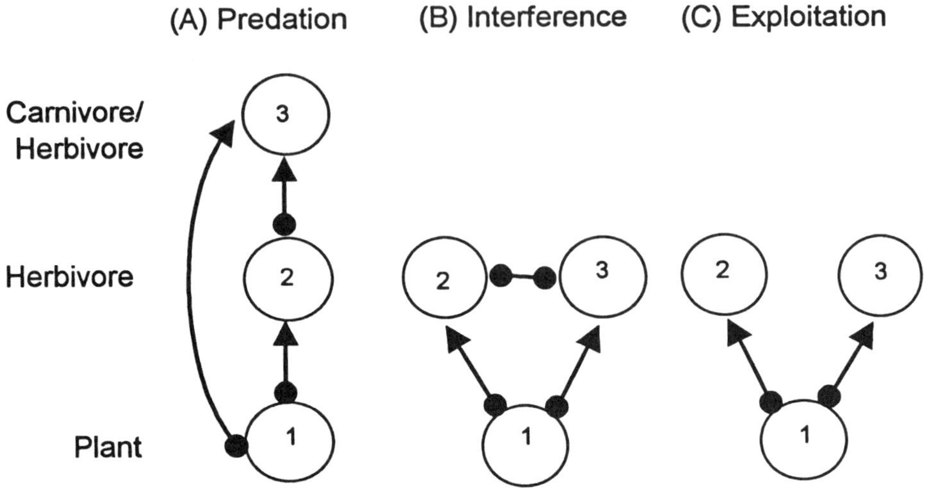

Figure 2. Interactions among populations in a weed biological control system including (A) predation, (B) interference competition, and (C) exploitation competition. Populations are represented as circles, a positive effect of one population on another is represented by an arrow while a negative effect is represented by a filled circle. No arrow means no direct interaction between two variables.

Put the insect species singly or in combination on plants in cages and the seed destruction by moth and fly larvae appears to be greater than with the moth or flies alone, despite the fact that *M. paucipunctella* larvae kill 19–67% of *U. affinis* larvae (Story et al. 1991). The evidence is equivocal because the artificial conditions of closed cages restrict movement, and the design confuses an increase the number of consumer species with an increase in the number of consumer individuals. The introduction of facultative carnivores illustrates how inadequate science and overly-optimistic management have contributed to growth in number, size, and complexity of biological weed control systems.

Third, some species used to control plant pests appear to be especially vulnerable to acquiring predators and parasites. High rates of parasitism rates are reported for gall midges (Diptera: Cecidomyiidae) including rates of 20–50% in Montana for the case of *Spurgia esula* Gagne introduced to control leafy spurge *Euphorbia esula* L. and rates of 20–90% in Idaho, Washington, and California for the case of *Cystiphora schmidti* Ruebsaamen introduced to control of rush skeletonweed *Chondrilla juncea* L. (Asteraceae) (Rees et al. 1996; Coombs, unpublished observations). The same is true for some Lepidoptera, including parsitism rate of 30–60% in Oregon populations of *Leucoptera spartifoliella* Huebner (Lyonetiidae) for control Scot's broom *Cytisus scoparius* (L.) (Fabaceae) and rates of 2–10% in Washington, Montana, and Oregon populations of *Metzneria paucipunctella* (Gelichiidae) for control several species of knapweeds *Centaurea* spp. (Asteraceae) (Rees et al. 1996; Coombs, unpublished observations).

One control organism species interferes with another's access to the shared resource (interference competition)

If biological control scientists can't agree about what it means for biological control when control organisms eat each other, it is small wonder they pay scant attention to more subtle forms of interference competition. According to the opinion survey cited earlier in this chapter (section 2), competition limits the effectiveness of weed-biological control organisms in only 13% of known cases (Crawley 1986). Interspecific competition for food is notoriously difficult to document in the field, even when it occurs; this is one reason why there is a persistent debate in ecology concerning the importance of this force (Denno et al. 1995, Stewart 1996). Several categories of interaction along a continuum are best lumped here: competition that involves a negative-negative interaction, asymmetric competition that implies unequal negative-negative interactions, and amensalism that involves negative-zero interaction.

One control organism species may interfere with another's access to shared resources such as hosts, mates, or microhabitats. Negative effects of one control organism on another's demography may include a decrease in reproduction or an increase in mortality or emigration. For example, weevils *Larinus minutus* Gyllenhal. (Coleoptera: Curculionidae) at high density interfere with access to the flower heads of diffuse knapweed (*Centaurea diffusa*) by the gall fly species *Urophora affinis* and *U. quadrifasciata* (Coombs, unpublished observations). The implications of these interactions for the dynamics of interacting populations in this system are unknown.

One control organism reduces the resources available to another (exploitation competition)

It is generally assumed that there nothing to fear if one control organism reduces host resources available to another; the likely outcome is a reduced host equilibrium. Most definitions of biological control assume that it is desirable to reduce pest equilibrium levels while retaining sufficient stability in the pest-enemy interaction to prevent the host from sporadically re-emerging as a pest. Contrary to conventional wisdom, there is mounting evidence that competition among control organisms can reduce control success. First, one control organism can undermine the establishment of another organism species. Briese (1997) reports that the leaf-defoliator *Chrysolina quadrigemina* (Suffrian) (Coleoptera: Chrysomelidae) may have inhibited the establishment of another, more effective control organism, root-borer *Agrilus hyperici* (Liro) (Coleoptera: Buprestidae), by inducing boom-and-bust fluctuations in populations of the target weed *Hypericum perforatum* L. (Clusiaceae) in Australia. If the order in which control agents are established is important for their eventual success, then we should try to predict the interactions that will occur between potential control agents, as well as their likely individual impact on the weed, in the country of introduction, as currently practiced in

New Zealand (Syrett et al. 1996). Second, one established control organism may undermine the effectiveness of another, more effective control organism. Woodburn (1996) reports that early, aggregated attack on thistle seed heads by a univoltine, seed-feeding weevil, *Rhinocyllus conicus* Frölich (Coleoptera: Curculionidae) hindered control of nodding thistle *Carduus nutans* L. (Asteraceae) in Australia by decreasing the numerical response of a more promising agent, a bivoltine tephritid seed fly *Urophora solstitialis* (Linnaeus) (Diptera: Tephritidae). Also, plant changes induced by one herbivore species may protect plants against a second, potentially more damaging species. An example can be found in recent studies involving plant-feeding mite species. Grape vines in the San Joaquin Valley of California are protected from the highly virulent Pacific mite *Tetranychus pacificus* McGregor by early inoculation with the more benign Willamette mite *Eotetranychus willametti* (McGregor) (Karban and English-Leob 1990, Karban et al. 1991, Karban et al. 1997). Plant pathologists have found similar cases in which "vaccination" with relatively benign pathogens reduces the negative impact of virulent viral, bacterial, and fungal pathogens (van Loon et al. 1998). In a similar way, releasing ineffective control organism species in weed control programs may induce plant resistance that reduces the impact of more virulent control organism species.

Precautionary measures for avoiding antagonistic interactions among control organisms have been offered but seldom followed. Almost 30 years ago, Zwölfer (1973) devised a prescription under the "balanced competition hypothesis" based on certain assumptions about the nature of competition and coexistence of control organisms operating side by side within the same seed head, and how the inferior competitor might respond if released from competition: If a trade-off exists between colonization and competition abilities (e.g. strong competitors are poor colonizers), then introduce the inferior intrinsic competitor first, knowing that if the result is unsatisfactory, the superior intrinsic competitor can be introduced later. The prescription was seldom followed (in part because it is not always easy to determine which species is likely to play which role in a given environment). Harris (1989, 1990) offers another recommendation based on metabolic source-sink relationships assumed to regulate resource allocation within plants. He suggests that there are seven guilds of phytophagous insects feeding in the seed-heads of knapweeds (*Centaurea* spp.) and other species in the Asteraceae subfamily Cynareae. The guilds in approximate order of attack on the flower-heads are: (1) the formation of woody galls, (2) receptacle feeding, (3) formation of non-woody ovary galls, (4) ovary feeding, (5) formation of achene galls, (6) feeding on well developed but still soft achenes, (7) feeding on mature achenes and preying on insect species in the previous guilds. He suggested that seed destruction is maximized by establishing a biological control systems with a single species from each of the following three guilds: (1) woody gall former, (4) ovary feeder, (6) feeder on well developed by still soft achenes. He assumes that (1) the primary control organism should be woody gall former because the gall that it induces acts as a strong metabolic sink capable of draining resources from the whole plant, (2) it is best to avoid adding competitors or facultative carnivores that might undermine the primary control-organism's effectiveness. Contrary to this parsimonious prescription of 3 species in 3 guilds, the actual biological control system for knapweeds in North America now contains 13 insect species in 6 guilds.

For the case of host-parasitoid interactions, momentum is building once again to obtain a mathematical understanding of what consequences antagonistic interactions among control organisms might have for biological control (May and Hassell 1981, Kakehashi et al. 1984, Briggs 1993, Briggs et al. 1993). These investigations are relevant to weed biological control because many insects that feed on plants (e.g. seed-slaying insects) have life styles analogous to parasitoids (Price 1980), in which a female lays an egg on a host and the larvae develops by feeding on the body of the host. Briggs (1993) develops two versions of a stage-structured, delay-differential equation model of two parasitoids attacking different developmental stages of a single host species. She found realistic situations in which the best competitor is not necessarily the parasitoid that produces the lowest adult host density and the combination of two parasitoid species may yield higher host density than that achieved by the single most effective parasitoid. This suggests that the lowest host density may be achieved in biological control by the release of only the most effective control organism rather than multiple control-organism species.

In summary, the tide of opinion seems to be turning against the long-standing, laissez-faire policy of introducing multiple control-organism species, leaving them to sort it out for themselves which single one or combination is best for biological control. Most studies to date are based on direct effects on individual plants. To weigh the implications of antagonistic interactions among control organisms for weed control, future studies should attempt to measure and model both direct and indirect effects on the plant responses at individual, population, and metapopulation levels.

One Pest Is Replaced by Another Pest That Can Be Even Harder to Control

If control of one weed leads to its replacement by another weed not susceptible to the control organism, then we are faced with a continuing cost of control and little benefit. This has happened in a number of cases. Control of water hyacinth *Eichhornia crassipes* (Martius) Soms-Laubach (Pontederiaceae) led to its replacement in many areas by water lettuce *Pistia stratiotes* Linnaeus (Araceae) (O'Brien 1995). Control of purple loosestrife *Lythrum salicaria* L. (Lythraceae) is leading to its replacement in many areas by reed canary grass *Phalaris arundinacea* L. (Poaceae) (Thompson et al. 1987). In Australia, control of a susceptible genotype of rush skeletonweed *Chondrilla juncea* Linnaeus (Asteraceae) led to its replacement by another, more resistant one (Burdon et al. 1981). Managing weed biological control often means managing plant succession (Luken 1990). There have been very few attempts to monitor plant succession following successful biological weed control (McEvoy et al. 1991) since the pioneering study of Klamath weed *Hypericum perforatum* L. that appeared nearly 40 years ago (Huffaker and Kennett 1959). There have been even fewer attempts to manage plant succession as part of a biological control program. Control of ragwort *S. jacobaea* in western Oregon led to replacement of ragwort by perennial grasses (McEvoy et al. 1991) and the reappearance of a rare native plant, the hairy-stemmed checker mallow *Sidalcea hirtipes* Hitchc. (Malvaceae) (Gruber and Whytemare 1997). In other cases, recolonization by the desired suite of species may be slow and unreliable because desired species are no longer in the vicinity and dispersal over long distances can be limited. Natural colonization may have to be supplemented by interventions such as removing plants or their parts, changing resource availability, changing propagule availability and using animals to remove weed biomass (Luken 1990). It is unfortunate that today applied ecologists concerned with managing ecological succession (Luken 1990) and those managing biological weed control (Harley and Forno 1992) take little notice of each other. Investigating the role of biological control in ecological restoration can provide important insights into the way that ecological communities are assembled and ecosystems function, and add another weapon to restoration-ecology's armory.

Control Organisms Introduced to Promote Environmental and Economic Health End Up Undermining It by Harming Nontarget Organisms

Releasing more and more control organisms challenges the capacity of monitoring efforts, increasing the likelihood that target and nontarget effects go undetected. There are several well documented examples where classical weed biological control has inadvertently harmed nontarget species. Recent studies document actual effects on nontarget species for the cases of both officially sanctioned control organisms—such as *C. quadrigemina* beetles on a groundcover in California (Andres 1985), *T. jacobaeae* caterpillars on ornamental and native plants in Oregon (Diehl and McEvoy 1990), *R. concicus* beetle larvae on native thistles in western North America (Turner et al. 1987, Louda et al. 1997)—as well as unsanctioned control organisms arriving accidentally in North America—*Cactoblastis cactorum* Bergroth (Lepidoptera: Pyralidae) caterpillars on native cacti in Florida (Bennett and Habeck 1995, Pemberton 1995, Simberloff and Stiling 1996a). Based on a worldwide review of recorded instances of damage to nontarget plants by biological control organisms (McFadyen 1998), about 3 in 95 control organism species in the western continental USA have attacked native nontarget plant species so far. These cases illustrate the dangers inherent in releasing oligophagous control organisms with relatively broad host ranges into environments with many potential nontarget species within the known host range. Oligophagous is used to refer to insects feeding on a number of plant species, usually in different genera within one plant family (Bernays and Chapman 1994). Faced with the possibility of harmful nontarget effects that already range in severity from transient to persistent attack, what is the probability of escalation to population suppression, local population extinction, species endangerment, species extinction, and community and ecosystem effects in the future? The most important uncertainties that remain have to do with forecasting control organism movement, evolution, and indirect effects, plus refining estimates of the severity, probability, and consequences of nontarget effects (McEvoy 1996, Secord and Kareiva 1996).

A rule of thumb in biological control favors target organisms that are taxonomically distinct from other plants (McClay 1989, Peschken and McClay 1995). In practice, this rule is often violated as indicated by the large number of other species, hybrids and infraspecific taxa within genera of target species for the western USA (Fig. 3). It would be prudent to anticipate and monitor potential effects in plant genera that contain native plant species in western North America, including *Centaurea**, *Cirsium**, *Cynoglosum**, *Euphorbia**, *Hieracium*, *Hypericum**, *Lepidium*, *Linaria**, *Lythrum**, *Potentilla*, *Salvia**, and *Senecio** for which control organisms are either already es-

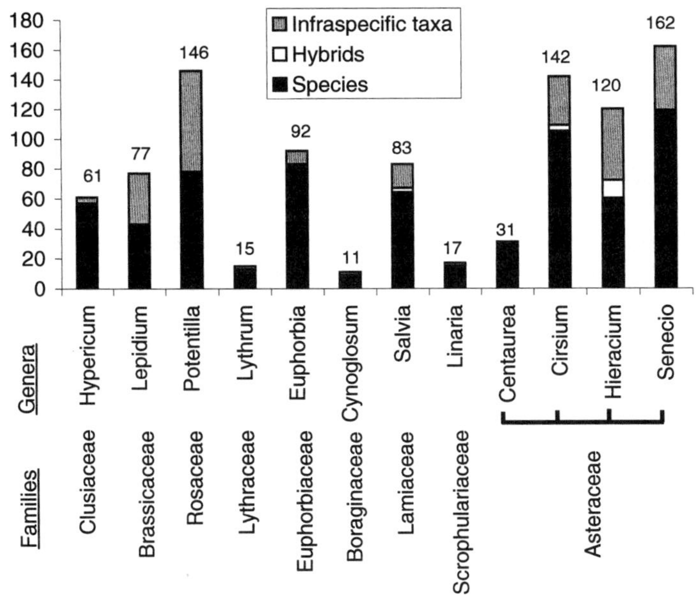

Figure 3. Taxa with high diversity indicated by the number of plant species, hydrids, and infraspecific taxa in target plant genera and families for biological weed control programs in western USA (CA, OR, WA, ID). It would be prudent to anticipate and monitor potential nontarget effects in these plant genera because they contain many native plant species.

tablished (*) or releases are planned in the near future (Fig. 3). This precautionary measure would improve the ability of researchers to predict severity and likelihood of harmful interactions between control organisms and nontarget species. It would increase the confidence of regulatory authorities that host range and environmental safety can be reliably predicted from protocols based on host plant records, phylogenetic analysis, genetic screening, plus host range tests in the lab and field.

Two case studies involve oligophagous control organisms imported and released in environments containing many native plants (potential nontargets) related to the target plant. The cinnabar moth *T. jacobaeae* was introduced to North America in 1959 to control ragwort *S. jacobaea* (Asteraceae). Host specificity testing prior to importing the cinnabar moth into Canada established the "physiological host range" by showing that the larvae of the cinnabar moth can feed and develop in the laboratory on plants species in the tribe Seneceonae (Bucher and Harris 1961), notably species in the genera *Senecio* and *Erechtites*. The genus *Senecio* is a treasure trove of native species containing 162 species, hybrids, and infraspecific taxa in North America (Fig. 3) (USDA 1997). Field observations 30 years after initial release of the cinnabar moth in North America confirm that "the ecological host range" includes other species in the genus *Senecio* including other exotic weeds (*S. vulgaris* L., *S. sylvaticus* L.), horticultural plants (*S. bicolor* ssp. *cineraria* (DC.) Cater), and at least one native species in genus *Senecio* (*S. triangularis* Hook, S.) (Diehl and McEvoy 1990). Turning to the second example, the weevil *R. conicus* was imported to control musk thistle *Carduus nutans*, Italian thistle *C. pycnocephalus* L., milk thistle *Silybum marianum* (L.) Gaertner in North America (Andres and Rees 1995, Goeden 1995b, Goeden 1995a). The larvae of *R. conicus* can feed and develop in seed heads of plants in four thistle genera (*Carduus*, *Cirsium*, *Silybum*, and *Onopordum*) found in North America with the tribe Cardueae (family Asteraceae) (Zwölfer and Harris 1984). The genus *Cirsium* contains many native plants among the 142 species, varieties and infraspecific taxa found in North America (Fig. 3). In a recent, influential paper, Louda et al. (Louda et al. 1997) claim that *Silybum* and *Onopordum* are native North American genera, but the sole North American representatives of these genera are of Eurasian origin (USDA 1997).

What are the lessons to be learned from this experience? First, the fundamentals of host specificity as a criterion for judging safety in biological weed control remain sound. Host specificity depends on attributes of the control organism, the recipient environment, and the organism by environment interaction. The cinnabar moth *T. jacobaeae* has a broad host range with the potential to feed and develop on species in the genus *Senecio* that contains many native species in North America. The weevil *R. conicus* has a broad host range with the potential to feed and develop on species in four genera of thistles (*Carduus*, *Silybum*, *Onopordum*, *Cirsium*) that occur in North America, and only *Cirsium* contains species native to North America. The same insect that is a hazard in North America is safely used in New Zealand, where there are no native thistles. Second, although the species is the operational unit of host specificity testing, new control organism genotypes with different potentials require independent screening. After screening one genotype of *R. conicus*, other genotypes of this species with different host ranges were introduced without additional screening (Peter Harris, personal communication). Third, impacts on nontargets should be monitored and assessments updated accordingly. A Precautionary Approach would recommend suspending redistribution and range expansion of *R. conicus* and *T. jacobaea* in North America while impacts of on target and nontarget organisms are being reassessed. This recommendation is being ignored: the cinnabar moth was recently transported from Oregon and released in Montana by the USDA Forest Service with few questions asked, and even fewer answers given, about risk to potential nontarget plants. The USDA Plants Data Base lists 42 taxa in the genus *Senecio* for Oregon and 41 taxa for Montana (USDA 1997). There should be a tacit moratorium barring redistribution of problem control organisms until scientists have assessed the damage and regulators can agree on new rules.

There is also the risk of introducing contaminating or misidentified species along with "approved" agents. Balciunas and Villegas (1999) reconstruct how the false peacock fly *Chaetorellia succinea* (Costa) (Diptera: Tephritidae) was accidentally released along with the true peacock fly *Ch. australis* Hering on yellow starthistle *Centaurea solstitialis* L. (Asteraceae) in southwest Oregon in 1991 and then redistributed to several states. At the time of release, *Ch. succinea* did not have approval for release granted by the regulatory authority U.S. Department of Agriculture's Animal and Plant Health

Inspection Service (USDA-APHIS). Brief host-range tests conducted on the false peacock fly in Europe showed that it would not attack safflower, *Carthamus tinctorius*, a close relative of yellow starthistle. However, the false peacock fly can hybridize with *Ch. carthami*, a minor pest of safflower in the Middle East. The false peacock fly is now more abundant, more widely distributed across California, and more damaging to seeds of the target host than the true peacock fly. Now, after the fact, the host range of the false peacock fly and the probability and severity of damage to commercial safflower varieties are being evaluated.

AVOIDING REVENGE EFFECTS

We earlier recommended that the best away to avoid revenge effects is to adopt a precautionary approach and to follow simple guidelines for decision making under uncertainty (McEvoy and Coombs 1999). The Precautionary Principle (O'Riordan and Cameron 1994) applied to Classical Biological Control decisions has four parts:

- First, potential harm to nontarget organisms can arise from the release of biological control organisms.
- Second, actual harm to nontarget organisms of sufficient magnitude and severity has occurred to warrant new principles for conducting biological control introductions.
- Third, the burden of proof for showing that new control organisms are necessary, safe, and effective rests with those proposing the activity.
- Fourth, the process of applying the precautionary principle must be open, informed and democratic and must include potentially affected parties. It must also involve an examination of the full range of alternatives, including no action.

Our foregoing review of revenge effects suggests that the practice of biological control currently falls short of general standards set by others (Ludwig et al. 1993) for decision making under uncertainty: "we must consider a variety of plausible hypotheses about the world; consider a variety of possible strategies; favor actions that are robust to uncertainties; hedge; favor actions that are informative; probe and experiment; monitor results; update assessments and modify policy accordingly; and favor actions that are reversible." The next efforts in weed biological control should combine prospective approaches, aimed at detailing the likely impacts that would arise from a proposed activity, and retrospective approaches, aimed a quantifying the actual impacts of the activity (Osenberg and Schmitt 1996).

Our prescription for avoiding revenge effects boils down to treating new control organisms as "guilty until proven innocent": presume each new control organism species is unnecessary, unsafe, and ineffective until it is shown, beyond a reasonable doubt, to be necessary, safe, and effective. Troubles arise when noble but narrowly focussed goals enter the complex interactions that make up nature and society. Unable to foresee which complexities need attention, we forge ahead, always expecting the best. In the USA, politics presently trumps policy: agency infighting, confusion about how to bring biosecurity policy in line with statutory authority, and regulatory uncertainty compound the problem of overseeing the growing global market for biological control organisms (OTA—U.S. Congress Office of Technology Assessment 1993,

OTA—U.S. Congress Office of Technology Assessment 1995). For biological control of weeds, it is too dramatic to claim, like Chicken Little, that harm caused to date by weed control organisms approaches ecological disaster. A tingid bug *Teleonemia scrupulosa* Stål (Hemiptera: Tingidae) caused local problems on a crop—but with effects localized by time, space, and crop variety (Harris 1988)—this is nothing like the ecological disaster that Miller and Aplet (1993) have alleged. It is equally misleading to brush aside concerns about possible environmental harm in the name of restoring public confidence, claiming like Voltaire's Dr. Pangloss that we already live in the best of all possible regulatory worlds. This adds one more version of a revenge effect, when official comments meant to shore up public confidence (Soper 1992) end up undermining it. Given the nature and magnitude of the actual risks and the inevitable costs of regulation, the challenge is to determine what kinds of oversight mechanisms are needed. Inadequate or misdirected oversight mechanisms may leave potentially serious risks unaddressed, while unwarranted regulatory burdens create disincentives to innovation that diminish the availability of important new techniques or products (Barton et al. 1997). Nevertheless, it should be possible to find a balance in biological control oversight mechanisms without the polarizing influence of Chicken Little or Dr. Pangloss.

CONCLUDING REMARKS

The practice of weed biological control currently resembles a lottery in which more and more control organisms are added to speed the discovery of the few that are effective. Practicing biological control as if it were a lottery fuels runaway importation rates; creates needless complexity, redundancy, and risk; distracts scientific attention from monitoring and evaluation; and stunts the development of more parsimonious alternatives. Biological control under the Lottery Model leads to four kinds of revenge effects: (1) Scarce resources are diverted from more profitable alternatives for managing pests, (2) One control organism undermines another, more effective control organism, leading to increase in pest density, (3) One pest is replaced by another pest that can be even harder to control, (4) Control organisms introduced to promote environmental and economic health end up undermining it by harming nontarget organisms.

Advances in understanding, prediction, and management of biological control could be made in two ways. The first way is through more timely and informative post-release evaluation of impacts on target and nontarget organisms. Currently biological control scientists often have very limited knowledge of the factors that limit effectiveness of control organisms, and much of that knowledge is subjective. For monitoring and evaluation of biological control programs, it is useful to recognize stages in the development of biological control (release, establishment, increase and spread of control organisms, suppression of the target organism, and plant succession replacing the weed with other plants), and then analyze progress step by step. The second way is through experimental screening of each control organism for effectiveness as well as safety prior to release and making educated guesses about effectiveness based on expert intuition, data bases, and experimental and mathematical models. Given the uncertainties involved in biological control, the best way to avoid revenge effects is to adopt a precautionary approach and to follow simple guidelines for decision making under uncertainty.

ACKNOWLEDGMENTS

We thank D. Briese, F. Grevstad, D. Isaacson, M. Lonsdale, S. Louda, H.-K. Luh, A.S. McClay, M. Pitcairn, D.K. Platt, M. Schat, S. Schooler, A. Sheppard, and T. Woodburn for stimulating discussions that improved this manuscript. Credit for the title goes to Edward Tenner, whose stimulating book inspired this chapter. We gratefully acknowledge support by National Science Foundation grants BSR-8516997 and BSR-9020483 and by a McMaster Fellowship to the first author from the Division of Entomology, CSIRO Australia. The opinions expressed are solely those of the authors and do not necessarily reflect the opinions, policies, or recommendations of the funding agencies. This chapter is Oregon State University Agricultural Experiment Station Technical Paper No. 11497.

REFERENCES

Andres, L. A. 1985. Interaction of *Chrysolina quadrigemina* and *Hypericum* spp. in California. *In* E. S. Delfosse (ed.), Proc. Sixth Intl. Symp. on Biological Control of Weeds, August 19–25, 1984, Agric. Canada, Ottawa. p. 235–239.

Andres, L. A., and N. E. Rees. 1995. Musk thistle. *In* J. R. Nechols, L. A. Andres, J. W. Beardsley, R. D. Goeden, and C. G. Jackson (eds.), Biological control in the western United States: accomplishments and benefits of regional research project W-84, 1964–1989. Univ. California Div. Agric. Nat. Res., Oakland, California. p. 248–251.

Balciunas, J., and B. Villegas. 1999. Two new seed head flies attack yellow starthistle. Calif. Agric. 53: 8–11.

Barton, J., J. Crandon, D. Kennedy, and H. Miller. 1997. A model protocol to assess the risks of agricultural introductions. Nature Biotech. 15: 845–848.

Beirne, B. P. 1985. Avoidable obstacles to colonization in classical biological control of insects. Can. J. Zool. 63: 743–747.

Bennett, F. D., and D. H. Habeck. 1995. *Cactoblastis cactorum*: A successful weed control agent in the Caribbean, now a pest in Florida? *In* E. S. Delfosse and R. R. Scott (eds.), Proc. 8th Intl. Symp. on Biological Control of Weeds, CSIRO Publishing, Melbourne, Australia. p. 21–26

Bergelson, J., and M. J. Crawley. 1989. Can we expect mathematical models to guide biological control programs: a comment based on case studies of weed control. Comments on Theoretical Biol. 1: 197–215.

Bernays, E. A., and R. F. Chapman. 1994. Host-plant selection by phytophagous insects. Chapman and Hall, New York. 312 p.

Blossey, B., and Nötzold, R. 1995. Evolution of increased competitive ability in invasive nonindigenous plants: a hypothesis. J. Ecol. 83: 887–889.

Blossey, B. 1996. What determines the increase competitive ability of invasive non-indigenous plants? *In* V. C. Moran and J. H. Hoffman (eds.), Proc. IX Intl. Symp. on Biological Control of Weeds, January 19–26, 1996, Stellenbosch, South Africa, Univ. Cape Town, South Africa. p. 3–9.

Briese, D. T. 1996. Biological control of weeds and fire management in protected natural areas: are they compatible strategies? Biol. Conserv. 77: 135–141.

Briese, D. T. 1997. Biological control of St. John's wort: past, present, and future. Plant Prot. Q. 12: 73–80.

Briggs, C. J. 1993. Competition among parasitoid species on a stage-structured host and its effect on host suppression. Am. Nat. 141: 372–397.

Briggs, C. J., R. M. Nisbet, and W. W. Murdoch. 1993. Coexistence of competing parasitoid species on a host with a variable life cycle. Theor. Popul. Biol. 44: 341–373.

Bucher, G. E., and P. Harris. 1961. Food-plant spectrum and elimination of disease of cinnabar moth larvae, *Hypocrita jacobaeae* (L.) (Lepidoptera: Arctiidae). Can. Entomol. 93: 931–936.

Burdon, J. J., R. H. Groves, and J. M. Cullen. 1981. The impact of biological control on the distribution and abundance of *Chondrilla juncea* in south-eastern Australia. J. Appl. Ecol. 18: 957–966.

Caswell, H. 1997. Matrix methods for population analysis. *In* S. Tuljapurkar and H. Caswell (eds.), Structured-population models in marine, terrestrial, and freshwater systems. Chapman and Hall, New York.

p. 19–58.
Coombs, E. M., H. Radtke, D. Isaacson, and S. Snyder. 1996. Economic and regional benefits from the biological control of tansy ragwart, *Senecio jacobea*, in Oregon. *In* V. C. Moran and J. H. Hoffmann (eds.), Proc. IX Intl. Symp. on the Biological Control of Weeds, Stellenbosch, South Africa. p. 489–494.
Coulson, J. R. 1992. Documentation of classical biological control introductions. Crop. Prot. 11: 195–205.
Coulson, J. R., and R. S. Soper. 1989. Protocols for the introduction of biological control agents in the U.S. *In* R. P. Kahn (ed.), Plant protection and quarantine. CRC Press, Boca Raton, Florida. p. 1–35.
Crawley, M. J. 1986. The population biology of invaders. Philos. Trans. Royal Soc. London, B, Biol. Sci. 314: 711–731.
Crawley, M. J. 1989a. Chance and timing in biological invasions. *In* J. A. Drake, H. A. Mooney, F. di Castri, R. H. Groves, F. J. Kruger, M. Rejmanek and M. Williamson (eds.), Biological invasions: a global perspective. John Wiley, New York. p. 407–423.
Crawley, M. J. 1989b. Insect herbivores and plant population dynamics. Annu. Rev. Entomol. 34: 531–564.
Crawley, M. J. 1989c. The successes and failures of weed biological control using insects. Biol. Control News Info. 10: 213–223.
Crawley, M. J. 1990. Plant life-history and the success of weed biological control projects. *In* E. S. Delfosse (ed.), Proc. VII international symposium on biological control of weeds, 6–11 March 1988, Ministero dell' Agricoltura e delle Foreste, Rome, and CSIRO, Melbourne. p. 17–26.
Cullen, J. M. 1990. Current problems in host specificity screening. *In* E. S. Delfosse (ed.), Proc. VII international symposium on biological control of weeds, 6–11 March 1988, Ministero dell'Agricoltura e delle Foreste, Rome, and CSIRO, Melbourne. p. 27–36.
Denno, R. F., M. S. McClure, and J. R. Ott. 1995. Interspecific interactions in phytophagous insects: Competition reexamined and resurrected. Annu. Rev. Entomol. 40: 297–331.
Diehl, J., and P. B. McEvoy. 1990. Impact of the cinnabar moth (*Tyria jacobaeae*) on *Senecio triangularis*, a nontarget native plant in Oregon. *In* E. S. Delfosse (ed.), Proc. VII Intl. Symp. on Biological Control of Weeds, 6–11 March 1988, Ministero dell'Agricoltura e delle Foreste, Rome, and CSIRO, Melbourne. p. 119–126.
Ehler, L. 1991. Planned introductions in biological control. *In* L. Ginzburg (ed.), Assessing Ecological Risks of Biotechnology. Butterworth–Heinemann, Boston. p. 21–39.
FAO. 1994. Code of conduct for the import and release of biological control agents. Draft version, FAO, Rome. 14 p.
Fowler, S. V., H. M. Harman, J. Memmott, Q. Paynter, R. Shaw, S. A.W., and P. Syrett. 1996. Comparing the population dynamics of broom, *Cytisus scoparius*, as a native plant in the United Kingdom and France and as an invasive alien weed in Australia and New Zealand. *In* V. C. Moran and J. H. Hoffman (eds.), Proc. IX Intl. Symp. on Biological Control of Weeds, January19–26, 1996, Stellenbosch, South Africa, Univ. of Cape Town, South Africa. p. 19–26.
Goeden, R. D. 1983. Critique and revision of Harris' scoring system for selection of insect agents in biological control of weeds. Prot. Ecol. 5: 287–301.
Goeden, R. D. 1995a. Italian thistle. *In* J. R. Nechols, L. A. Andres, J. W. Beardsley, R. D. Goeden and C. G. Jackson (eds.), Biological control in the western United States: accomplishments and benefits of regional research project W–84, 1964–1989. Univ. California Div. Agric. Nat. Res., Oakland, California. p. 242–244.
Goeden, R. D. 1995b. Milk thistle. *In* J. R. Nechols, L. A. Andres, J. W. Beardsley, R. D. Goeden and C. G. Jackson (eds.), Biological control in the western United States: accomplishments and benefits of regional research project W–84, 1964–1989. Univ. California Div. Agric. Nat. Res., Oakland, California. p. 245–247.
Goeden, R. D., and S. M. Louda. 1976. Biotic interference with insects imported for weed control. Annu. Rev. Entomol. 21: 325–342.
Grevstad, F. S. 1996. Establishment of weed control agents under the influences of demographic stochasticity, environmental variability and Allee effects. *In* V. C. Moran and J. H. Hoffman (eds.), Proc. IX Intl. Symp. on Biological Control of Weeds, January 19–26, 1996, Stellenbosch, South Africa, Univ. Cape Town, South Africa. p. 19–26.
Grevstad, F. S. 1999. Factors influencing the chance of population establishment: implications for release strategies in biological control. Ecol. Appl. In press:
Gruber, E., and A. Whytemare. 1997. The return of the native? *Sidalcea hirtipes* in coastal Oregon. *In* T. N. Kaye, A. Liston, R. M. Love, D. L. Luoma, R. J. Meinke and M. V. Wilson (eds.), Conservation and management of native plants and fungi. Native Plant Soc. Oregon, Corvallis, Oregon. p. 121–124.

Harley, K. L. S., and I. W. Forno. 1992. Biological control of weeds: a handbook for practitioners and students. Inkata Press, Melbourne, Australia. 74 p.

Harper, J. L. 1977. Population biology of plants. Academic Press, New York. 892 p.

Harper, J. L. 1982. After description. *In* E. I. Newman (ed.), The plant community as a working mechanism. Blackwell Scientific Publications, Boston, Massachusetts. p. 11–25.

Harris, P. 1973. The selection of effective agents for the biological control of weeds. Can. Entomol. 105: 1495–1503.

Harris, P. 1981. Stress as a strategy in the biological control of weeds. *In* G. C. Papavizas (ed.), Biological control and crop protection. Allahheld, Osmun, Totowa, New Jersey. p. 333–340.

Harris, P. 1988. Environmental impact of weed-control insects. BioScience 38: 542–548.

Harris, P. 1989. The use of Tephritidae for the biological control of weeds. Biol Control News Info. 10: 7–16.

Harris, P. 1990. Feeding strategy, coexistence and impact of insects in spotted knapweed capitula. *In* E. S. Delfosse (ed.), Proc. VII Intl. Symp. on Biological Control of Weeds, March 6–11, 1988, Ministero dell'Agricoltura e delle Foreste, Rome, and CSIRO, Melbourne. p. 39–47.

Harris, P. 1997. Monitoring and impact of weed biological control agents. *In* D. A. Andow, D. W. Ragsdale and R. F. Nyvall (eds.), Ecological interactions and biological control. Westview Press, Boulder, Colorado. p. 215–223.

Harris, P., and P. McEvoy. 1995. The predictability of insect host plant utilization from feeding tests and suggested improvements for screening weed biological control agents. *In* E. S. Delfosse and R. R. Scott (eds.), Proc. Eighth Intl. Symp. on Biological Control of Weeds, February 2–7, 1992, DSIR/CSIRO, Melbourne. p. 125–131.

Holt, R. D., and M. E. Hochberg. 1997. When is biological control evolutionarily stable (or is it)? Ecol. 78: 1673–1683.

Howarth, F. G. 1983. Classical biological control: panacea or Pandora's box. Proc. Hawaiian Entomol. Soc. 24: 239–244.

Howarth, F. G. 1991. Environmental impacts of classical biological control. Annu. Rev. Entomol. 36: 485–509.

Huffaker, C. B., and C. E. Kennett. 1959. A ten-year study of vegetational changes associated with biological control of Klamath Weed. J. Range Mgt. 12: 69–82.

Hughes, C. E. 1994. Risks of species introductions in tropical forestry. Commonwealth Forestry Rev. 73: 243–252.

James, R., J. C. Miller, and B. Lighthart. 1993. Bacillus thurlngiensis var. kurstaki affects a beneficial insect, the cinnabar moth (Lepidoptera: Arctiidae). J. Econ. Entomol. 86: 334–339.

Julien, M. H. 1989. Biological control of weeds worldwide: trends, rates of success and the future. Biol. Control News Info. 10: 299–306.

Julien, M. H., and M. W. Griffiths (eds.), 1998. Biological control of weeds. A world catalogue of agents and their target weeds. CABI Publishing, CAB International, Wallingford, UK. 240 p.

Kakehashi, N., Y. Suzuki, and Y. Iwasa. 1984. Niche overlap of parasitoids in host–parasitoid systems: its consequence to single versus multiple introduction controversy. J. Appl. Ecol. 21: 115–131.

Karban, R., and G. M. English-Leob. 1990. A "vaccination" of Willamette spider mites (Acari: Tetranychidae) to prevent large populations of Pacific spider mites on grapevines. J. Econ. Entomol. 83: 2252–2257.

Karban, R., G. English-Loeb, and D. Hougen–Eitzmann. 1997. Mite vaccinations for sustainable management of spider mites in vineyards. Ecol. Appl. 7: 183–193.

Karban, R., G. M. English-Loeb, and P. Verdigaal. 1991. Vaccinating grapevines against spider mites. Calif. Agric. 45: 18–21.

Klingman, D. L., and J. R. Coulson. 1983. Guidelines for introducing foreign organisms into the United States for the biological control of weeds. Bull. Entomol. Soc. Am., Fall 1983: 55–61.

Levin, S. A. 1989. Analysis of risk for invasions and control programs. *In* J. A. Drake, H. A. Mooney, F. di Castri, R. H. Groves, F. J. Kruger, M. Rejmanek and M. W. Williamson (eds.), Biological Invasions: a global perspective. John Wiley, New York. p. 425–435.

Levin, S. A. 1990. Ecological issues related to the release of genetically modified organisms into the environment. *In* H. A. Mooney and G. Bernardi (eds.), Introduction of genetically modified organisms into the environment. John Wiley, New York. p. 151–159.

Levins, R. 1986. Perspectives in integrated pest management: from an industrial to ecological model of pest management. *In* M. Kogan (ed.), Ecological theory and integrated pest management practice. John Wiley, New York. p. 1–18.

Lockwood, J. A. 1993. Environmental issues involved in biological control of rangeland grasshoppers (Orthoptera: Acrididae) with exotic agents. Environ. Entomol. 22: 505–518.

Lonsdale, W. M. 1994. Inviting trouble: introduced pasture species in northern Australia. Aust. J. Ecol. 19: 345–354.

Lonsdale, W. M., G. Farrell, and C. G. Wilson. 1995. Biological control of a tropical weed: a population model and an experiment for *Sida acuta*. J. Appl. Ecol. 32: 391–399.

Louda, S. M. 1998. Ecology of interactions needed in biological control practice and policy. Bull. Brit. Ecol. Soc. 29: 8–11.

Louda, S. M., D. Kendall, J. Connor, and D. Simberloff. 1997. Ecological effects of an insect introduced for the biological control of weeds. Science 277: 1088–1090.

Ludwig, D., R. Hilborn, and C. Walters. 1993. Uncertainty, resource exploitation, and conservation: lessons from history. Science 260: 17, 36.

Luken, J. O. 1990. Directing ecological succession. Chapman and Hall, New York. 251 p.

Marohasy, J. 1998. The design and interpretation of host–specificity tests for weed biological control with particular reference to insect behaviour. Biol. Control News Info. 19: 13N–20N.

May, R. M., and M. P. Hassell. 1981. The dynamics of multiparasitoid-host interactions. Am. Nat. 117: 234–261.

McClay, A. S. 1989. Selection of suitable target weeds for classical biological control in Alberta. Alberta Environmental Centre, Vegreville, Alberta, Canada. 97 p.

McClay, A. S. 1995. Beyond "before–and–after": experimental design and evaluation in classical weed biological control. *In* E. S. Delfosse and R. R. Scott (eds.), Proc. VIII international symposium on biological control of weeds, February 2–7, 1992, DSIR/CSIRO, Melbourne. p. 213–219.

McEvoy, P. B. 1996. Host specificity and biological pest control. BioScience 46: 401–405.

McEvoy, P. B., and E. M. Coombs. 1999. Biological control of plant invaders: Regional patterns, field experiments, and structured population models. Ecol. Appl. 9: 387–401.

McEvoy, P. B., C. Cox, and E. Coombs. 1991. Successful biological control of ragwort, *Senecio jacobaea*, by introduced insects in Oregon. Ecol. Appl. 1: 430–442.

McEvoy, P. B., and N. T. Rudd. 1993. Effects of vegetation disturbances on insect biological control of tansy ragwort *Senecio jacobaea*. Ecol. Appl. 3: 682–698.

McEvoy, P. B., N. T. Rudd, C. S. Cox, and M. Huso. 1993. Disturbance, competition, and herbivory effects on ragwort *Senecio jacobaea* populations. Ecol. Monogr. 63: 55–75.

McFadyen, R. E. C. 1998. Biological control of weeds. Annu. Rev. Entomol. 43: 369–393.

Memmott, J., S. V. Fowler, H. M. Harman, and L. M. Hayes. 1996. How best to release a biological control agent. *In* V. C. Moran and J. H. Hoffman (eds.), Proc. IX Intl. Symp. on Biological Control of Weeds, January19–26, 1996, Stellenbosch, South Africa, Univ. Cape Town, South Africa. p. 291–296.

Miller, M., and G. Aplet. 1993. Biological control: a little knowledge is a dangerous thing. Rutgers Law Rev. 45: 285–334.

Moran, V. C. 1985. The Silwood International Project on the biological control of weeds. *In* E. S. Delfosse (ed.), Proc. VI Intl. Symp. on Biological Control of Weeds, August 19–25, 1984, Agriculture Canada, Ottawa, Canada, Vancouver, British Columbia, Canada. p. 65–68.

Müller, H. 1990. An experimental and phytocentric approach for selecting effective biological control agents: insects on spotted and diffuse knapweed, *Centaurea maculosa* and *C. diffusa* (Compositae). *In* E. S. Delfosse (ed.), Proc. VII international symposium on biological control of weeds, 6–11 March 1988, Ministero dell'Agricoltura e delle Foreste, Rome, and CSIRO, Melbourne. p. 181–190,

Müller, H., and D. Schroeder. 1989. The biological control of diffuse and spotted knapweed in North America—what did we learn? *In* P. K. Fay and J. R. Lacey (eds.), Proc. 1989 knapweed symposium, Montana State Univ., Bozeman, Montana. p. 151–169.

Myers, J. H. 1985. How many insect species are necessary for successful biological control of weeds? *In* E. S. Delfosse (ed.), Proc. VI Intl. Symp. on Biological Control of Weeds, August 19–25, 1984, Agriculture Canada, Ottawa. p. 77–82.

Newman, R. M., D. C. Thompson, and D. B. Richman. 1998. Conservation strategies for biological control of weeds. *In* P. Barbosa (ed.), Conservation biological control. Academic Press, San Diego, California. p. 371–396.

Nowierski, R. M., C. B. Huffaker, D. L. Dahlsten, D. K. Letourneau, D. H. Janzen, and G. G. Kennedy. 1999. The influence of insects on plant populations and communities. *In* C. B. Huffaker and A. P. Gutierrez (eds.), Ecological Entomology. John Wiley, New York. p. 585–642.

O'Brien, C. W. 1995. Curculionidae premier biological control agents (Coleoptera: Curculionidae). Mem. Entomol. Soc. Washington 14: 119–128.

O'Riordan, T., and J. Cameron (eds.), 1994. Interpreting the precautionary principle. Earthscan Publica-

tions, London. 315 p.
Osenberg, C. W., and R. J. Schmitt. 1996. Detecting ecological impacts caused by human activities. *In* R. J. Schmitt and C. W. Osenberg (eds.), Detecting ecological impacts. Academic Press, New York. p. 3–16.
OTA—U.S. Congress Office of Technology Assessment. 1993. Harmful non-indigenous species in the United States, OTA–F–565. U.S. Govt. Printing Office, Washington, D.C. 391 p.
OTA—U.S. Congress Office of Technology Assessment. 1995. Biologically based technologies for pest control, OTA–ENV–636. U.S. Govt. Printing Office, Washington, D.C. 204 p.
Pemberton, R. W. 1995. *Cactoblastis cactorum* (Lepidoptera: Pyralidae) in the United States: An immigrant biological control agent or an introduction of the nursery industry? Am. Entomol. 41: 230–232.
Peschken, D. P., and A. S. McClay. 1995. Picking the target: a revision of McClay's scoring system to determine the suitability of a weed for classical biological control. *In* E. S. Delfosse and R. R. Scott (eds.), Proc. 8th Intl. Symp. on Biological Control of Weeds, CSIRO Publishing, Melbourne, Australia. p. 137–143.
Price, P. W. 1980. Evolutionary biology of parasites. Princeton Univ. Press, Princeton, New Jersey. 237 p.
Rees, M., and Q. Paynter. 1997. Biological control of Scotch broom: modeling the determinants of abundance and the potential impact of introduced insect herbivores. J. Appl. Ecol. 34: 1203–1221.
Rees, N. E., J. P.C. Quimby, G. L. Piper, E. M. Coombs, C. E. Turner, N. R. Spencer, and L. V. Knutson (eds.). 1996. Biological control of weeds in the West. Western Soc. Weed Sci., Bozeman, Montana.
Reichard, S. H., and C. W. Hamilton. 1997. Predicting invasions of woody plants introduced into North America. Conserv. Biol. 11: 193–203.
Rejmánek, M., and D. M. Richardson. 1996. What attributes make some plant species more invasive? Ecol. 77: 1655–1661.
Ruesink, J., I. M. Parker, M. J. Groom, and P. Kareiva. 1995. Guilty until proven innocent: reducing the risk of non–indigenous species introductions. BioScience 45: 465–477.
Secord, D., and P. Kareiva. 1996. Perils and pitfalls in the host specificity paradigm. Bioscience 46: 448–453.
Shea, K., and D. Kelly. 1998. Estimating biological control agent impact with matrix models: *Carduus nutans* in New Zealand. Ecol. Appl. 8: 824–832.
Sheldon, S. P. 1997. Ecological approaches for biological control of the aquatic weed Eurasian watermilfoil: resource and interference competition, exotic and endemic herbivores and pathogens. *In* D. A. Andow, D. W. Ragsdale and R. F. Nyvall (eds.), Ecological interactions and biological control. Westview Press, Boulder, Colorado. p. 53–70.
Sheppard, A. W. 1992. Predicting biological weed control. Trends Ecol. Evol. 7: 290–291.
Shigesada, N., and K. Kawasaki. 1997. Biological invasions: theory and practice. Oxford Univ. Press, New York. 205 p.
Simberloff, D. 1992. Conservation of pristine habitats and unintended effects of biological control. *In* W. C. Kauffman and J. E. Nechols (eds.), Selection criteria and ecological consequences of importing natural enemies. Entomol. Soc. Am., Lanham, Maryland. p. 103–117.
Simberloff, D., and P. Stiling. 1996a. How risky is biological control? Ecol. 77: 1965–1974.
Simberloff, D., and P. Stiling. 1996b. Risks of species introduced for biological control. Biol. Conserv. 78: 185–192.
Soper, R. S. 1992. USDA, Agricultural Research Service National Biological Control Program: program, policy, and constraints. *In* R. Charudattan and H. W. Browning (eds.), Regulations and guidelines: critical issues in biological control. Proc. of a USDA/CSRS National Workshop. Institute of Food and Agricultural Sciences, University of Florida, Gainesville, Florida. p. 49–52.
Stewart, A. J. A. 1996. Interspecific competition reinstated as an important force structuring insect herbivore communities. Trends Ecol. Evol. 11: 233–234.
Story, J. M., K. W. Boggs, W. R. Good, P. Harris, and R. M. Nowierski. 1991. *Metzneria paucipunctella* Zeller (Lepidoptera: Gelechiidae), a moth introduced against spotted knapweed: its feeding strategy and impact on two introduced *Urophora* spp. (Diptera: Tephritidae). Can. Entomol. 123: 1001–1007.
Syrett, P., S. V. Fowler, and R. M. Emberson. 1996. Are chrysomelid beetles effective agents for biological control of weeds? *In* V. C. Moran and J. H. Hoffman (eds.), Proc. IX Intl. Symp. on Biological Control of Weeds, January 19–26, 1996, Stellenbosch, South Africa, Univ. Cape Town, South Africa. p. 399–407.
Tenner, E. 1996. Why things bite back: Technology and the revenge of unintended consequences. Knopf, New York. 349 p.
Thomas, M. B., and A. J. Willis. 1998. Biological control—risky but necessary? Trends Ecol. Evol. 13:

325–329.

Thompson, D. Q., R. L. Stuckey, and E. B. Thompson. 1987. Spread, impact and control of purple loosestrife (*Lythrum salicaria*) in north american wetlands. U.S. Fish and Wildlife Res., Washington D.C. 55 p.

Tuljapurkar, S., and H. Caswell (eds.). 1997. Structured-population models in marine, terrestrial, and freshwater systems. Chapman and Hall, New York. 643 p.

Turner, C. E., R. W. Pemberton, and S. S. Rosenthal. 1987. Host utilization of native *Cirsium* thistles (Asteraceae) by the introduced weevil *Rhlnocyllus conicus* (Coleoptera: Curculionidae) in California. Environ. Entomol. 16: 111–115.

USDA. 1999. Reviewer's manual for the technical advisory group for biological control agents of weeds. Manuals Unit, Plant Protection and Quarantine, Animal and Plant Health Inspection Svc, U.S. Dept. Agric.

USDA. 1997. The PLANTS database (http: //plants.usda.gov). National Plant Data Center, Baton Rouge, Louisiana.

van Loon, L. C., P. A. H. M. Bakker, and C. M. J. Pieterse. 1998. Systemic resistance induced by rhizosphere bacteria. Annu. Rev. Phytopathol. 36: 453–483.

Wapshere, A. J. 1974. A strategy for evaluating the safety of organisms for biological weed control. Ann. Appl. Biol. 77: 201–211.

Wapshere, A. J. 1985. Effectiveness of biological control agents for weeds: present quandaries. Agric. Ecosys. Environ. 13: 261–280.

Wapshere, A. J., E. S. Delfosse, and J. M. Cullen. 1989. Recent developments in biological control of weeds. Crop. Prot. 8: 227–250.

Weidemann, G., and D. O. Tebeest. 1990. Biology of host range testing for biological control of weeds. Weed Technol. 4: 465–470.

Wilcove, D. S., D. Rothstein, J. Dubow, A. Phillips, and E. Losos. 1998. Quantifying threats to imperiled species in the United States. BioScience 48: 607–615.

Williamson, M. 1996. Biological invasions. Chapman and Hall, London. 244 p.

Wolpert, L. 1999. Is science dangerous? Nature 398: 281–282.

Woodburn, T. L. 1996. Interspecific competition between *Rhinocyllus conicus* and *Urophora solstitialis*, two biological control agents released in Australia against *Carduus nutans*. *In* V. C. Moran and J. H. Hoffman (eds.), Proc. IX Intl. Symp. on Biological Control of Weeds, January 19–26,1996, Stellenbosch, South Africa, Univ. Cape Town, South Africa. p. 409–415.

Zwölfer, H. 1973. Competition and coexistence in phytophagous insects attacking the heads of *Carduus nutans* L. *In* P. H. Dunn (ed.), Proc. Second Intl. Symp. on Biological Control of Weeds, October 4–7, 1971, Misc. Publ. no. 6, Commonwealth Inst. Biol. Control, Commonwealth Agric. Bur., Farnham Royal, Slough, UK. p. 74–77.

Zwölfer, H., and P. Harris. 1971. Host specificity determination of insects for biological control of weeds. Annu. Rev. Entomol. 16: 159–178.

Zwölfer, H., and P. Harris. 1984. Biology and host specificity of *Rhinocyllus conicus* (Froel.) (Col., Curculionidae), a successful agent for biological control of the thistle, *Carduus nutans* L. Z. Angew. Entomol. 97: 36–62.

12

IMPORTATION PROTOCOLS AND RISK ASSESSMENT OF WEED BIOLOGICAL CONTROL AGENTS IN AUSTRALIA: THE EXAMPLE OF *CARMENTA* NR *ITHACAE*

Toni Withers
CRC for Tropical Pest Management,
Alan Fletcher Research Station, Sherwood, Australia

Rachel McFadyen
Queensland Department of Natural Resources,
Alan Fletcher Research Station, Sherwood, Australia

Jennifer Marohasy
Canegrowers, Brisbane, Australia

IMPORTATION PROTOCOLS FOR BIOLOGICAL CONTROL AGENTS IN AUSTRALIA

In Australia there is a clear legislative process controlling the deliberate introduction of insects as biological control agents. The current system has been reviewed in Paton (1992) and described for weeds in McFadyen (1998). We will describe how the biological control introduction process currently runs in Australia, and demonstrate how the system provides an excellent legislative framework to ensure nontarget impacts are minimized. The process will be illustrated using the biological control agent *Carmenta* nr *ithacae* Beutenmuller (Lepidoptera: Sesiidae) as a case study. *C.* nr *ithacae* was recently approved for release against parthenium weed, *Parthenium hysterophorus* L., in Australia.

Two principal Acts apply: the Quarantine Act 1908 (Commonwealth of Australia 1985), designed to keep out agricultural pests and diseases of humans, and the Wildlife Protection Act 1982 (Commonwealth of Australia 1982), designed to control trade in endangered wildlife. The acts require permission first be sought to import a potential biological control agent into a registered quarantine insectary for host specificity testing, followed by the development of an acceptable list of nontarget species for testing, and finally, if the agent shows specificity, permission to release the organism. A third Act, the Biological Control Act 1984 (Commonwealth of Australia 1984), was designed to deal with conflicts of interest such as that which arose over the biological control of Paterson's curse, *Echium plantagineum*, which was regarded by some as a weed and by others as a useful plant (Cullen and Delfosse 1984). The Biological Control Act 1984 provides a legal basis for decisions, but it has seldom been called upon, suggesting that the current process effectively manages the decision-making process to minimize serious conflicts of interest.

The Quarantine Act 1908 and Wildlife Protection Act 1982 are administered by separate government departments and a permit to import and release a biological control agent is required under both Acts. Issue of permits is administered by the Australian Quarantine Inspection Service (AQIS) according to set protocols and with the assistance of reviewers from across Australia. The guidelines (currently available at www.aqis.gov.au/docs/appolicy/bcontrol.html) specify the information that applicants must provide on the target pest and the biological control agent, and are designed to ensure that reviewers are provided with all necessary information on which to base their decision. The system works as follows:

Permission to Import

The initial application to import a biological control agent into quarantine for detailed testing is submitted to AQIS in Canberra. This application includes basic information on the target pest, and on the biological control agent's biology and known host range. The application is considered by a panel of three experts (entomologists) who advise AQIS and consult with Environment Australia (EA) (Biodiversity Group). At this stage, the main consideration is whether the agent can be safely maintained in the quarantine insectary proposed by the applicant. If the application is approved, permits to import into quarantine are granted.

Host Testing List

Together with the original application or separately, the applicant submits to AQIS a proposed "Test List," a list of the nontarget arthropods or plants against which they propose to test the biological control agent. The list is usually based on the phylogeny, morphology, and chemical ecology of the target weed or arthropod (Forno and Heard 1997). "Test Lists" normally include all native genera and economically valuable species or genera closely related to the target in Australia. The proposed testing methodologies do not have to be specified. AQIS, in consultation with EA, sends the proposed "Test List" to the departments of agriculture and environment in each of the states, to equivalent Federal departments, to the Commonwealth Scientific and Industrial Research Organization (CSIRO), and to various other scientific or technical agencies—21 groups in all. In each of these agencies, the "Test List" is reviewed by scientists and their comments are submitted to AQIS for review. Guidance is now available to these reviewers in the form of a CD-ROM training manual (Univ. Queensland 1998). Any desired changes to the proposal are discussed with the applicant, and an agreed "Test List" is negotiated. Applicants can test additional nontarget species as they wish, but species on the list cannot be omitted, nor can other agencies later demand the addition of further species without justification.

Permission to Release

When host specificity testing has been completed, and if the applicant considers that the biological control agent would be safe to release, an "Application to Release" the agent is written. This application details the results of the host testing and includes all other relevant information on the agent's biology and host range. The application is submitted to AQIS along with information on the target species. AQIS sends the "Application to Release" to the same 21 groups in each state and territory as before for their consideration and comment. If all the 21 agencies are in favor of the application, both AQIS and EA issue a release permit. The review process, from submission of the "Application to Release" to granting the permits, usually takes 3 to 6 months.

Conflicts of Interest

If any of the 21 agencies have objections to the application, their comments are relayed back to the applicant for reply. Reviewers may be totally opposed to granting the permit, or may ask for extra tests or for more information. If most groups are in favor, while one or two are opposed, AQIS, in consultation with EA, may decide to grant the permit anyway. However, if there is a serious disagreement, the permit will be denied. In real conflicts of interest, such as where there is likely to be damage to the environment but real benefit to agriculture, the Biological Control Act 1984 allows for a public inquiry ending in a decision at the Ministerial level.

The important feature of this system that assures careful consideration of potential risks to nontarget organisms is the wide consultation that occurs within all the Australian states. There are clear avenues for government-facilitated resolution if minor conflicts should arise. In this way, plants or organisms which may be highly valued in localized regions of Australia or by different interest groups can be brought to the attention of the applicant, and the risks from the candidate biological control agent to those plants or organisms can be assessed and debated, both before and after host specificity testing.

A RISK ASSESSMENT FRAMEWORK APPLICABLE TO WEED BIOLOGICAL CONTROL

Applications to AQIS for the field release of biological control agents against weeds currently are not required to incorporate "International Best Practice in Environmental Risk Assessment" (Beer and Ziolkowski 1995). What such applications typically include are the biological/ecological attributes of the potential biological control agent indicating its likely host plant range and, in some cases, its likely geographic range with respect to the distribution of target and nontarget plants. As the potential risk to nontarget plants is an integral component of the "Application to Release," this application is an informal risk assessment, and we believe could be effectively treated within a risk assessment framework.

As biological control comes under increasing scrutiny for its environmental safety, it is important that the decision-making procedures in place in the future are compliant with international best practice in risk assessment. Formal risk assessment is comprised of four distinct stages: identification, analysis, decision making, and treatment of the risk (Table 1). Ecological-type risk assessments have been based upon this framework (Beer and Ziolkowski 1995), and a wider adoption for weed biological control will broaden the basis on which risk is assessed which should increase safety for nontargets.

Identification of the Risk

The identification of risk involves identifying the concerns of the community and the negative consequences of a potentially undesirable event occurring. In weed biological control in Australia, the gathering of relevant field data on nontarget and target impacts of biological control and effective communication of this to the public will aid in creating realistic public perceptions of risk.

Following the publication of an influential paper by Wapshere (1974), native plant species closely related to the target weed that grow within the target's geographic range were for the first time considered to be as important as exotic crop plants in terms of risk. The potential for an adverse effect on any native plant species by a weed biological control agent is now considered to be crucial. Therefore, identifying where the potential risk lies is integrally bound up with the first stage of the Australian protocols—the agreement on the test list. As mentioned above, the selection of plants species against which the proposed agent will be tested, is arrived at through a consultative process that exposes the proposed biological control project to the scrutiny of a wide-range of interest groups throughout Australia.

Analysis of the Risk

Risk analysis has two important components: the estimation of the probability of an event occurring, and the estimation of the probable consequences of that occurrence (Table 1). In weed biological control, the "event" is that the biological control agent establishes on a nontarget plant species, while the consequences are the damage the agent could cause to the nontarget plant, in the widest sense. Current submissions to

Table 1. Application of a generic risk assessment framework to weed biological control, compiled with reference to Beer and Ziolkowski (1995).

Generic framework	Ecological type assessment	Weed biological control type assessment
1. Identification of the risk		
Concerns	Stress on ecosystems	Attack on nontarget plant species, including native and economically important species
Consequences (impacts)	Ecological effects	Social, political, economic, environmental effects of attack on nontarget species
2. Analysis of the risk		
Certainty/uncertainty (probability)	Pathway analysis, Monte-Carlo modelling	Determine probability of establishing on nontarget plant species under range of conditions.
Calculations (consequences)	Quantify hazards	Quantify potential damage to nontarget plant species under range of conditions.
3. Deciding on the risk		
Compare with criteria	Ecosystem integrity	Agent should be approved for release if the harm caused by the agent would be significantly less than the harm caused, or likely to be caused, by failure to controltarget organisms (Biological Control Act 1984).
4. Treatment of the risk		
Control	Restrict hazardous activities	Do not approve field release of candidate biological control agent.
Communication	Document, publish and advise	If approve field release, monitor the impact of the control agent on nontarget plant species after release; document, publish, and advise.

AQIS for the release of biological control agents do not explicitly consider the probability of establishing on a plant separately from the harmful consequences that could result from establishment. We will show that the data gathered for current submissions can often be used to extrapolate these measures, at least qualitatively. The best methods for host specificity testing of different insects to accurately arrive at probabilities for these two components is currently under debate (McEvoy 1996, Heard and van Klinken 1998, Marohasy 1998, Withers 1998). There is, however, general agreement that unless field trials and cage assays incorporate behavioral and physiological concepts, the results could be either misleading or difficult to interpret (Marohasy 1998, Withers and Barton Browne 1998).

To arrive at realistic probabilities for establishment, we first need to understand that establishment is the result of various physiological and behavioral processes (McEvoy 1996) including habitat and host location, pre-landing and post-landing host discrimination, oviposition, larval acceptance, and host suitability. For risk analysis, a behaviorally realistic assay to determine the probability of a proposed agent locating, accepting, and feeding on a nontarget plant would be of considerable value.

The second part of risk analysis requires that the consequences of the "event" be quantified (Table 1). In most instances, quantifying consequences would involve determining the extent of damage to the nontarget plant species should that species be fed upon and/or oviposited on by the biological control agent. Preferably, this should involve the quantification of potential damage by the feeding stage(s), and determining the probability of reaching the feeding stage, under a range of conditions on the target and nontarget hosts. Plant damage could be measured in terms of the reproductive cost to the plant (e.g., the reduction in numbers of flowers and seeds produced) or in terms of the dollar value of lost production. Indirect negative effects on the environment, such as those discussed by Simberloff and Stiling (1996), may be important but are usually more difficult to quantify.

Further elaborating on the above points, we believe it apt to propose some guidelines on which to separate probability of establishment on a plant from its consequences, which we believe differ considerably between different insect guilds:

1. In passively dispersing mites or insects (e.g., mealybugs) the probability of establishment is the probability of arriving at the plant species in question, and is therefore related to the method of dispersal and the phenology of the plants. The consequences for the plant depend on the level of damage caused by the feeding or colonization behavior of various life stages, and this will be a function of survival of individuals of damaging life stages on the plant. The difficulty in obtaining probabilities for both of these aspects remains considerable. We recommend the use of continuing generation no-choice trials on all nontarget plants that support some colony development. Careful monitoring of populations at different life stages on target versus nontarget plants should generate some key data on both establishment and damage.
2. In many insects the adult female searches actively for host plants on which to oviposit, but larvae have a limited ability to disperse from the plant on which they hatch (e.g., Lepidoptera and Coleoptera with internal feeding larvae). In such insects the probability of establishment is the probability of the egg being laid on the plant in question. Female specificity in oviposition behavior is therefore a key factor for host specificity testing. The consequences of oviposition will be the product of egg viability and larval survival, and the damage inflicted by the feeding stages. In such insects oviposition trials should be compulsory to ascertain the probabilities of establishment, whereas consequences would best be ascertained from independent egg-transfer no-choice feeding trials. We propose such feeding trials need only be conducted on nontarget plants on which the female oviposited, however, if the true behavior of the female during oviposition assays is at all in doubt, no-choice feeding assays ought to be conducted on the complete test list.
3. In some insects the adult female may not search actively for host plants as much as for a suitable habitat, and the larvae retain the ability to move and discriminate between host plants on which they are attempting to feed (e.g., some Lepidoptera). In such insects the probability of establishment is the product of all the required steps leading to the larva feeding on the plant in question. In such an insect oviposition assays may be unnecessary although they will be helpful in obtaining a fuller

understanding of the biology of the agent. Larval feeding assays should form the basis of estimations of establishment. The consequences of attack are confined to assessment of plant damage following feeding by the damaging life stages.

In proposing such guidelines we acknowledge than none of the above are based upon a detailed comparative analysis, but will hopefully be a starting point to provoke further research in this area.

Deciding on the Risk

Permission to release biological control agents in Australia has sometimes been granted even when the agent has been known to be capable of completing development on a desirable nontarget plant species that grows sympatrically with the weed (McFadyen and Marohasy 1990). In such cases the proponents have had to argue for release of the agent on the basis of the criteria required for deciding on the risk in the Biological Control Act 1984 (Section 29). This section states that an agent should be approved for release if the harm caused by the agent "would be significantly less than the harm caused, or likely to be caused, by failure to control the target organisms" (Commonwealth of Australia 1984). Thus, the potential cost of releasing candidate biological control agents could be compared with the potential cost of not releasing them should the weed continue to spread. Pre- and post-target studies on weed ecology will assist significantly with assigning costs to these various events.

Treatment of the Risk

The fourth phase of a risk assessment process requires the "risk be treated." Risk of nontarget impacts is controlled most stringently by regulating authorities when they refuse to allow the release of a proposed agent. Refusal by the Australian Government bodies of permission to release weed biological control agents up until the present time has been rare. We believe this reflects that biological control practitioners in Australia do not request permission to release an agent unless their own assessments satisfy that the impact to nontarget plants is going to be negligible (McFadyen and Heard 1997). For instance, the potential biological control agent *Liothrips* sp. (Phlaeothripidae) was investigated for control of annual ragweed, *Ambrosia artemisiifolia* L. (Asteraceae) in Queensland. The host list for specificity testing was similar to that for agents against parthenium weed (see case study below). *Liothrips* sp. was thought to be a promising biological control agent because colonies showed significant oviposition and feeding preferences for annual ragweed in choice tests. In no-choice tests however, very young seedlings of the valuable crop plant sunflower (*Helianthus annuus*) were severely damaged and *H. annuus* proved to be an excellent host for *Liothrips* development (McFadyen and Weggler-Beaton 1999). After the host testing was completed the colony was destroyed because the probability of establishment on young sunflower plants in the field, and the potential for damage to them, were deemed unacceptably high.

When permission is given for an agent to be released, incorporation of risk assessment into weed biological control would require the applicant to quantify the impact of the biological control agent on nontarget plant species post-release, and to document and disseminate the results. Post-release monitoring of establishment and impacts on the target weed and reporting the results back to AQIS within 12 months is a formal requirement of all parties given permission to release biological control agents under the Australian protocols. As data are gathered from post-release monitoring studies and compared to pre-release predictions for a number of agent releases, the process and predictability of future risk assessments relative to nontarget impacts should improve.

We will now introduce a case study for a recent introduction of a biological control agent into Australia, and use this to demonstrate how the Australian protocols work, and how they allow for a degree of risk assessment to be incorporated into the current procedures.

CASE STUDY: *CARMENTA* NR *ITHACAE* FOR BIOLOGICAL CONTROL OF PARTHENIUM WEED

Background

Parthenium hysterophorus (parthenium weed) is native to the countries bordering the Gulf of Mexico (Navie et al. 1996). It is a short-lived ephemeral herb, germinating after rain at any season, flowering in 6 to 8 weeks, and senescing with drought or frost. Parthenium weed is an invasive annual of open land and pasture, being unpalatable to stock, and on suitable soils and following rainfall it will quickly dominate both native and planted pastures. Parthenium weed was first introduced into central Queensland from Texas, USA, in the late 1950s. It has spread rapidly to invade more than 10% of Queensland and parts of New South Wales. Parthenium weed, which is harmful to other plants, cattle and humans (McFadyen 1992) is a declared noxious weed in both States, with a legal requirement for control by all landholders (Navie et al. 1996). Once dominant, parthenium weed continues to persist as pure stands unless managed. Management options include de-stocking to maintain grass competition, fire, herbicides, and biological control (Navie et al. 1996).

Carmenta nr *ithacae* is the tenth biological control agent to be released against parthenium weed in Queensland (Dhileepan et al. 1996). This clear-wing moth is diurnal, emerging and mating in the morning, and ovipositing from mid-day to early afternoon. Eggs are laid singly on leaves and stalks, and the young larvae initiate feeding in the lower leaf axils and crown. Later instars feed in the crown and roots of the plant, covering external damage with a tube of frass and silk. Pupation occurs within the tunnel, from which the moth emerges, leaving the pupal exuviae protruding.

Carmenta nr *ithacae* has only been collected from *P. hysterophorus* in Mexico. *Carmenta ithacae* Beutenmuller is recorded from the roots and lower stems of *Helenium autumnale* and *Heliopsis helianthoides* (Asteraceae) in North America, where it is described as a wide-ranging species, occuring from New York state to northern Florida, and west to Manitoba, Wisconsin and Mississippi, Colorado and Arizona (Eichlin and Duckworth 1988). These two populations may be different species so we choose to refer to the former population as *C.* nr *ithacae*. *Carmenta* species are usually host-specific with most species being restricted to hosts in one or two closely related genera. Two *Carmenta* species have been used as biological control agents (Eichlin and Duckworth 1988).

Identifying the Risk of *C.* nr *ithacae* Attacking Nontarget Plants in Australia

In order to identify potential risks to nontarget plants a host test list was approved by AQIS and EA in 1996. Parthenium weed is in the subtribe Ambrosiinae of the tribe Heliantheae of the Asteraceae. There are no native or ornamental plants in Australia in this subtribe. The only plants in the subtribe occurring in Australia are introduced weeds in the genera *Iva, Xanthium*, and *Ambrosia. Parthenium argentatum*, guayule, is of potential economic importance as a source of rubber, and so was included in the test list despite not being currently utilized in Australia. The weed *Xanthium occidentale*, Noogoora burr, grows sympatrically with parthenium weed in parts of Queensland. Therefore from a risk perspective, all the plants most closely related to the target are exotic weeds.

Within the tribe Heliantheae there are 11 genera with species native to Australia. Three genera, *Chrysogonum, Blainvillea*, and *Moonia,* have limited distributions in far north Queensland, Western Australia, and the Northern Territory, and seed of species within these genera could not be obtained. Five other genera in the same subtribe (Coreopsidinae) were tested as the next best option. More widely occurring native genera as well as genera of ornamental or economic importance in this tribe, were included in the test list (Table 2). The major crop plant of significant economic importance to Australia in this tribe is sunflower, *Helianthus annuus*, and six cultivars of this in common use were tested, together with *Helianthus argophyllus*, used as a parent line in some cultivars. Sunflower is considered to be under particular risk from the biological control agents of parthenium weed, as the weed is commonly found growing within fields of sunflower in Central Queensland (McFadyen and Heard 1997) and sunflower shares a number of secondary chemicals with parthenium weed (Herz 1977). The recorded hosts of *C. ithacae*, *Helenium*, and *Heliopsis*, are also in the tribe Heliantheae, subtribes Gaillardinae and Zinniinae respectively. Neither *Helenium* nor *Heliopsis* occur in Australia, however *Heliopsis scabra* and several species of *Helenium* obtained from overseas seed suppliers were tested. In total, therefore, 27 species in 21 genera in the 8 subtribes of the Heliantheae were tested.

In the other 10 tribes of the Asteraceae represented in Australia, representatives of 20 genera, of which 11 occur as natives in Australia and the rest are ornamentals or crop plants, were included in testing. Attack by a parthenium biological control agent of these genera would therefore be viewed very seriously but is less likely because of their more distant relationship to the target weed.

As other species of *Carmenta* have hosts in the families Mimosaceae, Boraginaceae, Geraniaceae, and Solanaceae (Eichlin and Duckworth 1988), representatives of these families were also included. Two species of *Carmenta* feed in stems of *Phoradendron* spp. in the Loranthaceae (mistletoe), so a local representative, *Amyema* sp., was included in the oviposition tests.

Table 2. Host test list for *C.* nr *ithacae* with results of egg-transfer larval-feeding trials where the presence of larvae (L) was assessed after 30 days and for some plants also after 60 days.

Plant family, tribe (T), subtribe (sT)	Species, common name, (type[a])	Larvae found at (days) 30	60
ASTERACEAE			
T Astereae	*Aster novii-belgii* aster (O)	0	
	Brachycome multifida cutleaf daisy (N)	0	
	Olearia nernstii olearia (N)	L	1 adult
	Vittadinia sulcata fuzzweed (N)	0	
T Heliantheae	*Ambrosia artemisiifolia* annu. ragweed (W)	0	
sT Ambrosiinae	*Parthenium argentatum* guayule (PC)	0	0
	Parthenium hysterophorus	L	
	Xanthium occidentale Noogoora burr (W)	L	5–10/plant
sT Helianthiinae	*Helianthus argophyllus* silver-leafed sunflower (O)	0	
	H. annuus var. Suncross (C)	L	0
	H. annuus var. Hysun 33 (C)	L	0
	H. annuus var. Hysun 32 (C)	L	0
	H. annuus var. Double sunflower (C)	0	
	H. annus var. grey stripe (C)	0	
	H. annuus var. Russian giant (C)	0	
sT Gaillardinae	*Helenium montanum* (W, E)	L	
	Helenium amarum (W, E)	L	1 male
	Helenium bigelovii (E)	L	0
sT Gaillardinae	*Helenium hoopsi* (E)	0	
	Gaillardia grandiflora gaillardia (O)	L	
sT Galinsoginae	*Galinsoga parviflora* potato weed (W)	0	
	Tridax procumbens tridax (W)	0	
	Spilanthes grandiflora (N)	L	0
sT Zinniinae	*Zinnia elegans* zinnia (O)	0	
	Heliopsis scabra (E)	L	
sT Coreopsidinae	*Coreopsis lanceolata* coreopsis (O)	0	
	Cosmos bipinnatus comos (O)	L	
	Dahlia sp. (O)	L	
	Glossogyne tenuifolium cobbler's pegs (N, W)	0	
	Guizotia abyssinica niger seed (W)	0	
sT Melampodiinae	*Sigesbeckia orientalis* indian weed (N, W)	0	
sT Ecliptinae	*Eclipta platyglossa* yellow twinheads (N, W)	0	
	Eclipta prostrata white eclipta (N, W)	0	
	Enydra fluctuans (N)	L	
	Flaveria australasica speedy weed (N, W)	0	
	Wedelia biflora wedelia (N)	0	
T Inuleae	*Cassinia laevis* inland cassinia (N)	0	
	Gamochaeta pensylvanica (W)	L	
	Gamochaeta subfalcatum (W)	0	
	Helichrysum bracteatum golden everlasting (N)	0	
T Senecionideae	*Senecio lautus* fireweed (N, W)	0	
T Anthemideae	*Artemisia abrotanum* southernwood (W)	0	

Table 2. *(continued)*

	Chrysanthemum parthenium feverfew (O)	0	
	Cotula australis common cotula (N)	0	
	Chamomilla recutita German chamomile (O)	0	
T Eupatorieae	*Vernonia cinerea* veronia (N)	0	
T Cichorieae	*Cichorium intybus* chicory (C)	0	
	Lactuca sativa lettuce (C)	0	
	Picris evae hawkweed (N) (W)	0	
T Mutisae	*Gerbera jamesonii* gerbera (O)	0	
T Cynareae	*Carthamus tinctorius* safflower (O)	0	
T Calenduleae	*Calendula officinalis* calendula (O)	L	9 adults
MIMOSACEAE	*Mimosa pigra* mimosa (W)	0	
	Acacia fimbriata fringed wattle (N)	0	
FAGACEAE	*Nothofagus moorei* antarctic beech (N)	0	
BORAGINACEAE	*Heliotropium* sp. heliotrope (W)	0	
	Echium plantagineum paterson's curse (W)	0	
GERANIACEAE	*Geranium sp.* geranium (O)	0	
	Pelargonium stork's-bill (O)	0	
SOLANACEAE	*Lycopersicum esculentum* tomato (C)	0	

[a]Types: O = ornamental, N = native, W = weed, E = exotic, C = crop, PC = potential crop.

Analyzing the Risk of *C.* nr *ithacae* Attacking Nontarget Plants in Australia

Larval *C.* nr *ithacae* are believed to be incapable of moving off the plant on which they were laid as an egg, therefore establishment on a plant is a direct consequence of female oviposition behavior. Behaviorally based oviposition testing is therefore the most appropriate method of assessing the probability of establishment on nontargets, whereas no-choice larval feeding tests following egg transfer can independently assess the potential for damage by larval feeding on nontarget plants.

Method for larval feeding assay

When eggs laid on parthenium weed were 10–12 days old (almost ready to hatch) they were transferred to test plants on small pieces of leaf (1–2 mm) stuck with egg white onto the leaves of test plants. Ten eggs were placed onto each test plant, and replicated at least four times. Thirty days later the plants were examined, the stems, crowns, and roots cut open, and any larvae found were counted and their size in comparison to those in the control plant was recorded. With nontarget plants in which several larvae were found, further egg transfer tests were undertaken. Larvae were left until pupation on control plants (60 days), and any pupae found were held until adult emergence.

Results of larval feeding assays

Within the tribe Inuleae, small larvae were found in the weed species, *Gamochaeta pensylvanica.* Within the tribe Heliantheae, a few small larvae were found in at least one replicate of the genera *Gaillardia, Galinsoga, Enydra, Dahlia,* and *Helianthus*, and larvae were found in 50% or more of the replicates in the genera *Xanthium, Helenium, Spilanthes*, and *Heliopsis*. Pupation was recorded on the nontarget plants *Olearia nernstii* (Astereae), *Helenium amarum* (subtribe Gaillardinae), and *Calendula officinalis* (Calenduleae) (Table 2). Of these pupae, adults emerged normally only from *C. officinalis* plants. Emerging females called and were seen to mate successfully both with males from *C. officinalis* and with males reared from parthenium weed. These mated females were caged with *C. officinalis* plants, however they failed to oviposit upon them. A few eggs (15 and 21 per female, respectively) that were laid on the cage were shown to be fertile.

The results of the egg-transfer larval-feeding assays alone indicate a potential for negative impacts on nontarget plants by *Carmenta* nr *ithacae*. The physiological host range of larvae extends beyond the genus and subtribe of the target host, to include some plants in another subtribe of the Heliantheae, and even in two separate tribes. On the nontarget plants where there was significant larval feeding, plant vigour was reduced to the point of likely death or substantially reduced plant size and flowering. Despite the belief that development to the adult stage on *C. officinalis* is due to its inherent insect-susceptible characteristics, based on the larval feeding assay alone, the risk of negative consequences to native and ornamental plants in Australia from *C.* nr *ithacae* would be considered unacceptably high.

Adult oviposition tests

Oviposition tests were carried out with all plant species on which more than one healthy larva was found in the larval feeding assays, and on plants within the Ambrosiinae (the tribe containing parthenium weed) in order to assess the behavior of adult female moths. These tests were performed in large perspex and cotton mesh cages that were 1.2 m tall and between 0.5 m and 2 m long and contained the potted test plants in groups of two or three. These tests were designed to allow us to better predict the probability of establishment (oviposition) on those nontarget plants which sustain larval development, should females locate those plants in the field.

Three types of oviposition tests were used. First, "simple" choice oviposition tests were performed, with two or three specimens of each of three or four test plant species placed in a cage together with parthenium weed. Two to five mated female moths were introduced to the cage with plants for three days.

Secondly, "non-host choice tests" were performed to test the oviposition behavior of females at the time when egg laying is normally at its peak, in circumstances where parthenium weed plants were absent. One to three females emerging on the same day were mated in a cage containing no plants and then transferred to a cage containing three to four test plant species other than parthenium weed. The drawback with the method is that an absence of oviposition during the lifetime of the female might be due to an abnormality in the female, rather than to behavioral specificity.

Knowledge of the ephemeral nature of parthenium weed makes this type of test relevant. It is conceivable that under conditions such as drought following a successful rainy season, *C.* nr *ithacae* adults may emerge from pupae remaining in the root stock of senescent parthenium weed, only to find no parthenium weed plants available. Oviposition on nontarget plants in this situation will put them at risk. Additionally, the period of time that elapses before oviposition should any of these nontarget plants become acceptable, compared to the time at which oviposition would have commenced on parthenium weed, will tell us the approximate duration of the "discrimination phase" (Marohasy 1998).

Finally, "sequential multiple choice tests" (Marohasy 1998) were carried out, where mated female C. nr ithacae moth were placed in cages with initially two parthenium weed plants and groups of two to four other species of test plant. After >10 eggs per female had been oviposited on the parthenium weed plant (typically one or two days after mating), the parthenium weed plants were removed. Oviposition on the remaining test plant species was monitored daily, in order to ascertain which was next most stimulating for oviposition by the female in the absence of parthenium weed. After two complete days of exposure to test plants in the absence of parthenium weed, parthenium weed plants were returned to the test cage and a record kept of the eggs laid in the one to three days remaining before the females died. Eggs laid on the cage walls or floor were also recorded and removed at the completion of each test. Sequential choice tests were undertaken in order to assess the preference rank of various host plants by consistently removing the plant species that received the most oviposition, increasing the probability the female will oviposit on that which is the next most acceptable. Permitting oviposition on the target weed at the conclusion of the test ascertained whether or not the female was still capable of oviposition.

Oviposition test results

In the "simple" choice oviposition tests, no eggs were laid on any of the nontarget plants tested (data not shown). In each test, between 50 and 200 eggs were laid on the parthenium weed plants. In the "non-host choice tests" that excluded parthenium weed (data not shown), mated female moths rarely initiated oviposition on any test plants in the absence of parthenium weed (5 eggs on *H. bigelovii*, 4 eggs on *C. bipinnatus*; n = 9 mated females). In these assays the onset of oviposition was significantly delayed in the absence of parthenium weed, with the first eggs being laid (> 200 on the cage) 4-7 days after mating, compared to females with access to parthenium weed who invariably began oviposition 1–2 days after mating. Such temporal differences in oviposition indicate significant specificity (Withers 1997) to parthenium weed, as well as suggesting the duration of the discrimination phase (Marohasy 1998). The difference in time between acceptance of parthenium weed and acceptance of the cage surface or a less preferred plant may be as long as three days (i.e., half the normal adult lifespan). Results from the "sequential multiple choice tests" (14 tests using 19 females), were revealing (Table 3). A mean of 33 eggs per female were laid on parthenium weed when it was present at the beginning of the test and a mean of 33 eggs per female were laid on parthenium weed when it was reintroduced at the end of the test. Only three other plants received any eggs in the presence of parthenium weed; these were *X.*

occidentale with 35 eggs, and *H. bigelovii* and *H. annuus* cv. Hysun 33 with one egg each. In the case of *X. occidentale* and *H. bigelovii,* females died the day after laying eggs. Following the removal of parthenium weed, oviposition was always significantly reduced; for those two days a mean of 1 egg was laid per female. *Xanthium occidentale* was the only plant to receive eggs in the absence of parthenium weed (Table 3). We conclude that *X. occidentale* is ranked second in oviposition preference to parthenium weed, while cage netting or perspex walls are ranked third (cage walls received a mean of 8 eggs per female). All other nontarget plants rank below the cage in acceptability.

Deciding on the Risk of *C.* nr *ithacae* Attacking Nontarget Plants in Australia

On the basis of host specificity results with *C.* nr *ithacae,* particularly with the oviposition tests, permission was sought to release this agent. Primarily this was because the risk to nontarget native (*Olearia* spp.) and ornamental plants (*C. officinalis*) was considered to be low. We argued that despite the negative consequences of larval feeding on several nontargets, the probability of this occurrence is negligible because females did not oviposit upon any of the nontarget species tested. Indeed females in the presence of parthenium weed remained specific in oviposition behavior to parthenium weed, and in its absence did not initiate egg laying until just prior to dying. Furthermore, adult female *C.* nr *ithacae* reared from *C. officinalis* following egg transfer still showed no oviposition behavior on *C. officinalis* in a no-choice situation, which adds weight to the conclusion that a population would be unlikely to arise on *C. officinalis* in the field.

The plant that appeared to be second in ranking for oviposition (receiving eggs in two of four oviposition tests), *X. occidentale,* is closely related to parthenium weed and is believed to be most similar in plant chemistry (Herz 1977). It is also is the most likely of all the nontarget plants to be found growing in the same habitat as parthenium weed in Australia. However, not only was this weed a poor host for larval development, but any damage would be considered a positive benefit because of the exotic weedy status of *X. occidentale.* So in this case, although the probability of establishment is high, the consequences of any attempted larval feeding are either negligible or not considered harmful, meaning the risk is not significant.

The above nonquantified assessment of risks was used in the November 1997 "Application to Release" for *C.* nr *ithacae* in Australia (R. E. McFadyen and T. M. Withers, unpublished report, QDNR, Australia). In January 1998, AQIS relayed comments from one Australian State Department of Agriculture who was concerned with the interpretation of the results. The concerns were on the potential risk to nontarget plants particularly due to larval feeding ability on *Helianthus*, *Helenium*, and *Dahlia*, and more significantly, to the ability of *C. nr ithacae* to complete a life cycle on the native genus *Olearia* as well as the introduced *C. officinalis.* Both *Olearia* and *C. officinalis* are in different tribes of Asteraceae than the target parthenium weed. The ability of larvae to feed on a wider range of hosts than the female oviposited on was taken to suggest there was room for host shifts off the target weed in the future.

Our response to the criticisms included that weed biological control ought not to be prevented on the basis of the unknown potential for future host shifts. We reasoned

Table 3. Sequential choice oviposition tests with *C.* nr *ithacae* and various combinations of multiple nontarget hosts presented with then without parthenium weed.

No. females in test[a]	Days 1–2: choice test, parthenium present		Days 2/3 and 3/4: no parthenium	Days 3/4–7: choice test parthenium present		
	Eggs on:					
	Parthenium weed	Nontarget plant	Nontarget plant	Parthenium weed	Nontarget plant	Cage
1	21	0	0	39	0	
2	26	0	4 *X. o.*	19	0	145
3	177	0	18 *X. o.*	74	1 *X. o.*	
1	5	0	0	died	-	-
2	15	35 *X. o.*				
		1 *H. a.*	died	-	-	-
1	28	0	0	17	0	
1	9	1 *H. b.*	0	21	0	
1	73	0	0	49	0	
2	84	0	0	27	0	3
1	52	0	0	died	-	-
1	10	0	0	18	0	1
1	39	0	0	8	0	
1	45	0	0	36	0	
1	58	0	0	123	0	

[a]Number of female *Carmenta* nr *ithacae* tested against various nontarget plants: *Helenium bigelovii* (15), *Calendula officinalis* (8), *Xanthium occidentale* (7), *Cosmos bipinnatus* (6), *Dahlia* sp. (6), *Gamochaeta pensylvanica* (4), *Spilanthes grandifola* (4), *Gaillardia grandifolia* (4), *Helianthus annuus* (3), *Ambrosia artemisiifolia* (3).

that we can only judge a biological control agent on its current characteristics, which in this case included sufficiently specific oviposition behavior, with frequent ovipositions only on the target parthenium weed and a closely related exotic weed, *X. occidentale.* We also interpreted oviposition on cage walls as indicating that plastic or metal was a more acceptable oviposition site than any other of the nontarget plants presented, and that such behavior would put nontarget plants in the field at minimal risk.

Following this discourse in March 1998, it was agreed between the two parties involved that additional no-choice oviposition tests with *C.* nr *ithacae* adults on additional plants in the tribe Astereae would take place, using native species whereever possible. The results of these tests were sent to AQIS in May 1998 and forwarded to the state department that had requested the extra testing. The results (Table 4) confirmed that mated female *C. nr ithacae* will not oviposit on plants in the tribe Astereae of the family Asteraceae, even when confined from emergence to death on the plants without access to, or experience of, their normal host plant. The number of eggs laid on the cage walls and floor by females with only nontarget plants present (mean of 20 eggs laid per female) was a third of the normal oviposition on parthenium weed (mean of 61 eggs laid per female). Females with access to parthenium weed did not lay eggs on the cage walls and floor. We interpreted the willingness of females to lay on the cage walls or floor when confined without their host plant not as evidence of lack of discrimination, but rather of the preference ranking of the nontarget plants falling below even that of cage netting. We suggested that this behavior in the field would result in females continuing to search for parthenium weed until they die.

In this case, following further consultation with the department that originally requested the additional testing, permission to release this agent was given by AQIS in May 1998. The discourse required to reach this agreement, reinforces our belief that a risk assessment framework that is accepted within biological control will aid the dissemination and logical interpretation of results that attempt to predict consequences in the field using cage assays in the laboratory.

Should the negotiation between the parties have been unsuccessful in the case of *C.* nr *ithacae*, a more formal risk assessment may have been required under the Biological Control Act 1984. It is likely that more experiments to increase sample sizes would have been required, as well as additional information such as quantifying the impact of larval feeding on flowering and seed set of *Olearia* and *C. officinalis*. The services of a risk assessment modeler also would have been requested.

Below, we present a brief example of how a more open application of generic risk assessment coupled with an increased understanding of this approach could have made this application and its approval even more straightforward. A similar method has been applied to another weed biological control agent (Wan and Harris 1997). We fully acknowledge the limitations of estimating probabilities given the small sample sizes used in our tests, and we make the assumption that no adaptation or genetic change occurs in the agent's oviposition preference or larval host range over time.

Deciding on the risk of attack by C. *nr* ithacae *on* P. hysterophorus *(the target weed)*

- Probability of co-existing with *P. hysterophorus* in the field (worst case scenario) is 1.0
- Probability of ovipositing on the plant (all females oviposited in laboratory assays) is 1.0
- Probability of egg to larval survival to inflict damage (larvae in 42/ 44 plants) is 0.95

Risk = probability of establishing * probability of damage = (1.0 *1.0)*0.95 = 0.95
Consequences of damage include significant reduction in plant height, and significant reduction in number of mature capitula and viable seeds produced (based on observations that impact on *P. hysterophorus* plants is similar to that when attacked by stem-galling moth larvae of *Epiblema strenuana* Walker, (Tortricidae) although *E. strenuana* occupies a feeding niche further up the stem; Navie et al. 1998).

Table 4. Lifetime oviposition of mated *C.* nr *ithacae* females placed in cages with either parthenium weed or a nontarget host (no choice test, N = 4 females per test).

	Eggs		
Plant species	On plant	On cage walls/floor	Total
Brachycome multifida (native)	0	82	82
parthenium control	322	0	322
Olearia nernstii (native)	0	187	187
parthenium control	167	0	167
Olearia microphylla (native)	0	30	30
parthenium control	180	0	180
Aster novii-belgii (introduced)	0	19	19
parthenium control	322	0	322
Conyza canadensis (introduced)	3	85	88
parthenium control	386	0	386
Conyza bonariensis (introduced)	0	71	71
parthenium control	134	0	134
Calotis cuneata (native)	0	150	150
parthenium control	191	0	191
Vittadinia dissecta (native)	0	26	26
parthenium control	335	0	335

Deciding on the risk of attack by C. *nr* ithacae *on* X. occidentale *(Noogoora burr)*

- Probability of co-existing with *X. occidentale* in the field (take 1.0 as a worst case scenario) is 1.0
- Probability of ovipositing on the plant (5 females out of 9 oviposited in cage assays) is 0.6
- Probability of egg to larval survival to inflict damage (3 small larvae/ 30 eggs) is 0.1

Risk = probability of establishing * probability of damage = (1.0 *0.6)*0.1 = 0.06
Consequences of damage to *X. occidentale* remain unidentified at this stage, but any reduction in viable seed produced would be viewed as a positive consequence due to the exotic weed status of this plant.

Deciding on the risk of attack by C. *nr* ithacae *on* C. officinalis *(ornamental calendula)*

- Probability of co-existing with *C. officinalis* in the field (base on inferred probability of a likely frequent event with unanimous community concern, Beer and Ziolkowski 1995) is 0.5
- Probability of ovipositing on the plant (0 females out of 13 oviposited in cage assays) is 0
- Probability of egg to larval survival to inflict damage (28 larvae/ 180 eggs) is 0.16

Risk = probability of establishing * probability of damage = (0.5 *0)*0.16 = 0.0
Consequences of damage to *C. officinalis* remain unquantified, although a significant impact on seed set and plant survival is expected due to the small size of this plant relative to parthenium weed. However, the risk analysis indicates the probability of attack is zero, therefore the negative consequences are unlikely to ever arise in the field in Australia.

Treating the Risk of *C.* nr *ithacae* Attacking Nontarget Plants in Australia

The approval to release *C.* nr *ithacae* in Australia (releases began in 1998) was fully conditional on the party responsible monitoring the effect of the biological control agent on species of native flora growing in the vicinity of release sites, and providing reports to EA on the effectiveness of the biological control agent in controlling parthenium weed. Post-release monitoring is the current way of treating the risks of weed biological control. Pre- and post-release data will be compared in order to refine the decision-making processes for future releases. We believe that the current Australian system is assessing nontarget risks of biological control pre-release in a way that controls the process without preventing the useful application of the science of biological control.

CONCLUSIONS

The Australian legislation governing the approval of agents for biological control encourages a step-wise approach to gathering and disseminating data important to making educated and consultative decisions on nontarget impacts. Although guidelines are provided to assist in the preparation of the requests for import and release permits, the methods required to demonstrate safety of potential biological control agents remain at the discretion of the researcher. This gives researchers an excellent opportunity to incorporate behaviorally and physiologically based tests to reveal the degree of specificity of the proposed agent. In this way, researchers can begin to approach quantified risk assessments through the generation of data that lend themselves to analysis, not only of the probabilities of establishment occurring on nontargets, but of the consequences of any damage. If the quality of information supplied to the governing bodies in such applications remains high, formal risk assessments may rarely be required. Finally, through post-release monitoring, public perceptions and treatments of any risks of biological control can be improved.

ACKNOWLEDGMENTS

We are grateful to F. Ambatali, M. Rogers and M. Trevino for providing essential technical assistance to the *C. ithacae* project, to A. McClay for collecting and rearing the initial colony, to the staff at QDNR for helpful discussions, and to our anonymous referees for helpful comments on this manuscript.

REFERENCES

Beer, T., and F. Ziolkowski. 1995. Environmental risk assessment: an Australian perspective, Supervising Scientist Report 102, Commonwealth of Australia, Darwin.

Commonwealth of Australia. 1982. Wildlife Protection (Regulation of Exports and Imports) act 1982. Commonwealth Government Printer, Canberra.

Commonwealth of Australia. 1984. Biological Control Act 1984. Commonwealth Government Printer, Canberra.

Commonwealth of Australia. 1985. Quarantine Act 1908. Commonwealth Government Printer, Canberra.

Cullen, J. M., and E. S. Delfosse. 1984. *Echium plantagineum*: catalyst for conflict and change in Australia, 249–292. *In* E. S. Delfosse (ed.), Proc. VI Intl. Sym. on Biological Control of Weeds. Agriculture Canada, Vancouver.

Dhileepan, K., B. Madigan, M. Vitelli, R. E. McFadyen, K. Webster, and M. Trevino. 1996. A new initiative in the biological control of Parthenium. *In* R. C. H. Shepherd (ed.), Proc. 11th Australian Weeds Conf.. Weed Sci. Soc. Victoria, Melbourne. p. 309–213.

Eichlin, T. D., and W. D. Duckworth. 1988. Sesoidea: Sesiidae, *In* R. B. Dominick et. al. (eds.), The moths of America north of Mexico including Greenland. The Wedge Entomological Research Foundation, Washington.

Forno, I. W., and T. Heard. 1997. Compiling a plant list for testing the host range of agents. *In* M. Julien, and G. White (eds.), Biological control of weeds; theory and practical application. ACIAR Monograph Series, GPO Box 1571, Canberra, ACT 2601, Australia. p. 71–75.

Heard, T. A., and R. D. van Klinken. 1998. An analysis of test designs for host range determination of insects for biological control of weeds. *In* M. P. Zalucki, R. A. I. Drew, and G. G. White (eds.), Proceedings of the Sixth Australasian Applied Entomological Research Conference. Univ. Queensland, Brisbane. 1: 539–546.

Herz, W. 1977. Sesquiterpene lactones in the Compositae. *In* V. H. Heywood, J. B. Harborne, and B. L. Turner (eds.), The biology and chemistry of the Compositae. Academic Press, London. p. 337–357.

Marohasy, J. 1998. The design and interpretation of host specificity tests for weed biological control with particular reference to insect behavior. Biocontr. News Info. 19: 13–20.

McEvoy. P. B. 1996. Host specificity and biological pest control. BioSci. 46: 401–405.

McFadyen, R. E. 1992. Biological control against parthenium weed in Australia. Crop Prot. 11: 400–407.

McFadyen, R. E. 1998. Biological control of weeds. Annu. Rev. Entomol. 43: 369–393.

McFadyen, R. E., and T. A. Heard. 1997. Decision making based on host range tests. *In* M. Julien, and G. White (eds.), Biological control of weeds; theory and practical application. ACIAR Monograph Series, Canberra, Australia. p. 83–88.

McFadyen, R. E., and J. Marohasy. 1990. A leaf feeding moth, *Euclasta whalleyi* (Lep.: Pyralidae) for the biological control of *Cryptostegia grandiflora* (Asclepiadaceae) in Queensland, Australia. Entomoph. 35: 431–435.

McFadyen, R. E., and K. Weggler-Beaton. 1999. The biology and host specificity of *Liothrips* sp., rejected as a potential biological control agent for annual ragweed. BioControl. in press.

Navie, S. C., R. E. McFadyen, F. D. Panetta, and S. W. Adkins. 1996. The biology of Australian weeds. 27. *Parthenium hysterophorus* L. Plant Prot. Quart. 11: 76–88.

Navie, S. C., T. E. Priest, R. E. McFadyen, and S. W. Adkins. 1998. Efficacy of the stem–galling moth *Epiblema strenuana* Walk. as a biological control agent for ragweed parthenium (*Parthenium hysterophorus* L.). Biol. Control 13: 1–8.

Paton, R. 1992. Legislation and its administration in the approval of agents for biological control in Australia. *In* E. S. Delfosse, and R. R. Scott (eds.), Proc. Eighth Intl. Sym. on Biological Control of Weeds.

DSIR/CSIRO. p. 653–658.
Simberloff, D., and P. Stiling. 1996. How risky is biological control ? Ecology 77: 1965–1974.
University of Queensland. 1998. Reviewing applications to import and/or release biological control agents into Australia. Brisbane.
Wan, F.–H., and P. Harris. 1997. Use of risk analysis for screening weed biological control agents: *Altica carduorum* Guer. (Coleoptera: Chrysomelidae) from China as a biological control agent of *Cirsium arvense* (L.) Scop. in North America. Bio. Sci. Tech. 7: 299–308.
Wapshere, A. J. 1974. A strategy for evaluating the safety of organisms for biological weed control. Ann. Appl. Biol. 77: 201–211.
Withers, T. M. 1997. Changes in plant attack over time in no–choice tests: an indicator of specificity. N.Z. Plant Prot. Soc. Bull. 50: 214–217.
Withers, T. M. 1998. Influence of plant species on host acceptance behavior of the biological control agent *Zygogramma bicolorata* (Coleoptera: Chrysomelidae). Biol. Control 13: 55–62.
Withers, T. M., and L. Barton Browne. 1998. Possible causes of apparently indiscriminate oviposition in host specificity tests using phytophagous insects. *In* M. P. Zalucki, R. A. I. Drew, and G. G. White (eds.), Proc. Sixth Australasian Applied Entomological Res. Conf.. Univ. Queensland, Brisbane. 1: 539–546.

13

NEGATIVE ECOLOGICAL EFFECTS OF THE MUSK THISTLE BIOLOGICAL CONTROL AGENT, *RHINOCYLLUS CONICUS*

Svata M. Louda
School of Biological Sciences,
University of Nebraska, Lincoln, Nebraska

INTRODUCTION

Concern over the economic and environmental effects of invasive species is widespread (see McKnight 1993, OTA 1993, 1995). Clearly some plant species, when released from limiting factors in their indigenous environments and spread by their interaction with humans, expand to form large, dense, undesirable stands, e.g., *Opuntia* spp. cacti in Australia (Dodd 1940), as well as *Carduus* spp. thistles (Dunn 1976) and *Rosa multiflora* Thunb. (Amrine and Stasny 1993) in the middle and eastern USA. Some of these weedy species are considered threats to the integrity of protected plant communities in preserves. Examples include *Centauria solstitialis* (yellow star thistle) in California grassland, *Tamarix ramosissima* (tamarisk) in southwestern deserts, *Vinca major* (periwinkle) in Arizona, and *Casuarina equisetifolia* (Australian pine) in Florida coastal habitats (Randall 1993).

Biological control agents have been released, or are being studied for release, for the control of many weeds. Classical biological control of weeds involves the establishment of exotic natural enemies to limit the exotic weedy plant. The strategy involves finding, testing, and introducing relatively host-specific, coevolved phytophagous insects or pathogens (DeBach and Rosen 1991). In a simple case, this means reestablishing a critical limiting interaction from the indigenous region. This appears to have been the case in the biological control of *Opuntia* spp. cacti in Australia (Dodd 1940) and in the Channel Islands (Goeden 1968), although it did not work as well in South Africa (Pettey 1950). In more complicated cases, it has meant releasing secondary natural enemies from the indigenous area without their predators, parasitoids and competitors in hopes that they will limit the pest plant. This appears to have been the case for Musk thistle, since the flowerhead weevil, *Rhinocyllus_conicus* Froelich, is out-competed by *Larinus* spp. in flowerheads of *Carduus nutans* L. in its indigenous region of central Europe (Zwoelfer 1978). Underlying all cases is the assumption that we can gather enough information prior to the deliberate importation and release of an exotic species to prevent undesirable side effects over time. The recent evidence of *R. conicus* impact on native thistles and their insects illustrates how difficult it is to do this, suggesting increased caution in future introductions of nonnative species.

Native thistles have a long history of both economic use and environmental importance. Thistle honey is a lucrative commodity, and at least ten native thistles have medicinal value (Moerman 1986). Trophic interconnections are extensive, and many are intimate. For example, thistle seed and down are used for food and nest building by many small mammals and birds, including the American goldfinch (*Carduelis tristis*). Also, thistle nectar and pollen are important to a wide array of native insects, including several named after their host thistle, such as the swamp metalmark (*Calephelis muticum* McAlpine) named after the northeastern swamp thistle, *Cirsium muticum* Michaux (Louda et al. 1998). Such interconnections, as well as concerns for the maintenance of natural communities and their biodiversity, contribute both economic and environmental incentives for maintaining native thistle populations.

In this chapter, I summarize key findings from our ongoing research on several native thistles, review the evidence on the direct and indirect numerical consequences to date of host range expansion of *Rhinocyllus conicus* onto native thistle species, and suggest seven lessons that emerge from this case history with hindsight. Our long term research on thistle-insect interactions and their dynamics in native prairie serendipitously provides the background data to quantify the influence of the host range expansion by this biological control agent on the reproduction, fitness, and population dynamics of several native plants and their dependent native insect fauna. Our data and the economic and environmental considerations led me to investigate the information that was used to justify the release of *R. conicus*. This investigation provides the basis for a list of lessons that I believe can be learned from this case history. This list includes the need for (1) better documentation of the weed problem, (2) improved incorporation of ecological criteria into pre-release testing, (3) increased investigation of plausible ecological interconnections, (4) quantitative evaluation of the efficacy of the proposed agent, (5) more research on alternative control methods, (6) enhanced public review prior to release, and (7) an overall revision in our thinking on which situations qualify for the use of biological control releases of exotic organisms.

NATURAL HISTORY

Plants and Study Sites

Wavyleaf and Platte thistles co-occur in sand prairie in the upper Great Plains. Wavyleaf thistle (*Cirsium undulatum* (Nutt.) Spreng.; Fig. 1A) has a broad edaphic range and geographic distribution within the plains, from Texas to southern Canada. Platte thistle (*C. canescens* Nutt.; Fig. 1B), however, is more restricted. It is characteristic of prairie grasslands on sand and gravel soils in the upper Missouri River drainage, specifically in central and western Nebraska, southeast Wyoming and northeast Colorado (Great Plains Flora Assoc. 1986). The center of Platte thistle's quantitative distribution is in the Sandhills of Nebraska (McCarty et al. 1967, Great Plains Flora Assoc. 1977, 1986), the largest continental dune system in the Western Hemisphere (Bleed and Flowerday 1989). The local distribution of each of these native thistles is patchy (Great Plains Flora Assoc. 1986, Pavlovic et al. 1992), and neither is considered a serious weed (McCarty et al. 1967, McCarty and Scifres 1967).

Our research on wavyleaf and Platte thistles summarized here was performed between 1984 and 1997 at two preserves owned by The Nature Conservancy, 330 km apart in the Sandhills formation of Nebraska: 1984–1997 at Arapaho Prairie (Arthur Co.), and 1990–1997 at the Niobrara Valley Preserve (Brown Co.). This work builds on that by W.O. Lamp (1976–1979) at Arapaho Prairie. Both preserves contain characteristic sandhills prairie vegetation (Keeler et al. 1980, Kaul 1989), but represent different dune formations, geographic positions, and climatic regimes within the Sandhills (Bleed and Flowerday 1989). Arapaho has not been grazed since 1978, whereas our sites at the Niobrara were moderately grazed by cattle.

The federally listed, threatened Pitcher's thistle (*Cirsium pitcheri* (Torr.) Torrey and Gray); Fig. 1C), a close relative of Platte thistle, occurs in the sand dune grasslands around the Great Lakes along the Canadian-USA border (Pavlovic et al. 1992). Not only are Platte and Pitcher's thistles considered sister species (Johnson and Iltis 1963, Loveless and Hamrick 1988), the rare Pitcher's thistle is ecologically similar to Platte thistle (Louda 1994). Striking parallels include key habitat requirements, early season phenology, monocarpic life history strategy, and genetic similarity (Keddy and Keddy 1984, Loveless 1984, Loveless and Hamrick 1988, McEachern 1992, Pavlovic et al. 1992, Louda 1994). For example, most flowerheads of both species are initiated in May and June, making them vulnerable during the feeding and oviposition period of *Rhinocyllus conicus*. The close, derivative relationship of Pitcher's thistle to Platte thistle, the experimental data showing the importance of insect herbivory for Platte thistle fitness and dynamics (Louda and Potvin 1995), and the reports of herbivory on Pitcher's thistle by native insects (Keddy and Keddy 1984, Loveless 1984) suggest *R. conicus* eventually may also impact this protected species (Louda et al. 1997). Our data for Pitcher's thistle were collected 1993–1996 at Sleeping Bear Dunes and Indiana Dunes National Lakeshores on Lake Michigan (see Pavlovic et al. 1992).

All three of these native thistles are tap-rooted, short-lived perennials that grow as rosettes for some years before they bolt, flower, and set seed. Wavyleaf thistle is iterocarpic in our region, flowering several times before dying (McCarty et al. 1967). Flowerheads are initiated in Nebraska in June. Flowering peaks in July, and it is com-

Figure 1. Native thistles studied (A, B, C) and *Rhinocyllus conicus* (D): A, wavyleaf thistle (*Cirsium undulatum*); B, Platte thistle (*C. canescens*); C, Pitcher's thistle (*C. pitcheri*); D, *R. conicus* adult and egg cases (near base) on a Platte thistle flowerhead (photographs by author).

pleted in August. Platte thistle, the earliest flowering native *Cirsium* spp. in the Sandhills, is monocarpic; it dies after flowering once (Lamp and McCarty 1981). Juvenile rosettes of Platte thistle grow for 1–5 years, then bolt, flower, set seed, and die (Louda and Potvin 1995). Flowerhead development and stem growth in Nebraska start early in May, peaks in mid June, and is completed by the end of June. Pitcher's thistle is also monocarpic. Flowerhead development and stem growth in Indiana and Michigan start in early May. Flowering peaks in mid July and it is usually completed by the end of August. In the field, juvenile rosettes of Pitcher's thistle grow for 3–11 years before maturing, setting seed, and dying (Loveless 1984, McEachern 1992, Pavlovic et al. 1992).

Native Insects

A characteristic set of insects feeds on the inflorescences of thistles (Zwolfer 1965, 1988; Redfern 1983). At least five native insects commonly feed on or within the developing inflorescences of the two native thistles at our sites. On wavyleaf and Platte thistles in sandhills prairie, these include: (1) two picture-winged flies (*Paracantha culta* Wiedeman, *Orellia occidentale* [Snow]: Tephritidae); (2) two pyralid moths, *Pyrausta subsequalis plagialis* Haim. (Subfamily Pyraustinae) and *Homeosoma* [*stypticellum*] *eremophasma* Goodson & Neunzig 1993 (Subfamily Phycitinae); plus (3) adults of a native weevil, *Baris subsimilis* Casey (Curculionidae), whose larvae feed internally on stems and roots (Lamp 1980, Louda and Potvin 1995, S.M. Louda, unpub. data). The most common insects feeding on the inflorescences of Pitcher's thistle at Sleeping Bear Dunes National Lakeshore in Empire, Michigan include (1) the same curculionid weevil (*B. subsimilis*) found on Nebraska thistles (C.W. O'Brien, personal communication), and (2) two moths, the artichoke plume moth (*Platyptilia carduidactyla*: Pterophoridae) and a new species of *Homeosoma* (Pyralidae; G. Balogh, personal communication). There are no published records of these native insects developing on flowerheads of the exotic *Carduus* spp. in Nebraska (Lamp 1980, Lamp and McCarty 1982a, b,c).

On Platte thistle, two insects damage the very small inflorescences early in the season. Adults of the native weevil (*B. subsimilis*) feed at the base of small, unopened flowerheads, especially on the small buds on the lower half of the plant, before they oviposit into the leaf axils on the stem. Larvae of the first pyralid moth (*P. subsequalis plagialis*) occur abundantly in early and mid-season in some years; they feed on the smaller flowerheads along the middle of the flowering stem, as well as within the upper rootcrowns of juvenile rosettes later in the season (S. M. Louda, unpublished data). Three insects feed sequentially over the season within flowerheads as the inflorescences and seeds develop (Lamp and McCarty 1979, 1982a, c). Each oviposits on successively larger, more developed flowerheads. The tephritid *P. culta* is the earliest feeder; it prefers unopened heads that are 10–15 mm in diameter (Lamp and McCarty 1982a, c), heads that *R. conicus* also uses (S. M. Louda and A. E. Arnett, in preparation). This fly has a second generation each season in Nebraska that develops in flowerheads of later-flowering native thistles (Lamp 1980). The second, less common fly (*O. occidentalis*) occurs next, preferring larger flowerheads (18–22 mm) that are beginning to open (Lamp and McCarty 1979, 1982a, c). Finally, feeding by the second pyralid moth (*H. eremophasma*) occurs later in the season. Eggs are placed among the florets in large, opened heads (Lamp and McCarty 1979, 1982a, c). Late instar larvae of *H. eremophasma* are sometimes seen moving from one head, in which the seeds have been completely consumed, to another head. Feeding damage by this pyralid is most severe on medium to large flowerheads.

On wavyleaf thistle, which flowers 2–3 weeks after Platte thistle in our region, the native weevil (*B. subsimilis*) often causes abortion of the earliest flowerhead, the apical terminal (S. M. Louda, unpublished data). The most common tephritid on wavyleaf thistle is *O. occidentalis*, since it is more temporally synchronous with the larger, just opening flowerheads on wavyleaf than with those on the earlier flowering Platte thistle. Although *Orellia* can be important, the most damage to flowerheads of wavyleaf thistle is done by larvae of the late-occurring pyralid moth, *H. eremophasma*. In some years, many of the flowerheads that develop, most of which look undamaged externally, contain only seed fragments and frass, unambiguous evidence of feeding by this moth.

On Pitcher's thistle, feeding by the native weevil (*B. subsimilis*) damages the receptacles of a large number of flowerheads early (S.M. Louda and A.K. McEachern, unpub. data). Stem mining by the larvae continues after adults disappear, often causing head mortality and sometimes plant mortality in severe cases (Bevill 1998, Bevill et al. 1999, R.L. Bevill and S.M. Louda, unpub. data). As the season develops, both the pterophorid (*P. carduidactyla*) and the pyralid moth (*Homeosoma* sp. nov.) feed inside the flowerheads, causing a major reduction in seed set and maturation by the heads that flowered (Louda and McEachern 1995, and unpub. data). Both *B. subsimilis* and *P. carduidactyla* also feed within the upper rootcrown of the juvenile rosettes (Stanforth et al. 1997, Bevill 1998).

Introduced Biological Control Agent

The flowerhead weevil, *Rhinocyllus conicus* (Fig. 1D), was introduced from Europe into the United States in 1969 to limit seed production by invasive Eurasian *Carduus* spp. thistles, especially Musk thistle (*C. nutans* L. complex) (Kok and Surles 1975, Rees 1977, 1991, Surles and Kok 1978). It was introduced into Nebraska in 1972 and actively distributed within the state (McCarty and Lamp 1982). Redistribution of *R. conicus* in the United States continues (e.g. Boldt and Jackman 1993). Although *R. conicus* is common in inflorescences of thistles in Europe, it is out-competed by other weevils (Zwoelfer 1978). Variation occurs among individuals within *R. conicus*. These variants have been considered host plant biotypes (Goeden 1978, Goeden et al. 1984, Zwoelfer and Preiss 1983) or, more recently, geographic subspecies (Klein 1986, Klein and Seitz 1995, Zwoelfer and Romstock-Volkl 1991). In spring, overwintered adults of *R. conicus* congregate on early flowerheads, and eggs are deposited on the bracts (Rees 1982). After 6–8 days the larvae hatch, burrow through the bract into the receptacle, feed for 25–40 days, and then pupate in hardened cells within the base of the flowerhead. The pupal stage lasts 8–14 days. A newly emerged adult may remain within the pupal chamber for up to 2 weeks before dispersing to an overwintering site (Rees 1982).

No significant damage to North American *Cirsium* spp. by *R. conicus* was ex-

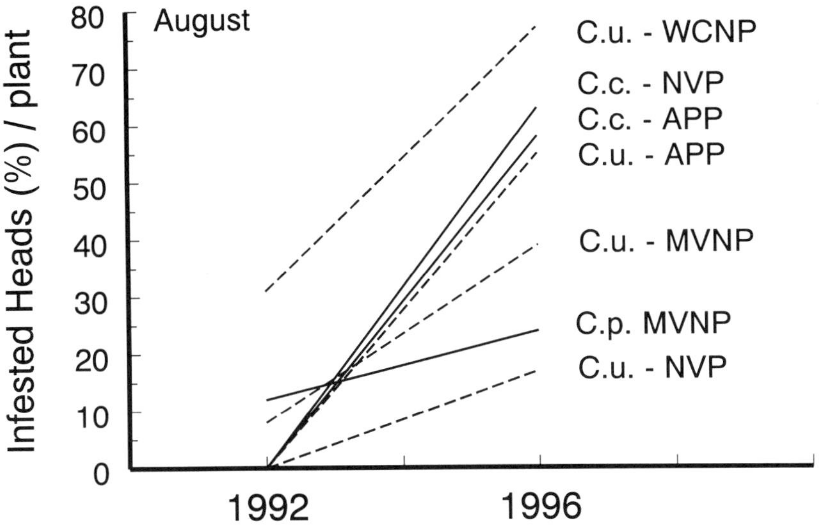

Figure 2. The proportion of flowerheads infested with *Rhinocyllus conicus* for native thistles in the upper Great Plains measured in both 1992 and 1996 (from Louda et al. 1997). The symbols represent species: C. u. = *Cirsium undulatum*, C. c. = *C. canescens*, C. p. = *C. pulchellum*; and sites: WCNP = Wind Cave National Park, NVP = Niobrara Valley Preserve, APP = Arapaho Prairie Preserve, and MVNP = Mesa Verde National Park.

pected, based on extensive pre- and early post-release studies of *R. conicus* feeding, oviposition, and larval development (e.g., Ward et al. 1974; Rees 1977, 1978, 1991; Surles and Kok 1977; Kok 1979; Zwoelfer and Harris 1984). Although *R. conicus* fed slightly on some *Cirsium* spp., stronger preference for *Carduus* spp., and slower larval development to smaller adult sizes on *Cirsium* spp., were expected to select against any significant use of native North American *Cirsium* spp (Zwoelfer and Harris 1984). Unfortunately, subsequent evidence of *R. conicus* feeding on native thistles in the USA is not consistent with this expectation.

After a lag of 20+ years, *R. conicus* now occurs in flowerheads of multiple native thistle preserves and national parks in the west and in the central plains and mountains, where population growth is occurring (Fig. 2; and Louda et al. 1997). This weevil has been reared from flowerheads of a third of the native *Cirsium* spp. in California (Pemberton et al. 1985; Goeden and Ricker 1986a, b, 1987a, b; Turner et al. 1987; Turner and Herr 1996; Palmisano and Fox 1997), from three of the six native thistle species in Rocky Mountain National Park, Colorado (*C. centaureae* (Rydb.) K. Schum., *C. scopulorum* (Greene) Cockll., and *C. undulatum*) (Louda, et al. 1997), from *C. undulatum* in Colorado, Montana, Nebraska and South Dakota (Rees 1977, 1991; Louda et al. 1997), and from *C. canescens* in Colorado, Nebraska and Wyoming (Louda et al. 1997, S. M. Louda and A. E. Arnett, unpublished data).

No *R. conicus* were reported on either native thistle (*C. undulatum*, *C. canescens*) in Sandhills prairie until 1993, in spite of longterm data collection from 1976 at Arapaho Prairie site (Lamp and McCarty 1979, 1981, 1982a, b, c; Lamp 1980; Louda et al. 1990, 1992; Louda and Potvin 1995) and from 1990 at Niobrara (Louda et al. 1997, Louda 1998b). At our sites *R. conicus* has increased dramatically on Platte thistle (Louda 1998b), reducing the number of viable seeds 5-fold by 1996 (Louda et al. 1997). Fortuitously, our ongoing studies of insect floral herbivory in the population dynamics of native thistles, especially Platte thistle (Louda et al. 1990, 1992; Louda and Potvin 1995), provide an unusual opportunity. We can compare plant reproductive success and population dynamics before versus after the host range expansion by this deliberately introduced biological control insect. Furthermore, the similarities between Platte thistle and its rare close relative, Pitcher's thistle (*C. pitcheri* [Torr.] Torrey & Gray), a federally-listed, threatened species in the dunes around the Great Lakes, provide the basis for inference that similar impacts could occur for Pitcher's thistle as documented for Platte thistle if the species come into contact.

SEED PRODUCTION AND POPULATION DYNAMICS OF PRAIRIE THISTLES

In our studies over the last 15 years, we have quantified herbivory by native insects on the native thistles in grasslands, evaluating these interactions for influence on individual plant performance and population dynamics under natural conditions. Both observations and controlled exclusion experiments have been employed. For example, using a stratified random sampling procedure, we have recorded the number of viable seeds produced per head and per plant in relation to frequency and intensity of insect herbivore feeding on the developing flowerheads for: Platte thistle (1984–1997), wavyleaf thistle (1990–1997), and Pitcher's thistle (1993–1996). Additionally, seed production and seedling establishment in insecticide and control experimental treatments have been completed for all three thistle species (see below). Our results show that inflorescence herbivory by native insects is an important factor in thistle seed production, and that seed limits population recruitment and density of all three of these native thistles in dune grasslands. Specific results relevant to the assessment of the nontarget effects by *Rhinocyllus conicus* can be summarized as follows.

Platte Thistle (*Cirsium canescens*)

Long-term observations show that floral and seed herbivory by native insects severely reduces the number of viable seeds matured (Lamp and McCarty 1979, 1981, 1982c; Louda et al. 1990, 1992). Experimental reduction of floral and seed herbivory with insecticide (Louda et al. 1990, 1992, Louda and Potvin 1995) verified the causal nature of the negative relationship between insect feeding and production of viable seed, and it demonstrated that the feeding by inflorescence insects has important demographic consequences. First, seed output increased when herbivory was decreased experimentally, indicating that insects rather than resources limit the total number of viable seeds produced. Second, more of the later flowerheads survived, flowered, and produced viable seed when protected from insect herbivores, suggesting that feeding by native insects restricts flowering period. Third, the density of seedlings around plants that had been treated with insecticide increased from 2- to 6-fold, showing that safe sites do not limit recruitment and local density. Finally, fourth, the increase in seedlings subsequently led to higher numbers of flowering adult progeny, demonstrating that native inflorescence-feeding insects limit both population density and lifetime maternal fitness of Platte thistle under natural conditions. In sum, this accumulated evidence shows that prior to 1993 native insects limited seed, seedling, juvenile, and adult numbers of this characteristic, relatively sparse native thistle in sand prairie of the upper Great Plains.

Wavyleaf Thistle (*Cirsium undulatum*)

Parallel studies of wavyleaf thistle, an iterocarpic native species that also occurs in sand prairie in our region, have also been done at our sites. These studies include quantification of seed production and herbivory, plant demography, and experimental reductions in insect floral herbivory. Initial analyses show that inflorescence-feeding insects significantly reduce seed production of *C. undulatum*. From 1990–1997, for example, 87.1% of the flowerheads on the sampled wavyleaf plants (N = 364) had unambiguous external evidence of insect feeding. A flowerhead with such evidence of insect feeding averaged 89.2% fewer viable seeds than did a flowerhead with little or no such evidence, 21.2 viable seeds (S.E. 2.26, N = 317) versus 197.0 viable seeds (S.E. 6.82, N = 47). Furthermore, the reduction in seed caused by inflorescence-feeding insects lowered seedling establishment in the native grassland habitat. In an exclusion experiment, we found that wavyleaf plants sprayed with insecticide in 1992 established 0.7 seedlings per plant (S.E. 0.5, N = 12) in 1993, compared of no seedlings established by control plants (S.M. Louda, T. Huettner Tesar and J.A. Burger, in preparation). Herbivore pressure by native insects is also important to understanding reproduction and density of wavyleaf thistle in sandhills prairie grasslands.

Pitcher's Thistle (*Cirsium pitcheri*)

Our studies of insect herbivory on the federally-listed Pitcher's thistle examined the hypothesis that a closely related, ecologically similar sister species could be used as a predictive experimental model for understanding factors limiting the population of the rare species (Louda 1994). Observational data, like those collected for Platte thistle, document significant levels of insect herbivory on Pitcher's thistle, both on juvenile plants (Stanforth et al. 1997, Bevill 1998, Bevill et al. 1999) and on inflorescences of adult plants (Keddy and Keddy 1984, Loveless 1984, Louda and McEachern 1995). For example, on juvenile plants at Sleeping Bear Dunes National Lakeshore in 1993 and 1994, insects damaged 10.9–36.0% of the leaves, as well as 6.3% of all rootcrowns and 29.9% of the rootcrowns of the large juveniles (Stanforth et al. 1997). Insecticide treatment and caging of juvenile rosettes reduced herbivory by 15–40%, and this reduction increased survival by 67%, growth by 25%, and subsequent seed reproduction by the survivors 2- to 4-fold ($p < 0.01$) wherever evidence of insect herbivory was significant (Bevill 1998, Bevill et al. 1999). Also, among the flowerheads sampled on adults in 1993 and 1994, 40–45% were unambiguously insect-damaged by the end of the season (Louda and McEachern 1995, and unpub. data). Experimental reduction of insect herbivory on inflorescences with insecticide reduced external evidence of insect damage by 10% and subsequently increased seedling recruitment by 10% (Louda and McEachern 1995, and unpub. data). In sum, although differences in plant-insect interactions exist between the two closely related thistle species, our results show that the basic patterns predicted using Platte thistle as a model held for Pitcher's thistle. Native insect herbivores also play a significant role in the determination of successful seed reproduction and seedling recruitment by the related, rare Pitcher's thistle.

HOST RANGE EXPANSION BY *RHINOCYLLUS*

Population Growth of *R. conicus*

The first unambiguous evidence of successful host range expansion by *R. conicus* onto both native thistle species was found in 1993, and it occurred simultaneously at both sites, over 300 km apart (Louda et al. 1997, Louda 1998b). Population growth of the weevil on the native thistles at these sites has been rapid and destructive, especially on Platte thistle. Three types of evidence support these conclusions.

First, no weevils were ever recorded from either native thistle prior to 1993 (Figs. 3, 4, 5A), in spite of ongoing annual data collection from 1976 at Arapaho (Lamp and McCarty 1979, 1981, 1982a, b, c; Lamp 1980; Louda et al. 1990; Louda and Potvin 1995; Louda 1998b, and unpub. data) and from 1990 at Niobrara Preserves (Louda et al. 1997; Louda 1998b, and unpub. data). In 1993, a few pupae and new adults were dissected out of the flowerheads on sampled plants (Fig. 3). The number of *R. conicus* developing within the flowerheads has increased consistently on both thistles at both sites since that time (Fig. 3), especially on Platte thistle (Fig. 3A).

Second, the number of adult weevils feeding and ovipositing on Platte thistle in late May increased significantly from the first adult observation recorded at each site, May 1994 at Arapaho and May 1995 at Niobrara (Fig. 4). By 1996, adult weevil densities on Platte thistle had increased significantly at both sites (ANOVA on square root-transformed counts per plant: Arapaho Prairie, $F_{6,122} = 2.21$, $P < 0.05$; and Niobrara Valley Preserve, $F_{6,134} = 3.49$, $P < 0.003$). This increase continued in 1997. The number of adult weevils observed on wavyleaf thistle peak later, between our regular sampling dates. However, the build-up in adult weevil activity on wavyleaf thistle at Arapaho is documented by the change in the proportion of flowerheads with *R. conicus* eggs. The number of heads with egg cases increased dramatically between 1993 and 1997 (Fig. 5A). The increase in *R. conicus* on native thistles has been slower at the Niobrara, in the north, than at Arapaho Prairie (Figs. 3, 4, 5A), in the southwest part of the Sandhills.

Third, evidence of successful weevil development within the flowerheads of sampled plants increased on both native thistles from 1993 to 1997 (Fig. 3), dramatically so on Platte thistle (Fig. 3A). The number of weevils developing in sampled flowerheads of both species tended to be lower at the northern sandhills site (Niobrara) than at the southwestern one (Arapaho) (Fig. 3). Increasing frequency of weevil occurrence and activity (Fig. 3, 4, 5A) suggest that the population buildup has continued. However, significantly fewer weevils were found within flowerheads of wavyleaf thistle than those of Platte thistle each year (ANOVA, $P < 0.001$; Fig. 3). Flowering

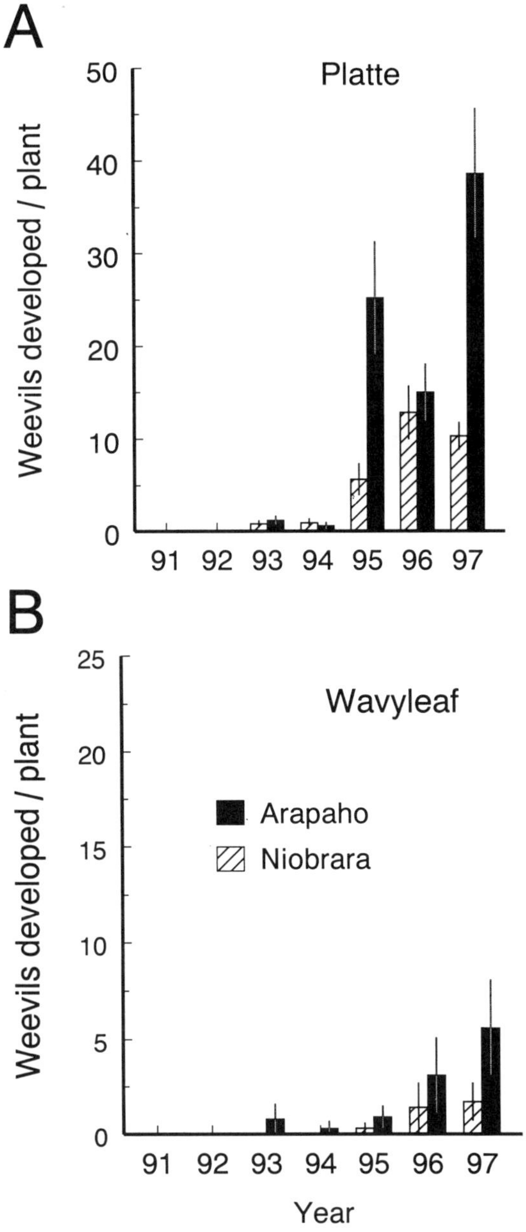

Figure 3. Average number (± S.E.) of *Rhinocyllus conicus* (late larvae, pupae, new adults or chambers) found within all of the flowerheads initiated and dissected for each plant sampled, at the end of the season, at Arapaho Prairie (none: 1984–1992) and Niobrara Valley (none: 1990- 1992): A, for Platte thistle, *Cirsium canescens*; and B, for wavyleaf thistle, *C. undulatum* (based on Louda 1998b).

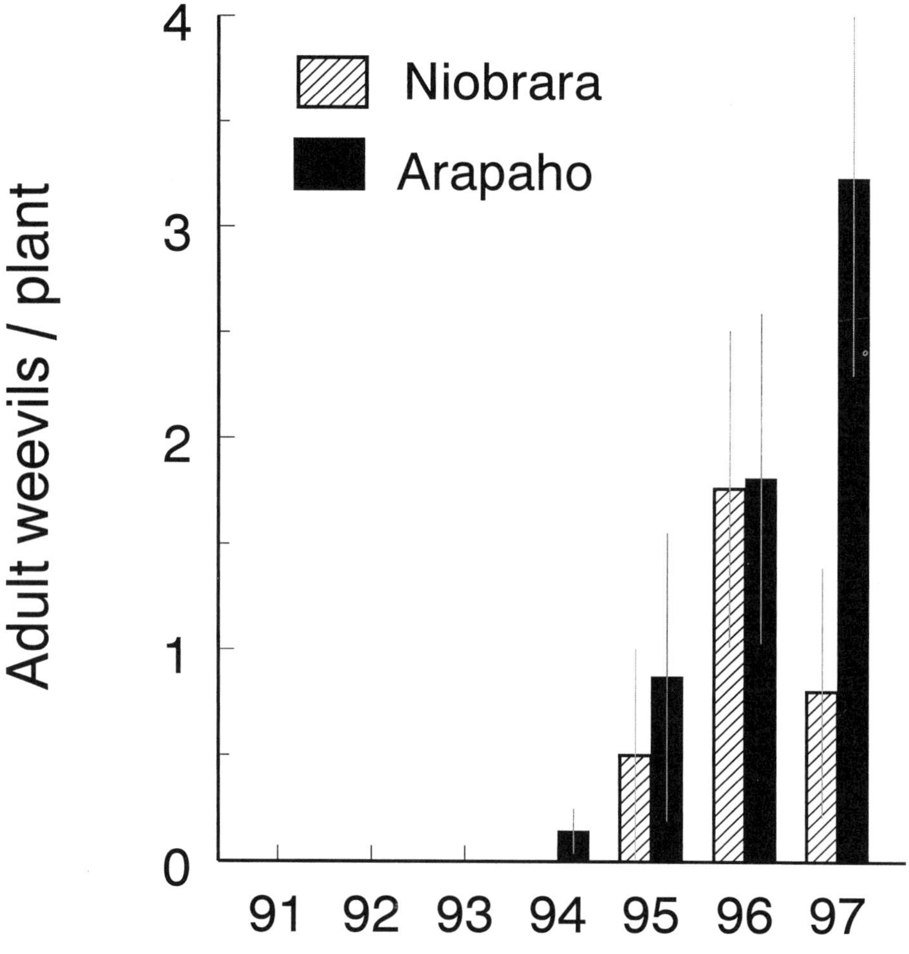

Figure 4. Average number (± S.E.) of *Rhinocyllus conicus* adults observed per Platte thistle (*Cirsium canescens*) plant in late May, 1990–1998 (zero observed 1990–1992), at both Niobrara Valley Preserve in the northcentral region and Arapaho Prairie Preserve in the southwestern region of the Sand Hills formation, Nebraska (based on Louda 1998b).

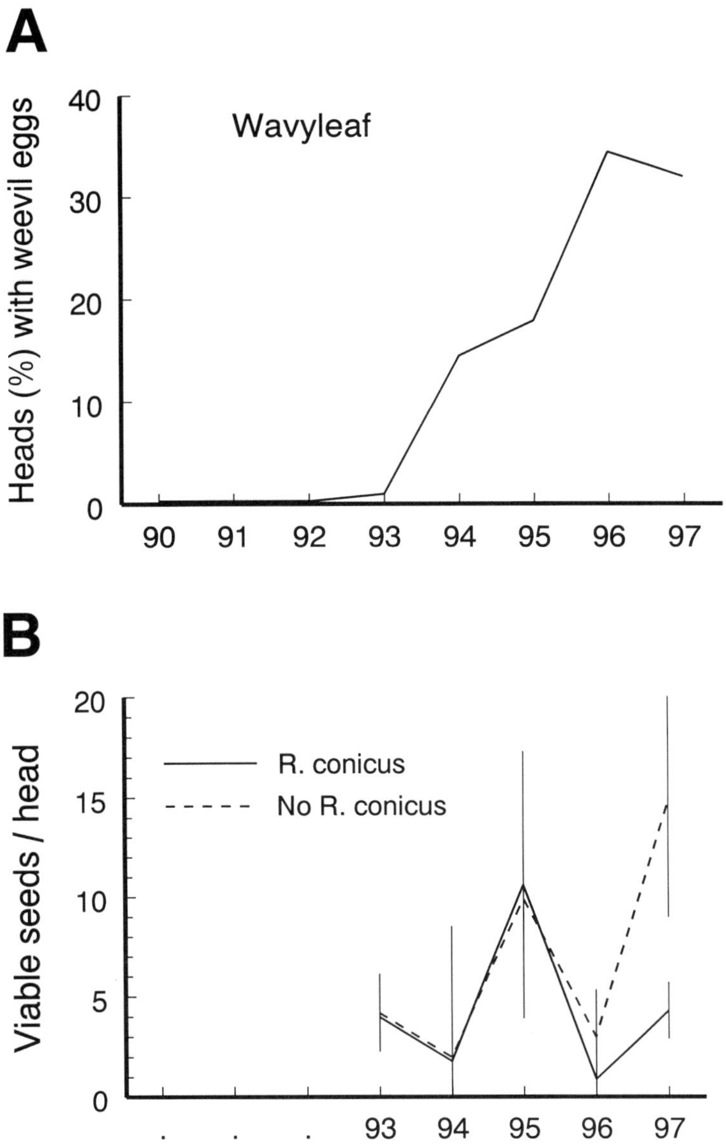

Figure 5. *Rhinocyllus conicus* oviposition on wavyleaf thistle (*Cirsium undulatum*) and average number (± S.E.) of viable seeds produced by wavyleaf at Arapaho Prairie Preserve in the Sandhills, Nebraska: A, the percent of all wavyleaf flowerheads on census plants with *R. conicus* eggs in late May (N = 78, 89, 93, 108, 74, 117, 93, and 97 heads, 1990–1997 respectively); and B, the average number (± S.E.) of viable seeds per head from wavyleaf census plants for heads without evidence of *R. conicus* versus heads with evidence of weevil oviposition from when *R. conicus* first observed (1993).

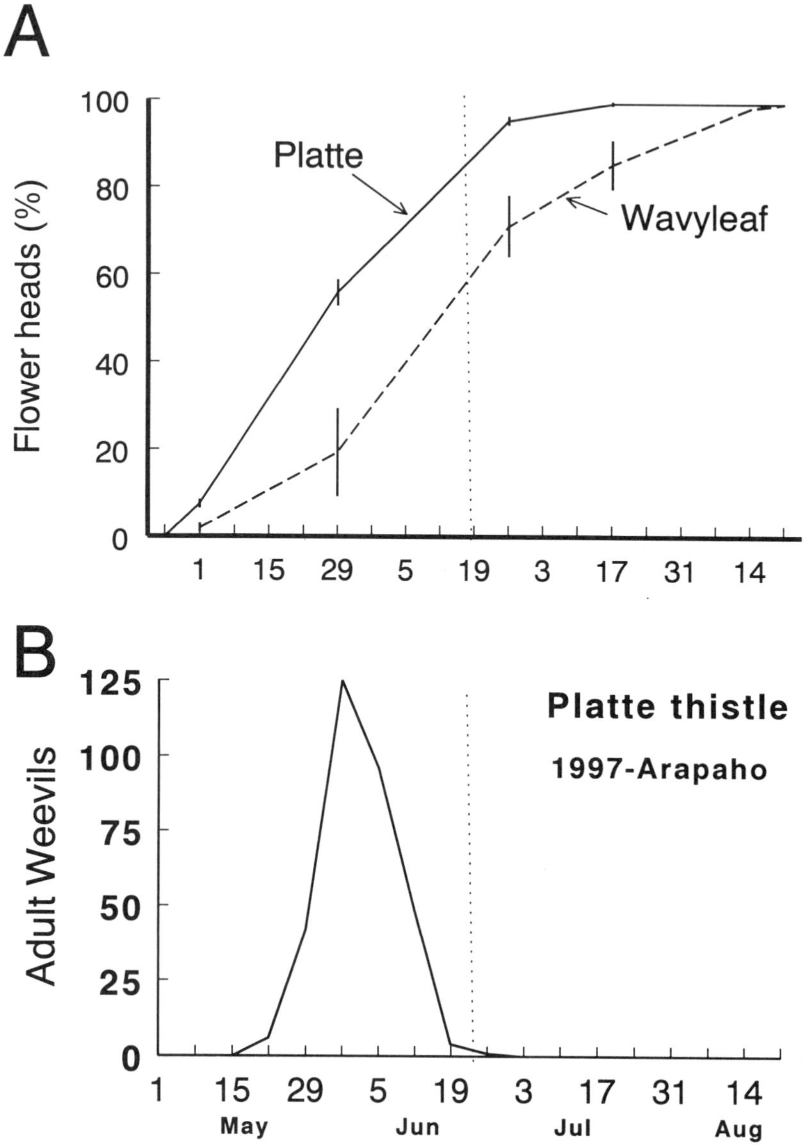

Figure 6. Phenology of Platte and wavyleaf thistles and *Rhinocyllus conicus* adult feeding and oviposition in sandhills prairie at Arapaho Prairie: A, percent of final number of heads initiated by each census data for Platte thistle (*Cirsium canescens*) and wavyleaf thistle (*C. undulatum*) (1996); and B, total number of adult *R. conicus* observed on Platte thistle census plants (1997).

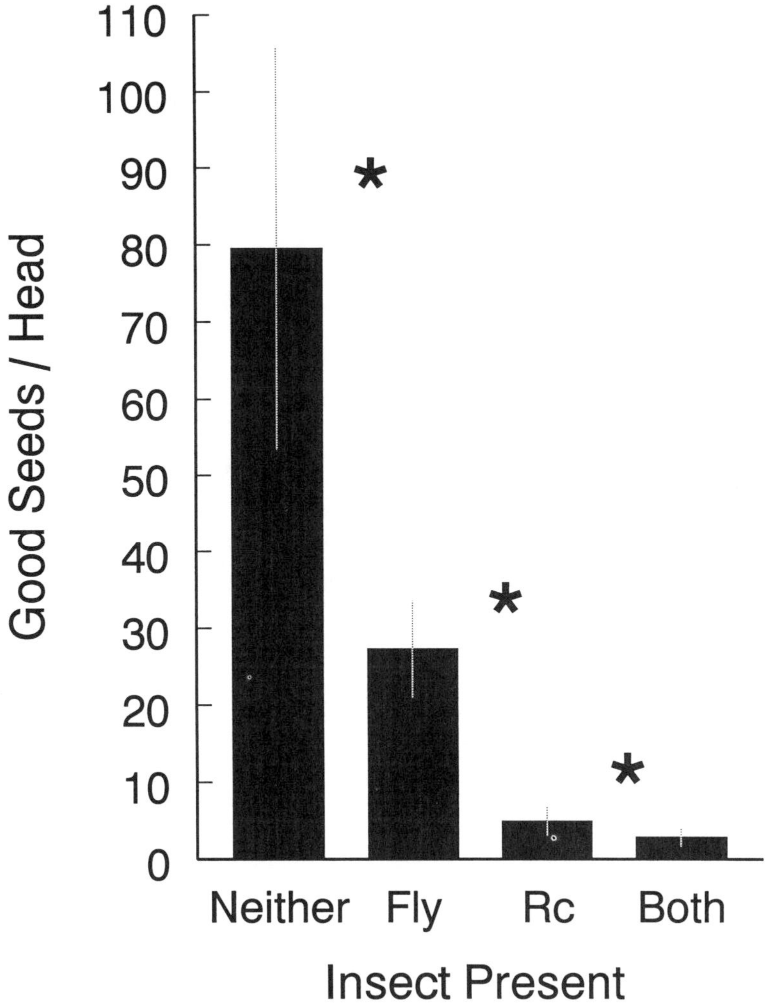

Figure 7. Average number of viable seeds (± S.E.) produced per head for Platte thistle (*Cirsium canescens*) in sandhills prairie for flowerheads with neither flies nor weevils, or native flies, or *Rhinocyllus conicus*, or both (ANOVA: $F_{3,208} = 30.89$, $P < 0.001$).

phenology may help to explain the difference (Louda 1998b). Platte thistle typically initiates flowerheads about 3 weeks earlier than does wavyleaf thistle (Fig. 6A). At Arapaho, adult *R. conicus* densities on Platte thistle peak in early June (Fig. 6B). At that time, the majority of Platte flowerheads (64% in 1997) were already initiated compared to only a third for wavyleaf (30% in 1997: Fig. 6A). By the end of the adult weevil oviposition period near the end of June, 86% of Platte heads were initiated versus 64% for wavyleaf (Fig. 6A). These data suggest that a higher proportion of Platte thistle than wavyleaf thistle flowerheads are available when *R. conicus* is ovipositing, making Platte the more vulnerable native species initially.

In summary, host range expansion by *R. conicus* succeeded at our study sites in 1993, over 20 years after initial release in Nebraska in 1972. In addition, there has been a steady increase in the numbers of weevils feeding, ovipositing, and developing on both native species since that time. The increase has been especially significant on the earlier flowering Platte thistle.

Impact of *R. conicus* on Seed Reproduction by Wavyleaf Thistle

The frequency of damage by *R. conicus* to flowerheads of wavyleaf thistle increased at all of the sites in central U.S.A for which we had data in both 1992 and 1996 (Fig. 1; Louda et al. 1997). Within Sandhills prairie, the evidence of weevil development on wavyleaf was higher at Arapaho than at Niobrara (Fig. 3B). By 1997, weevil eggs occurred on 33% (N = 135) of the heads on plants censused at Arapaho Prairie (Fig. 5A), but only 24% (N = 100) of those at Niobrara Valley Preserve. And by 1997, it was clear that the increase in *R. conicus* use of wavyleaf flowerheads at Arapaho (Fig. 6A) was associated with a significant decrease in seed production ($P < 0.05$; Fig. 5B). So, flowering phenology may delay the use of wavyleaf relative to Platte thistle (Louda 1998b), but not preclude a major negative effect of *R. conicus* on reproduction by wavyleaf thistle at Arapaho Prairie. The longterm effect of adding *R. conicus* to the inflorescence-feeding guild of wavyleaf thistle population dynamics is not yet clear. Successful seed set in wavyleaf is variable (5B), as is seedling establishment (S.M. Louda, unpub. data), increasing reliance of wavyleaf on persistence and iterocarpy. However, we do know that seedling establishment increased when the number of viable seeds was increased experimentally. So, it is likely that R. *conicus* has the potential of further limiting the density, and altering the population dynamics, of wavyleaf thistle in sand prairie.

Impact of *R. conicus* on Reproduction and Dynamics of Platte Thistle

The direct effect of the weevil on Platte thistle has been to reduce further the number of viable seeds released (Fig. 7). One of the best estimates of the weevil's influence on seed reproduction is the reduction in number of viable seeds in heads with the weevil compared to those without the weevil. In 1996, the average number of viable seeds produced by Platte thistle flowerheads infested with weevils averaged 4.9 per head (SE 1.85) versus 35.4 (SE 6.00) without weevils (t-test, $P < 0.01$), an 86% reduction. Early phenology and a short flowering season (Fig. 6A) preclude compensatory flowering by Platte thistle (Louda and Potvin 1995), contributing to this high numerical impact of *R. conicus* on Platte thistle.

These data can also be examined to evaluate the effect of the weevil relative to those of the native insects (Louda et al. 1998). In 1996, flowerheads that had no insects averaged 79.5 viable seeds (Fig. 7). Flowerheads infested only with native tephritid flies averaged 27.3 viable seeds (SE 6.20), while flowerheads only infested with the introduced weevil produced 4.9 (SE 1.85), a 5-fold further reduction in viable seed by

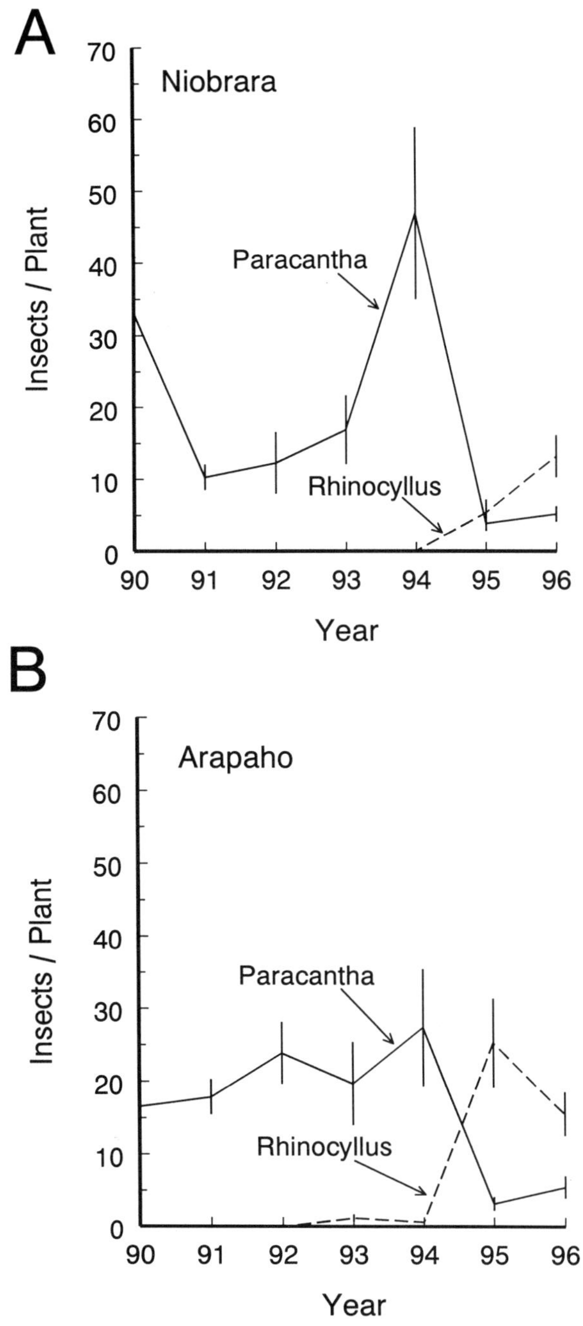

Figure 8. Number of *Rhinocyllus conicus* and the native tephritid fly, *Paracantha culta*, within flowerheads dissected for each Platte thistle (*Cirsium canescens*) census plant, 1990–1996 at: A, Niobrara Valley Preserve; and B, Arapaho Prairie Preserve.

the weevil (Fig. 7). And, when weevils and flies co-occurred in flowerheads, the number of viable seeds fell from 27.3 with flies only to 2.8 (SE 1.13) with weevils in addition to flies, a further 89.7% reduction! These data suggest that the weevil is causing a major additive reduction in viable seeds per attacked head (ANOVA: $F_{3,208} = 30.89$, $P < 0.001$). In addition, 36% more heads were attacked by weevils (N = 165) than by the native fly (N = 121), suggesting the weevil had a higher frequency of attack as well as a higher impact per attacked head. In sum, *R. conicus* is significantly reducing the number of viable seeds produced by Platte thistle in native prairie.

Since recruitment, density, and lifetime fitness of Platte thistle are proportional to the number of viable seeds (Louda and Potvin 1995), the added 5-fold or more reduction in seed caused by *R. conicus* leads to the prediction of a severe (potentially 5-fold) reduction in plant population density for this sparse native species. Small population sizes increase the chance of extinction. The striking, well-documented ecological parallels between Platte thistle and its rare congener, Pitcher's thistle, also lead to the inference that the same thing could happen to Pitcher's thistle, should *R. conicus* become established in the dune ecosystem around the Great Lakes. Early season initiation of flowering by Pitcher's thistle, like Platte thistle, would make a majority of its heads vulnerable to egg deposition and feeding by *R. conicus*. The weevil is not yet introduced into Michigan. Based on the evidence presented here, such introduction should be prevented if possible.

Indirect Effects of *R. conicus* on Platte Thistle-feeding Insects

Among the native inflorescence insects, picture-winged flies (Tephritidae) typically exploit flowerheads of the same stage and size as *R. conicus* (Mellini 1951, Harris and Zwoelfer 1968, Frick 1978, Lamp and McCarty 1982a, c, Zwoelfer and Preiss 1983, Zwoelfer and Harris 1984). Thus, the potential for indirect interactions exists. In fact, our observations are consistent with this hypothesis. From 1994 to 1996, as the average number of *R. conicus* feeding within Platte heads increased dramatically, from 0.1 to 3.1 per head ($P < 0.001$; Fig. 3A), the number of *Paracantha culta* decreased (Fig. 8). The negative correlation was greatest at Arapaho Prairie (Fig. 8B). Preliminary analysis of an experimental test of the relationship between weevil attack and fly numbers suggests the weevil interferes with fly oviposition and development (S.M. Louda and A.E. Arnett, unpub. data).

SUMMARY

The evidence to date shows five unexpected consequences. First, post-release ecological studies show that use of native thistles by *R. conicus* is geographically widespread (Goeden and Ricker 1986a, b, Turner et al. 1987, Turner and Herr 1996, Louda et al. 1997, Palmisano and Fox 1997). Second, use of native thistles by *R. conicus* in the central USA is increasing (Fig. 1; and Louda et al. 1997). Third, near exponential population growth of *R. conicus* on native thistles in Nebraska followed a 20+ yr. lag after introduction (Fig. 1; and Louda 1998b). Fourth, the weevil reduces seed production of both native species (Fig. 5B, 7), especially Platte thistle (Fig. 7) which flowers early and is more phenologically synchronized with the weevil's oviposition period

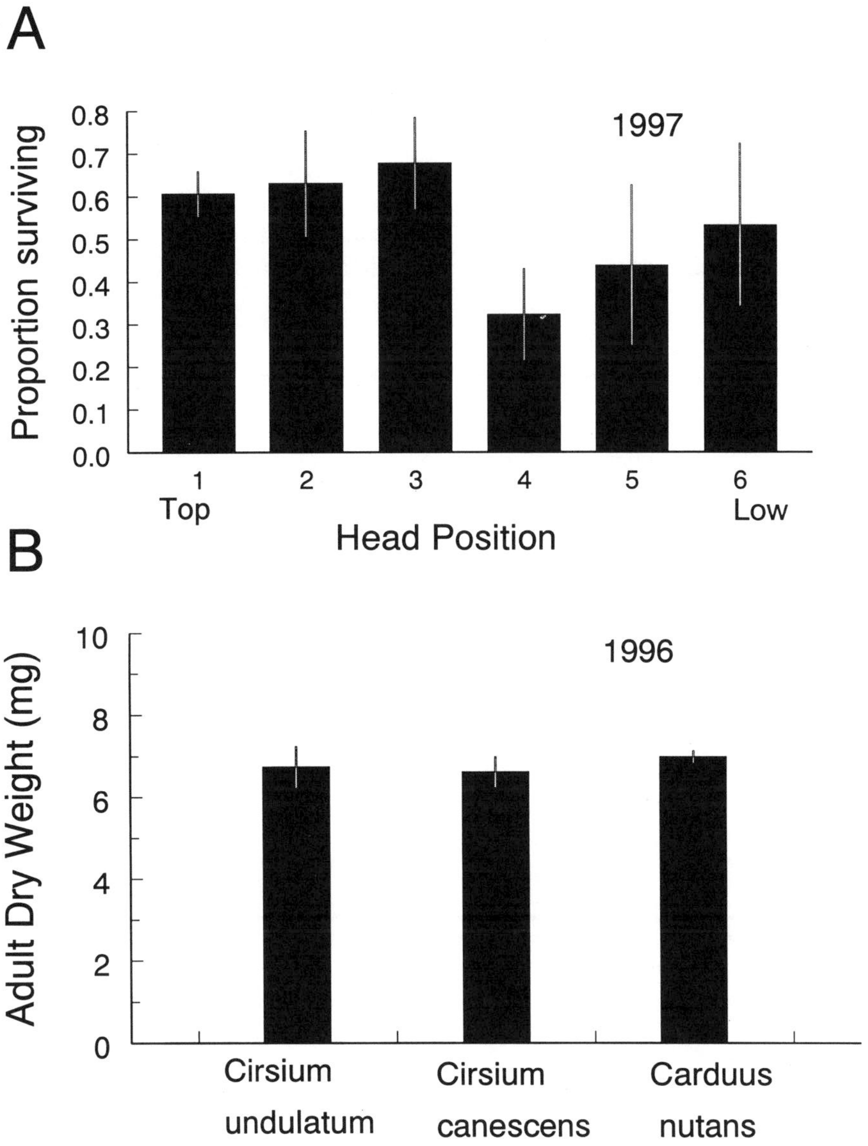

Figure 9. Performance of *Rhinocyllus conicus* on thistles in Nebraska: A, proportion of successful first instar larvae (mean = 54%), estimated as number of egg cases with a tiny penetration hole ("pinhole") under it, surviving to produce a late instar larva, or pupa or adult with chamber in flowerheads of Platte thistle (*Cirsium canescens*) census plants, from Arapaho Prairie Preserve (1997); and B, average adult dry weight of *R. conicus* emerging from heads of the two native species (wavyleaf thistle, *Cirsium undulatum*; Platte thistle, *C. canescens*), and from the exotic Musk thistle (*Carduus nutans* ssp. *leiophyllus* [Kartesz 1998]) in western Nebraska, 1996 (ANOVA, $P > 0.10$).

(Fig. 6). Since Platte thistle has previously been shown to be seed-limited (Louda and Potvin 1995), the population consequences of the further reduction of seed are predictable. And, fifth, the data so far suggest an indirect effect occurred as well, since a decrease in numbers of native picture-winged flies was correlated with the rapid increase in weevils especially at Arapaho Prairie (Fig. 8). Some nontarget feeding and oviposition by *R. conicus* on native *Cirsium* spp. may have been expected (Boldt 1997), but clearly the extent of that feeding and oviposition were not predictable based on the pre-release and early post-release studies (see Zwoelfer and Harris 1984). Thus, the character and the magnitude of the ecological effects now quantified represent an unexpected, undesired outcome of a deliberate release of an exotic species for biological control.

Observed vs. Expected Performance of *Rhinocyllus* on Native Thistles

The expectation, based on pre- and early post-release studies, was that *R. conicus* would not have any quantitative effects on nontarget North American species. For example, although Zwoelfer and Harris (1984: 59) noted there was "considerable opposition to the introduction to North America of biological control agents for thistles on the grounds that they may also damage native *Cirsium* spp.," they also argued that four factors would prevent any major impact. These factors were: 1) adult survival and feeding were significantly greater on *Carduus* spp. than *Cirsium* spp. (p. 53), including one North American species tested (wavyleaf thistle, *C. undulatum*); 2) *R. conicus* had a strong oviposition preference for *Carduus* spp. over European *Cirsium* spp. (p. 57); 3) larval survival was much higher on *Carduus* spp. than on *Cirsium* spp., including one North American species (wavyleaf thistle); and, 4) average adult weights were higher on *Carduus* spp. than on *Cirsium* spp. (p. 51), including on one North American species (*Cirsium flodmanii*). Zwoelfer and Harris (1984: 59) concluded: "...to overcome the negative selection pressures of adult oviposition preferences and high larval mortality, the thistle involved should be abundant. Thus, it is unlikely that the normally scattered native thistles will be adopted as a prime host."

Unfortunately, however, these factors have not precluded the significant effects on Platte thistle that we have documented. In this case, adult feeding and oviposition preferences were not relevant, since no *Carduus* spp. occurred at our sites. Larval survival on Platte thistle in 1997 averaged 54% (Fig. 9A), lower than reported for *Carduus* spp.(97.4% in Rees 1978) as expected, but were higher than that previously reported on wavyleaf thistle (46% in Rees 1978). However, the lower larval survival rate was still sufficient to lead to rapid population growth in the new environment (Louda 1998b). Finally, we found no differences in adult size related to host for *R. conicus* from the field in western Nebraska (Fig. 9B). Therefore, the obvious conclusion is that adult preference and larval performance were insufficient bases on which to predict population growth and ecological impact with adoption of a new native host by an insect introduced to control an exotic plant species when it had even marginally acceptable related host plants (Arnett and Louda 1999). Relative preference and performance, however, still are the basic criteria used to predict ecological risk.

Lessons for Future Biological Control Projects

The data on the impact of *Rhinocyllus conicus* on Platte thistle (*Cirsium canescens*) may be used to help protect rare native species. For example, Pitcher's thistle (*Cirsium pitcheri*), a close relative of Platte thistle, is already threatened by habitat destruction (Pavlovic et al. 1992) and limited by native insects (Louda and McEachern 1995, Bevill et al. 1999). Based on the data available, it is likely that Pitcher's thistle reproductive success and population density will decline further if *R. conicus* is added to its inflorescence-feeding guild. And, even smaller populations would increase its chance of extinction. The phylogenetic and ecological similarities between these thistles argue *R. conicus* should not be distributed to the states and provinces surrounding the Great Lakes. This may or may not be possible now. Insects can disperse on their own, and no national regulations control the distribution of biological control agents among states or regions once the organism has been approved for release within the USA. Guidelines for such deliberate movement, plus periodic reviews of agent effectiveness and side effects (Thomas and Willis 1998), are needed to manage the risk associated with deliberate importation, introduction and spread of exotic organisms.

What can be learned from this case for future biological control projects? A review of the documentation supporting the release *R. conicus* for musk thistle weed control leads me to make or support seven suggestions for the future. All involve a more rigorous examination of the specific applicability of the key assumptions underlying biological control. Biological control rests on several central assumptions, including (1) the target species poses major economic and environmental problems, (2) no less risky alternatives exist, (3) control by introduced natural enemies is predicted, (4) significant harm to native species is unlikely, and (5) release involves known risks acceptable to the public (Louda et al. 1998).

Although these assumptions underlie every responsible biological control project, examination of the *Rhinocyllus* introduction for musk thistle control suggests that the validity of each of these must be better quantified in the future. Briefly, I suggest we need: (1) better *a priori* quantification of the problem; (2) improved incorporation of ecological criteria into pre-release testing; (3) increased assessment of plausible direct and indirect ecological interactions before release; 4) quantitative evaluation of the likely efficacy of the proposed biological solution; 5) more evidence on alternative control methods; (6) enhanced peer and public review prior to release; and, finally, (7) redefinition of the situations in which releases of exotic species for biological control are considered a viable option.

First, the need for better *a priori* quantification of the problem is obvious with hindsight (Louda et al. 1998, Thomas and Willis 1998, Simberloff and Stiling 1998). For musk thistle, the data that were taken initially are not adequate to determine current or cumulative costs and benefits of the biological control program. Economic costs were not quantified before *Rhinocyllus* was introduced. The best study to quantify economic threat was done after release (Dunn 1976). Unfortunately, this study relied completely on anecdotal information, and on an inappropriate county-by-county spatial scale. Counties are large, hundreds km^2, and heterogeneous in terrain, land use, and weed densities. Using presence or absence on this scale to quantify the density of a patchy weed, such as musk thistle, overestimates the scale of the problem (Louda et

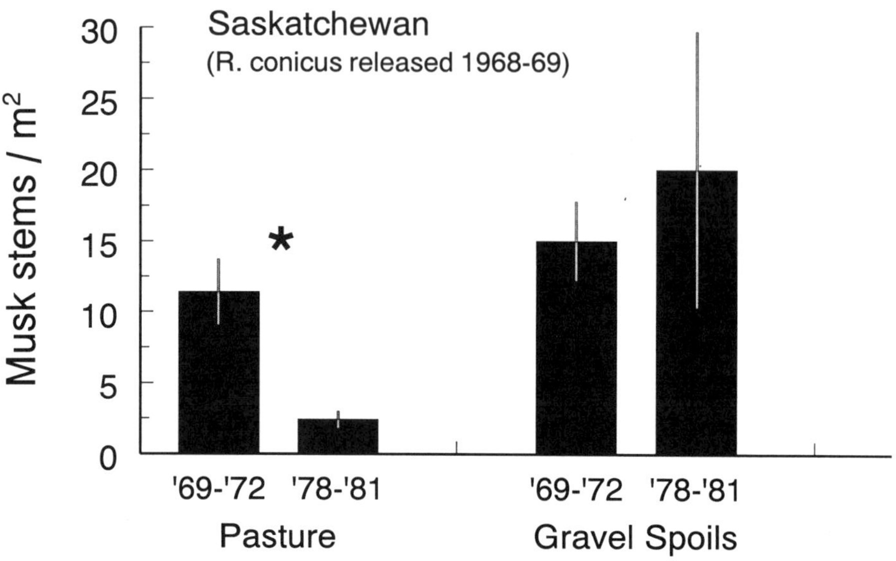

Figure 10. Density of musk thistle (*Carduus nutans*) in Saskatchewan, Canada, at the time of the initial release (1968–1969) and after about 10 years (1978–1981) in plots in two types of habitats, pasture grassland and gravel pit spoils, showing importance of grass competition for reduction in musk thistle density (data from Zwoelfer and Harris 1984; graph from Louda 1998a).

al. 1998). The economic threat in each county was also scored. Dunn (1976) defined a county as having a *serious economic threat* if "one or more pesticide applications had been used or would have been used if funds were available," and as having a *potential economic threat* if "the weed occurred but was not considered a problem." This method of evaluating the magnitude of the problem seriously overestimates economic impact as well. However, such estimates continue to be used even today (e.g., Boldt 1997, Frank 1998). Better economic evaluations can be made and should be required (Louda et al. 1998, Simberloff and Stiling 1998). Furthermore, environmental costs from the spread of musk thistle were unlikely. Musk thistle is a fugitive species that becomes a weed in disturbed areas and overgrazed pasture. It loses in competition with grasses, such as observed in normal plant succession or good pasture management (Austin et al. 1985, Hamrick and Lee 1987, Popay and Medd 1990). Thus, I concur with the suggestion (Thomas and Willis 1998) that more precise evaluation of a pest problem should now be required to justify the deliberate introduction of an exotic species.

Second, this case also supports the suggestions that ecological criteria based on conceptual considerations must become an integral part of pre-release testing (Louda et al. 1998, Thomas and Willis 1998). Host preference and host range tests are necessary, but insufficient, evidence of safety by themselves (Second and Kareiva 1996,

McEvoy 1996, Thomas and Willis 1998, Arnett and Louda 1999). For example in this case, the magnitude of direct nontarget effects on related natives would have been better predicted if more than one native species had been tested thoroughly for potential nontarget effects. In addition, ecological criteria, such as flowering phenology, should have been used to select those native species. Phenological synchrony is known to be an important factor in host plant use, even by *R. conicus* (Klein 1986, Zwoelfer and Romstock-Volkl 1991, Louda 1998b). Thus, it is not surprising that tests that focused primarily on one later-flowering North American thistle, *C. undulatum* (Zwoelfer and Harris 1984), did not make a strong prediction of nontarget ecological impacts. Studying species in which flowering was phenologically synchronized with *Rhinocyllus* oviposition would have had a high probability of identifying the potential for serious nontarget effects, such as observed for Platte thistle (Louda 1998b).

Third, the case lends direct support to the pleas for increased assessment of plausible direct and indirect ecological interactions before release (Simberloff and Stiling 1996, Thomas and Willis 1998). Better information on the population dynamics of interactions and food web interconnections associated with plausible nontarget effects would have provided data critical to the evaluation of environmental side effects. Such information is necessary for a realistic discussion of cost/benefit ratios. In this case, pre-release data suggested flowerheads of *Cirsium* spp. could be used by *Rhinocyllus*. Yet, no data were collected on the conditions under which this might be expected, or on the outcome if various degrees of nontarget host use occurred. We should be asking not only will the proposed agent utilize native North American species, but also which species could be vulnerable and what would happen if they were utilized? I appreciate such data collection is labor-intensive. However, it is feasible, if the number of proposed releases is more limited, such as to the most promising candidates for control of the most invasive, damaging weed species. Without a thorough, ecologically sound investigation, we have no basis on which to predict the consequences of deliberate introductions for indigenous biodiversity.

Fourth, better quantitative evaluation of the efficacy of the proposed biological solution prior to release would help justify the risk associated with deliberate introductions of exotic species (Thomas and Willis 1998, Louda 1998a), and it would decrease the chance of interference among agents released (McEvoy and Coombs 1999). The most rational biological control strategy would entail introducing the fewest, most effective agents that had the lowest probability of nontarget effects (McEvoy and Coombs 1999). To do this, we need better prediction of control efficacy. For musk thistle, even the early evidence for control by *R. conicus* was equivocal. Decreases in musk thistle density after the release of *R. conicus* in test plots occurred only where musk thistle was in a competitive context (Fig. 10; and Zwoelfer and Harris 1984). Subsequent studies have confirmed the importance of grass competition for thistle control (e.g. Austin et al. 1985, Hamrick and Lee 1987, Popay and Medd 1990). Every introduction of an exotic species entails some risk (Simberloff 1981, 1992; Howarth 1983, 1990; Secord and Kareiva 1996; Simberloff and Stiling 1996, 1998). Thus, common sense suggests that the pre-release data should demonstrate a high probability of control to justify taking that risk. Pre-release studies of effectiveness could be initiated as field experiments in the indigenous region of the most likely agents for our worst weeds. Such tests would likely improve effectiveness of biological control efforts as well as decrease the potential of serious side effects.

Fifth, this case also supports the call for greater consideration of alternative, potentially more manageable, control methods (Barbosa 1998, Newman et al. 1998, Thomas and Willis 1998). Solid evidence on options is needed to justify use of the environmentally riskier strategies, including both use of broad-spectrum chemicals and deliberate releases of exotic organisms. In this case, the local and regional scales of variation in a patchy weed suggest a wide range of possible options for control strategies. In general, multiple smaller-scale alternative tactics can be employed to manage a pest problem (e.g., Newman et al. 1998, Thomas and Willis 1998). However, biological control is often presented as the only alternative to broad-spectrum chemical application (e.g., Boldt 1997, Frank 1998). Several alternative strategies have been reported to control local infestations of musk thistle control, including mechanical control by hand weeding, localized spraying, and intensified grass management (Randall 1993). Augmentation of indigenous natural enemies (Newman et al. 1998) remains unexplored in this case. Each of these options entails less risk to other organisms than either biological control or broad-spectrum pesticides. Logic argues for the implementation of the least risky options first. However, rational decisions are difficult if all the options are not equally evaluated. If petitions for release of an exotic organism included a requirement for equivalent evaluation of all alternatives, then optimization of weed management strategies could provide adequate weed control with minimal environmental risk.

Sixth, the musk thistle story also suggests the need for enhanced peer and public review in pest management, before targeting organisms for biological control, before releasing exotic species, and before moving species between eco-regions within the USA. External review prior to targeting should help focus resources on the most significant problems (Louda 1998a). More discussion prior to deliberate releases of exotics could go a long way to preventing a single perspective from dominating pest management decisions. In the case of musk thistle, such reviews could have focused more consideration on the facts that appear to have been overlooked in the pressure "to solve the problem." These include the local scale of the problems caused by a fugitive, poor competitor weed the close similarity of native species to the targeted weed making host range expansion plausible, and the trophic dependence of other organisms upon potential nontarget species. It seems clear to me that such discussions would improve decisions and reduce the chances of making undesirable, irreparable ecological alterations in natural communities.

Finally, the data and these considerations raise the issue of which species are sufficiently damaging to merit increased potential threats to our indigenous species and biodiversity by the deliberate importation and release of exotic organisms. I think we need an examination and revision of the criteria that qualify situations for the use of biological control releases. A biological risk is inherent with any release. "Recalling" harmful species is next to impossible (Simberloff 1992). Given these facts and our data, I suggest that the use of biological control should be restricted to only the most invasive species (particularly those with no known, even marginally acceptable native relatives) where no clear evidence of serious environmental effects exists and for which equal investigation has produced no shorter term, smaller scale alternative methods for weed management. The wisdom of an old adage, "better safe than sorry," seems particularly applicable to deliberate "non-recallable" species additions to biological communities.

The effects of *R. conicus* on native plants and dependent fauna also suggest that it is difficult to anticipate the outcome of new interactions and the long-term consequences of introductions. Doing so requires more information on the dynamics of interactions, activity of already introduced agents, and consequences of invasions. Use of such information, plus more rigorous economic and environmental assessments, should increase environmental safety in future biological control efforts (Louda 1998a, Louda et al. 1998, Thomas and Willis 1998). The recent findings and permanence of "the solution" with classical biological control suggest that (i) past biological control agents should be reviewed periodically and (ii) new efforts should be reserved for only the most invasive, harmful species with no native relatives, if and only if intensive efforts have shown that no reasonably effective, more local alternatives exist.

ACKNOWLEDGMENTS

These studies could not have been done without the input of many people, and I am grateful to each. Dr. Charles W. O'Brien identified all of the weevils (Curculionidae). Dr. George Balogh identified the moths (Pyralidae). Dr. R. Foote identified the flies (Tephritidae). Student field assistants not only did a competent, cheerful job in helping collect the data, they also buoyed my spirits by sharing their discovery of the prairie with me, especially S. Jedlicka Arevalo, R. Bevill, N. Bandt, J. Burger, K. Farrell, T. Huettner Tesar, K. Jackson, C. Murphy, L. Stanforth, and K. Stormburg. I appreciate my discussions of these studies and data with: A. E. Arnett, A. Basolo, L. R. Fox, A. Joern, W. O. Lamp, S. A. Louda, J. Mulroy, J. Nechols, D. O'Dowd, R. W. Otley, R. Pemberton, D. Pilson, M. A. Potvin, and A. Steuter. I am grateful for the support and encouragement of the staffs of both the Niobrara Valley Preserve of The Nature Conservancy and the University of Nebraska Cedar Point Biological Station. The funding that made the research possible was provided by the University of Nebraska Research Council, The National Science Foundation (DEB92-21065, DEB96-15299), and The Nature Conservancy through the Rodney Johnson and Katharine Ordway Stewardship Endowments.

REFERENCES

Amrine, J. W., Jr., and T. A. Stasny. 1993. Biological control of multiflora rose. *In* B. N. McKnight (ed.), Biological pollution: the control and impact of invasive exotic species. Indiana Acad. Sci., Indianapolis. p. 9–21.

Arnett, A. E., and S. M. Louda. 1999. Criteria for biological control introductions: Evidence preference and performance are insufficient to predict ecological risk. In review.

Austin, M. P., R. H. Groves, L. M. F. Fresco, and P. E. Kaye. 1985. Relative growth of six thistle species along a nutrient gradient with multispecies competition. J. Ecol. 73: 667–684.

Barbosa, P., ed. 1998. Conservation biological control. Academic Press, San Diego.

Bevill, R. L. 1998. The study and management of rare plant species with special attention to the threatened Pitcher's thistle and insect herbivory. MS Thesis. Univ. Nebraska, Lincoln.

Bevill, R. L., S. M. Louda, and L. M. Stanforth. 1999. Protection from natural enemies in managing rare species. Conserv. Biol. In press.

Bleed, A., and C. Flowerday. 1989. An atlas of the Sand Hills. Resource Atlas No. 5. Univ. Nebraska, Inst. Agric. Nat. Res., Lincoln.

Boldt, P. E. 1997. Response of a *Rhinocyllus* researcher. Biocontr. News Info. 18: 100N.

Boldt, P. E., and J. A. Jackman. 1993. Establishment of *Rhinocyllus conicus* Froelich on *Carduus macrocephalus* in Texas. Southwest. Entomol. 18: 173–181.

Debach, P., and D. Rosen. 1991. Biological control by natural enemies, 2nd Edition. Cambridge Univ. Press, Cambridge.

Dodd, A. P. 1940. The biological campaign against prickly pear. Commonwealth Prickly Pear Board, Brisbane.

Dunn, P. H. 1976. Distribution of *Carduus nutans*, *C. acanthoides*, *C. pycnocephalus*, and *C. crispus*, in the United States. Weed Sci. 24: 518–524.

Frank, J. H. 1998. How risky is biological control? comment. Ecol. 79: 1829–1834.

Frick, K. E., editor. 1978. Biological control of thistles in the genus *Carduus* in the United States. Sci. and Educ. Admin., U.S. Dept. Agric., Stoneville, Mississippi.

Goeden, R. D. 1978. Initial analysis of *Rhinocyllus conicus* (Froelich)(Coleoptera: Curculionidae) as an introduced natural enemy of milk thistle (*Silybum marianum* [L.] Gaertner) and Italian thistle (*Carduus pycnocephalus* L.) in southern California.. *In* Biological control of the genus *Carduus* in the United States. U.S. Dept. Agric., Stoneville, Mississippi. p. 39–50.

Goeden, R. D., and D. W. Ricker. 1986a. Phytophagous insect faunas of the two most common native *Cirsium* thistles, *C. californicum* and *C. proteanum*, in southern California. Ann. Entomol. Soc. Amer. 79: 953–962.

Goeden, R. D., and D. W. Ricker. 1986b. Phytophagous insect faunas of two introduced *Cirsium* thistles, C. *ochrocentrum* and *C. vulgare*, in southern California. Ann. Entomol. Soc. Amer. 79: 945–952.

Goeden, R. D., and D. W. Ricker. 1987a. Phytophagous insect faunas of native *Cirsium* thistles, *C. mohavense*, *C. neomexicanum*, and *C. nidulum*, in the Mojave Desert of southern California. Ann. Entomol. Soc. Amer. 80: 161–175.

Goeden, R. D., and D. W. Ricker. 1987b. Phytophagous insect faunas of the native thistles, *Cirsium brevistylum, Cirsium congdonii, Cirsium occidentale*, and *Cirsium tioganum* in southern California. Ann. Entomol. Soc. Amer. 80: 152–160.

Goeden, R. D., C. A. Fleschner, and D. W. Ricker. 1967. Biological control of Prickly Pear cacti on Santa Cruz Island, California. Hilgardia 38: 579–606.

Goeden, R. D., D. W. Ricker, and B. A. Hawkins. 1984. Ethological and genetic differences among three biotypes of *Rhinocyllus conicus* (Coleoptera: Curculionidae) introduced into North America for the biological control of Asteraceous thistles. *In* E. S. Delfosse (ed.), Proceedings of the VI international symposium on the biological control of weeds. Agric. Canada, Vancouver, Canada. p. 181–189.

Great Plains Flora Association. 1977. Atlas of the flora of the Great Plains. Iowa State Univ. Press, Ames.

Great Plains Flora Association. 1986. Flora of the Great Plains. Univ. Press of Kansas, Lawrence.

Hamrick, J. L., and J. M. Lee. 1987. Effect of soil surface topography and litter cover on the germination, survival, and growth of musk thistle (*Carduus nutans*). Amer. J. Bot. 74: 451–457.

Harris, P., and H. Zwoelfer. 1968. Screening of phytophagous insects for biological control of weeds. Can. Entomol. 100: 295–303.

Howarth, F. G. 1983. Classical biological control: Panacea or Pandora's box? Proc. Hawaiian Entomol. Soc. 24: 239–244.

Howarth, F. G. 1991. Environmental impacts of classical biological control. Ann. Rev. Entomol. 36: 485–509.

Johnson, M. F., and H. H. Iltis. 1963. Preliminary reports on the flora of Wisconsin. No. 48. Compositae I–Composite family. Trans. Wisconsin Acad. Arts and Sci.

Kartesz, J. 1998. A synonymized checklist of the vascular flora of the United States, Puerto Rico, and the Virgin Islands, Full text index–July, 1998 (http://www.csdl.tamu.edu/FLORA/b98/check98.htm).

Kaul, R. B. 1989. Plants. *In* Bleed, A., and C. Flowerday (eds.), An atlas of the Sand Hills. Univ. Nebraska, Lincoln. p. 127–142.

Keddy, C. J., and P. A. Keddy. 1984. Reproductive biology and habitat of *Cirsium pitcheri*. Michigan Bot. 23: 57–67.

Keeler, K. H., A. T. Harrison, and L. Vescio. 1980. The flora and Sand Hills prairie communities of Arapaho Prairie, Arthur County, Nebraska. Prairie Natur. 12: 65–78.

Klein, M. 1986. Anpassungen von *Rhinocyllus conicus* Froel. (Col.: Curculionidae) an allochrone wirtspflanzensituationen. Verhandlungen der Deutschen Zoologischen Gesellschaft, 19 Jahresversammlung 1986, uberreicht im Namen der Deutschen Zoologischen Gesellschaft. Gustav Fischer

Verlag, Stuttgart.
Klein, M., and A. Seitz. 1994. Geographic differentiation between populations of *Rhinocyllus conicus* Froelich (Coleoptera: Curculionidae): concordance of allozyme and morphometric analysis. Zool. J. Linn. Soc. 110:181–191.
Kok, L. T. 1979. Influence of temperature and photoperiod on the mortality of diapausing *Rhinocyllus conicus* (Col.: Curculionidae). Ann. Entomol. Soc. Am. 72: 206–208.
Kok, L. T., and W. W. Surles. 1975. Successful biological control of Musk thistle by an introduced weevil, *Rhinocyllus conicus*. Environ. Entomol. 4:1025–1027.
Lamp, W. O. 1980. Predispersal seed predation of the Platte thistle and its effect on seed production. Ph.D. Dissertation. Univ. Nebraska, Lincoln.
Lamp, W. O., and M. K. McCarty. 1979. A preliminary study of seed predators of Platte thistle. Trans. Nebraska Acad. Sci. 7: 71–74.
Lamp, W. O., and M. K. McCarty. 1981. Biology and ecology of Platte thistle (*Cirsium canescens*). Weed Sci. 29: 686–692.
Lamp, W. O., and M. K. McCarty. 1982a. Biology of predispersal seed predators of the Platte thistle (*Cirsium canescens*). J. Kansas Entomol. Soc. 55: 305–316.
Lamp, W. O., and M. K. McCarty. 1982b. Observations of *Corythucha distincta* (Hemiptera: Tingidae) on Platte thistle in Nebraska. J. Kansas Entomol. Soc. 55: 34–36.
Lamp, W. O., and M. K. McCarty. 1982c. Predispersal seed predation of a native thistle, *Cirsium canescens*. Environ. Entomol. 11: 847–851.
Louda, S. M. 1994. Experimental evidence for insect impact on populations of short–lived, perennial plants, and its application in restoration ecology. M. L. Bowles and C. J. Whelan (eds.), Restoration of endangered species. Cambridge Univ. Press, Cambridge. p. 118–138.
Louda, S. M. 1998a. Ecology of interactions needed in biological control practice and policy. Bull. Brit. Ecol. Soc. 29(4): 8–11.
Louda, S. M. 1998b. Population growth of *Rhinocyllus conicus* (Coleoptera: Curculionidae) on two species of native thistles in prairie. Environ. Entomol. 27: 834–841.
Louda, S. M., and A. K. McEachern. 1995. Insect damage to inflorescences of the threatened dune thistle *Cirsium pitcheri*. Bull. Ecol. Soc. Am. 76: 358.
Louda, S. M., and M. A. Potvin. 1995. Effect of inflorescence–feeding insects in the demography and lifetime fitness of a native plant. Ecol. 76: 229–245.
Louda, S. M., M. A.. Potvin, and S. K. Collinge. 1990. Predispersal seed predation, postdispersal seed predation and competition in the recruitment of seedlings of a native thistle in sandhills prairie. Am. Midl. Natural. 124: 105–113.
Louda, S. M., D. Kendall, J. Connor, and D. Simberloff. 1997. Ecological effects of an insect introduced for the biological control of weeds. Science 277: 1088–1090.
Louda, S. M., D. Simberloff, G. Boettner, J. Connor, D. Kendall, and A. E. Arnett. 1998. Insights from data on the nontarget effects of the flowerhead weevil. Biocontr. News Info. 19: 70N–71N.
Loveless, M. D. 1984. Population biology and genetic organization in *Cirsium pitcheri*, an endemic thistle. Ph.D. Dissertation. Univ. Kansas, Lawrence.
Loveless, M. D., and J. L. Hamrick. 1988. Genetic organization and evolutionary history in two North American species of *Cirsium*. Evol. 42: 254–265.
McCarty, M. K., and W. O. Lamp. 1982. Effect of a weevil, *Rhinocyllus conicus*, on Musk Thistle (*Carduus thoermeri*) seed production. Weed Sci. 30: 136–140.
McCarty, M. K., and C. J. Scifres. 1969. Life cycle studies with musk thistle. Nebraska Agric. Exp. Sta. Res. Bull. 230: 1–15.
McCarty, M. K., C. J. Scifres, and L. R. Robison. 1967. A descriptive guide for major Nebraska thistles. Nebraska Agric. Exp. Sta. Bull. 493: 1–24.
McEachern, A. K. 1992. Disturbance dynamics of Pitcher's thistle (*Cirsium pitcheri*) populations in Great Lakes sand dune landscapes. Ph.D. Dissertation. Univ. Wisconsin, Madison.
McEvoy, P. B. 1996. Host specificity and biological pest control. BioSci. 46: 401–405.
McEvoy, P. B., and E. M. Coombs. 1999. A parsimonious approach to biological control of plant invaders. Ecological Applications: in press.
McKnight, B. N., editor. 1993. Biological pollution: The control and impact of invasive exotic species. Indiana Acad. Sci., Indianapolis.
Mellini, E. 1951. Insetti del *Carduus nutans* L. II. *Rhinocyllus conicus* Froelich e *Larinus jaceae* F. (Coleoptera: Curculionidae). Boll. Ist. Ent. 18:319–349.

Moerman, D. E. 1986. Medicinal plants of native America, Tech. Rep. 19 ed., vol. 1. Univ. Michigan Mus. Anthropol., Ann Arbor.

Newman, R. M., D. C. Thompson, and D. B. Richman. 1998. Conservation strategies for the biological control of weeds. *In* P. Barbosa (ed.), Conservation biological control. Academic Press, San Diego. p. 371–396.

OTA. Office of Technology Assessment, Congress of the United States. 1995. Biologically based technologies for pest control, OTA-ENV-636 Edition. U.S. Govt. Printing Office, Washington, D.C.

OTA. Office of Technology Assessment, Congress of the United States. 1993. Harmful non-indigenous species in the United States, OTA-F-565 Edition. U.S. Govt. Printing Office, Washington, D.C.

Palmisano, S., and L. R. Fox. 1997. Effects of mammal and insect herbivory on population dynamics of a native Californian thistle, *Cirsium occidentale*. Oecol. (Berl.) 111: 413–421.

Pavlovic, N. B., M. Bowles, S. R. Crispin, T. C. Gibson, K. D. Herman, R. T. Kavetsky, A. K. McEachern, and M. R. Penskar. 1992. Pitcher's thistle (*Cirsium pitcheri*) recovery plan (draft). U.S. Fish and Wildlife Svc. (Region 3), Dept. Interior, Minneapolis, Minnesota.

Pemberton, R. W., C. E. Turner, and S. S. Rosenthal. 1985. New host records for tephritid flies (Diptera) from *Cirsium* and *Saussurea* thistles (Asteraceae) in California. Proc.Entomol. Soc. Washington 87: 790–794.

Pettey, F. W. 1950. The cochineal insect (*Dactylopius opuntiae*), an the problem of its control in spineless cactus plantations. Part 1. Its history, distribution, biology, and achievements in the control of prickly pear in South Africa, Bull. 196. Dept. Agric., Union of South Africa, Cape Town.

Popay, A. I., and R. W. Medd. 1990. The biology of Australian weeds 21. *Carduus nutans* L. Plant Prot. Quart. 5: 3–13.

Randall, J. M. 1993. Exotic weeds in North American and Hawaiian natural areas: The Nature Conservancy's plan of attack. *In* B. N. McKnight (ed.), Biological pollution: the control and impact of invasive exotic species. Indiana Acad. Sci., Indianapolis. p. 159–172.

Redfern, M. 1983. Insects and Thistles. Cambridge Univ. Press, Cambridge.

Rees, N. E. 1977. Impact of *Rhinocyllus conicus* on thistles in southwestern Montana. Environ. Entomol. 6: 839–842.

Rees, N. E. 1978. Interactions of *Rhinocyllus conicus* and thistles in the Gallatin Valley. *In* Biological control of thistles in the genus *Carduus* in the United States. U.S. Dept. Agric., Agric. Res. Svc., Stoneville, Mississippi. p. 31–38.

Rees, N. E.. 1982. Collecting, handling and releasing *Rhinocyllus conicus*, a biological control agent of musk thistle. U.S. Dept. Agric., Agric. Res. Svc. 579: 1–7.

Rees, N. E.. 1991. Biological control of thistles. *In* L. F. James, J. O. Evans, M. H. Ralphs and R. D. Child (eds.), Noxious range weeds. Westview Press, Boulder, Colorado. p. 264–273.

Secord, D., and P. Kareiva. 1996. Perils and pitfalls in the host specificity paradigm. BioSci. 46: 448–453.

Simberloff, D. 1981. Community effects of introduced species. *In* T. H. Nitecki (ed.), Biotic crises in ecological and evolutionary time. Academic Press, New York. p. 53–81.

Simberloff, D. 1992. Conservation of pristine habitats and unintended effects of biological control. *In* W. C. Kauffman and J. E. Nechols (eds.), Selection criteria and ecological consequences of importing natural enemies. Entomol. Soc. Am., Lanham. p. 103–117.

Simberloff, D., and P. Stiling. 1996. How risky is biological control? Ecol. 77: 1965–1974.

Simberloff, D., and P. Stiling. 1998. How risky is biological control?; reply. Ecol. 79: 1834–1836.

Stanforth, L. M., S. M. Louda, and R. L. Bevill. 1997. Insect herbivory on juveniles of a threatened plant, *Cirsium pitcheri*, in relation to plant size, density and distribution. EcoSci. 4: 57–66.

Surles, W. W., and L. T. Kok. 1977. Ovipositional preference and synchronization of *Rhinocyllus conicus* with *Carduus nutans* and *C. acanthoides*. Environ. Entomol. 6: 222–224.

Surles, W. W., and L. T. Kok. 1978. *Carduus* thistle seed destruction by *Rhinocyllus conicus*. J. Weed Sci. Soc. Am. 26: 264–269.

Thomas, M. B., and A. J. Willis. 1998. Biological control–risky but necessary? Tr. Ecol. Evol. 13: 325–329.

Turner, C. E., and J. C. Herr. 1996. Impact of *Rhinocyllus conicus* on a non–target, rare, native thistle (*Cirsium fontinale*) in California. *In* V. C. Moran and J. H. Hoffman (eds.), Proc. IX international symposium on biological control of weeds. Univ. Cape Town, Stellenbosch, South Africa. p. 103.

Turner, C. E., R. W. Pemberton, and S. S. Rosenthal. 1987. Host utilization of native *Cirsium* thistles (Asteraceae) by the introduced weevil *Rhinocyllus conicus* (Coleoptera: Curculionidae) in California. Environ. Entomol. 16: 111–115.

Ward, R. H., R. L. Pienkowski, and L. T. Kok. 1974. Host specificity of the first–instar of *Ceuthorhynchidius horridus*, a weevil for biological control of thistles. J. Econ. Entomol. 67: 735–737.

Zwoelfer, H. 1965. Preliminary list of phytophagous insects attacking wild Cynareae (Compositae) in Europe. Commonw. Inst. Biol. Contr. Tech. Bull. 6: 81–154.

Zwoelfer, H. 1978. Mechanismen und ergebnisse der co-evolution von phytophagen und entomophagen insekten und hoheren pflanzen. Sonderband naturwissenschaften Verein im Hamburg 2: 7–50.

Zwoelfer, H. 1988. Evolutionary and ecological relationships of the insect fauna of thistles. Annu. Rev. Entomol. 33: 103–229.

Zwoelfer, H., and P. Harris. 1971. Host specificity determination of insects for biological control of weeds. Annu. Rev. Entomol. 16: 159–178.

Zwoelfer, H., and P. Harris. 1984. Biology and host specificity of *Rhinocyllus conicus* (Froel.) (Col., Curculionidae), a successful agent for biological control of the thistle, *Carduus nutans* L. Z. ang. Entomol. 97: 36–62.

Zwoelfer, H., and M. Preiss. 1983. Host selection and oviposition behaviour in west–European ecotypes of *Rhinocyllus conicus* Froel. (Col.: Curculionidae). Z. ang. Entomol. 95: 113–122.

Zwoelfer, H., and M. Romstock-Volkl. 1991. Biotypes and the evolution of niches in phytophagous insects on Cardueae hosts. *In* P. W. Price, T. M. Lewinsohn, G. Wilson Fernandes, and W. W. Benson (eds.), Plant-animal interactions: evolutionary ecology in tropical and temperate regions. John Wiley, New York. p. 487–507.

14

BIOLOGICAL CONTROL OF MUSK THISTLE: A REASSESSMENT

James R. Nechols
*Department of Entomology,
Kansas State University, Manhattan, Kansas*

HISTORICAL SYNOPSIS

The Weed

Musk thistle and nodding thistle are names commonly used to describe Old World thistles in the *Carduus nutans* species group (family Asteraceae). These include *Carduus macrocephalus* Desfontaines (big-headed thistle), *C. nutans* ssp. *leiophyllus* [Petrovic]) Stoj.and Stef. (nodding thistle) a.k.a. *C. thoermeri* , and *C. nutans* L. (musk thistle proper). According to McGregor (1986), *C. macrocephalus* is a subspecies of *C. nutans* along with *C. nutans* ssp. *leiophyllus.* Other classifications exist (Kazmi 1964, McCarty 1982, Desrochers et al. 1988). In this chapter, the term "musk thistle" will be used generically to refer to any and all of the names used in the various classification schemes. Musk thistle became established in North America sometime during the mid-1800s (Stuckey and Forsythe 1971). Introduction appears to have been accidental, possibly as seed contaminants of imported grain. By the mid-1900s, musk thistle had proliferated throughout much of the eastern and north central parts of United States (Dunn 1976) and into parts of southern Canada (Harris and Zwölfer 1971). In 1981, this weed occupied an estimated 730,000 hectares in 40 of the United States (Batra et al. 1981). In Kansas, *C. nutans* occurred on more than 400 hectares in 75 of the 105 counties (Robinson 1976). Musk thistle's weed status derives from a combination of structural, developmental and ecological characteristics. The presence of sharp spines on stems, leaf margins and floral bracts of plants, and the fact that plants may reach two meters or more in height, represent a nuisance in urban and suburban landscapes. Encounters with stands of musk thistle by livestock also are reported to interrupt grazing (Batra 1978, Trumble and Kok 1982, Rees 1992). Economic losses also have resulted from the thistle's competition with forage crops, particularly in overgrazed or disturbed pastures and rangeland (Trumble and Kok 1982, Reece and Wilson 1983, Moore et al. 1989, Fick 1995, Thompson et al. 1995). Musk thistle has been categorized as a noxious weed in many states, where control is mandated by law.

The Biological Control Program

In response to increases in the distribution and density of musk thistle (Dunn 1976, Robinson 1976, Harris 1984), searches began in Europe during the mid 1960s for insects that could be introduced and established as biological control agents of musk thistle in North America. The first insect to be imported was the head weevil, *Rhinocyllus conicus* Froelich. Adults collected in France were released in Canada in 1968 (Harris and Zwölfer 1971) and in Virginia and Montana in 1969 (Surles et al. 1974, Kok and Surles 1975, Hodgson and Rees 1976). Deliberate redistribution of this weevil, along with natural range expansion, have resulted in the establishment of *R. conicus* in more than 20 states and at least six Canadian provinces (Möller-Joop and Schroeder 1986, Julien 1992, Andres and Rees 1995, Harris and Shorthouse 1996).

Between 1977 and 1981, a second weevil, the rosette-feeder *Trichosirocalus horridus* (Panzer), was released in Virginia and in seven other states in the eastern, midwestern and western United States (Kok and Trumble 1979, Andres and Rees 1995). Initial establishment occurred in Virginia, and subsequent redistribution has resulted in the establishment of *T. horridus* in Kansas, Missouri and Wyoming (Andres and Rees 1995). Releases in Canada have led to the establishment the rosette weevil in British Columbia, Alberta, Saskatchewan and Manitoba (Moller-Joop and Schroeder 1986, Julien 1992).

Comparatively, *R. conicus* has a much wider North American range than *T. horridus*, and it occurs at higher densities where distributions of the two weevils overlap (Andres and Rees 1995). However, no experiments have been done to evaluate their relative (individual) impact on *Carduus* thistles. In fact, experimental data are generally lacking (Kok et al. 1986). On the other hand, a considerable amount of indirect, correlative data are available. For example, a three-year survey in Montana at sites where *R. conicus* had been released showed that the amount of viable musk thistle seed decreased markedly as infestations of *R. conicus* increased (Rees 1977). Kok and Surles (1975) made similar findings in Virginia. Workers in Virginia also noted that a steady decline in musk thistle density over an 11-year period at two widely separated sites corresponded with an increase in head weevil infestation (Kok and Pienkowski 1985). These authors also observed a brief resurgence of thistles in areas of low plant density from which weevils had dispersed, followed by a precipitous decline in thistle populations a year later. On the basis of these results, they speculated that weevils might have responded in a density-dependent manner. However, they stated that other suppressive factors, like plant competition, could have contributed to the population trend. The authors further noted that resurgent plants were smaller, and had significantly fewer buds, than plants present before the weevil was released. But it is unclear whether a functional relationship existed between plant productivity and previous head weevil infestation.

The role of the rosette weevil, *T. horridus*, as a suppressive agent of musk and other *Carduus* thistles also is equivocal, both because controlled experiments are lacking and because of conflicting results. For example, Cartwright and Kok (1985) found that the reproductive response of musk thistle and plumeless thistle (*Carduus acanthoides* L.) to infestations of *T. horridus* depended on the size of the rosette at the time of infestation. Plants with small-to-medium-sized rosettes had smaller heads and fewer seeds than uninfested plants, whereas infestation of large rosettes stimulated plants to produce larger heads than uninfested plants. A 15-year survey at multiple sites in Virginia showed a progressively large decline in the density of plumeless thistle that was correlated with increases in rosette weevil populations (Kok and Mays 1991). Although this pattern was consistent across sites, no data from uninfested fields were available for comparison. However, findings from a six-year study in Virginia showed that sites where *T. horridus* failed to establish experienced increases in thistle density, whereas sites where rosette weevils were present had large (up to 96%) declines in thistle density (Kok 1986). The largest percentage changes were found at sites where *T. horridus* had been established the longest. Kok (1986) also noted that populations of musk thistle — a host plant that is preferred by *T. horridus* (Sieburth and Kok 1982) — declined first, followed by plumeless thistle. He suggested that population shifts by *T. horridus* from *C. nutans* to *C. acanthoides* could have accounted for the temporal difference in decline of the two thistle species.

Circumstantial evidence suggests that the introduced weevils are capable of exerting some impact on *Carduus* thistle populations despite the paucity of experimental data, and the fact that existing data do not support this hypothesis. However, short-term experiments like that of Kok et al. (1986) potentially can misrepresent long-term impacts of weevils. An untested hypothesis is that *R. conicus* has an impact on musk thistle recruitment by reducing the amount of viable seed in seed banks. However, the counter-argument is that, even after seed reduction, sufficient viable seeds may be left to allow thistle populations to replace themselves. Long-term experiments are needed, with and without weevils, in which disturbance is imposed at different intervals. Ideally, these studies should be done under a range of field conditions representing different levels of plant competition. Where both the head and rosette weevils are present, treatments also should be established to test for their individual and combined impact on thistle demography. The duration of seed viability in different soil strata, and under different environmental conditions (McCarty et al. 1969, Hamrick and Lee 1987), also represents an important component in the analysis.

The gall-forming tephritid fly, *Urophora solstitialis* (L.), was first released in Canada in 1986. Subsequently, it established on nodding thistle (*C. nutans* complex) in Saskatchewan, and on plumeless thistle (*C. acanthoides*) in Ontario (Möller-Joop and Schroeder 1986). On *C. acanthoides*, up to 70% of the flower heads are infested (P. Harris. Agriculture Canada, personal communicatioon). Releases of *U. solstitialis* on plumeless thistle in Maryland between 1992 and 1995 have been unsuccessful despite the availability of large numbers of flies under apparently suitable conditions (Tipping and Holko 1995; P. Tipping, Maryland Department of Agriculture, personal communication). Because of the broad range of *U. solstitialis* in Europe, biotypes probably exist (P. Harris, Agriculture Canada, personal communication). These may be differentially adapted to *C. nutans, C. acanthoides,* and other *Carduus* species, either through differences in host plant affinities or phenology (i.e., synchronization with the flowering periods of the various *Carduus* species) (P. Harris, Agriculture Canada, personal communication).

The stem-feeding syrphid fly, *Cheilosia corydon* (Harris), was approved for release in the United States in 1989. Between 1990 and 1995, releases were made in Maryland (P. Tipping, Maryland Department of Agriculture, personal communication), Texas (De Quattro 1997), Montana (P. Boldt, USDA-ARS, Temple, TX, personal communication) and Oregon (Coombs 1996). Recoveries were made in Texas a year after its release, but it is unclear whether this fly is established (De Quattro 1997).

The chrysomelid flea beetle, *Psylliodes chalcomera* (Illiger), was approved for release in 1997 following extensive host specificity testing. Adults collected near Rome, Italy were released on musk thistle in Texas in 1997 (P. Boldt, USDA-ARS, Temple, TX, personal communication). A small number of adults also were liberated in Kansas. Larvae of *P. chalcomera* feed primarily on leaf and floral buds. The outcome of these releases is unknown, but to date no recoveries of this beetle have been made in either state (J. Nechols, personal observations, T. Robbins, USDA-ARS, Temple, TX, personal communication).

Nontarget Feeding

Besides musk thistle and plumeless thistle, *R. conicus* feeds on other introduced thistles that have attained weed status in the New World. These include Italian thistle (*Carduus pycnocephalus*), bull thistle (*Cirsium vulgare*), Canada thistle (*Cirsium arvense*), milk thistle (*Silybum marianum),* and Scotch thistle (*Onopordum acanthium*) (see Batra et al. 1981; chapters in Nechols et al. 1995 and references cited therein). However, *R. conicus'* relatively broad host range has allowed it to feed and develop on at least 16 native North American species of *Cirsium,* some of which are listed as threatened or endangered (Turner et al. 1987, Turner and Herr 1996, Louda et al. 1997, Louda 1998). In some cases, heavy infestation of native thistles has resulted in significant seed reduction (Louda et al. 1997, Louda 1998). Seed reduction on *C. canescens* has demographic consequences because experiments have shown that populations of this thistle already are seed-limited (Louda and Potvin 1995). These authors also have expressed concern that native insect species that live exclusively on or in native thistles may be in jeopardy.

In the light of available evidence to date, two related questions need to be considered: Should we proceed with an active biological control program for musk thistle? And, if so, how? In the sections that follow, I present a comparative review of the insects approved for release on musk thistle, cite options for biological control of musk thistle (including the possible consequences of these options), make recommendations for managing musk thistle, and offer some general challenges for resolving future conflicts involving classical biological control of weeds.

COMPARISON OF INSECTS APPROVED FOR RELEASE

Host Specificity

Of the five insects approved for release, two have relatively broad host ranges: the head weevil, *R. conicus* (Zwölfer 1967, Batra et al. 1981), and the rosette weevil, *T. horridus* (Ward et al. 1974, Kok 1975, Boldt et al. 1980). From the perspective of biological control, the ability of these weevils to feed and develop on more than one exotic thistle may be beneficial since availability of alternate host plants during periods of low musk thistle abundance could lead to greater population stability of *R. conicus* and *T. horridus* and, thus, potentially more persistent weed suppression. The broader host ranges of *R. conicus* and *T. horridus* compared with the other agents also may contribute to the suppression of exotic weeds other than musk thistle, though experimental evidence is lacking. However, increases in the density of *T. horridus* have been correlated with the reduction of musk thistle and plumeless thistle (see Kok 1986, Kok and Mays 1991). The temporal pattern of decline in the densities of these two weeds corresponds with the switch of rosette weevils from musk thistle, a preferred host, to plumeless thistle (Kok 1986). In contrast, in cases where *R. conicus* has been released on or moved to other introduced thistles, its populations have been lower than those observed on musk thistle (see Goeden and Ricker 1985, Goeden 1995a,b). Evaluation of the effectiveness of *R. conicus* on exotic thistles is complicated by the existence of known biotypes (Goeden 1978, Goeden et al. 1985) and the different degrees of seasonal synchrony that exist between the weevils and host thistles (Rees 1977, Surles and Kok 1977, Andres and Rees 1995).

From a conservation standpoint, the fact that *R. conicus* exploits native *Cirsium*, and the likelihood that *T. horridus* will do the same once it becomes established in areas occupied by these plants, represents a direct threat to native thistles. Thus, it poses a serious conflict for biological control. The other three insect species that have been imported have a diet breadth that is considerably narrower than that of *R. conicus* and *T. horridus*. However, none of the three species is absolutely specific to the genus *Carduus*. For example, open field tests in Italy showed that the syrphid fly, *Cheilosia corydon*, oviposited only on selected *Carduus* species, including musk thistle (Rizza et al. 1988). Additional starvation tests by these investigators showed no survival of first instars when placed on Canada thistle (*Cirsium arvense*) or six *Cirsium* species that are native to the United States [*C., andrewsii* (Gray) Jeps., *C. discolor* (Muhl.) Spreng., *C. occidentale* (Nutt.) Jeps., *C. crassicaule* (Greene) Jeps., *C. douglasii* DC., *C. foliosum* (Hook) DC, *and C. flodmani* (Rydb.) Arthur]. However, one larva was recovered on 2 of 5 plants of *C. crassicaule* (Rizza et al. 1988). Larvae also have been recovered at very low densities from three European species—bull thistle, *Cirsium vulgare*, *C. palustre*, and *C. eriophorum* (Nurse 1910, van Leeuwen 1983, Rizza et al. 1988).

The gall fly, *Urophora solstitialis*, has been recorded mainly from thistles in the genus *Carduus*. However, it has been reported in Europe on *Centaurea* and *Carlina*. Its affiliation with two additional genera, *Arctium* and *Carthamus*, is open to question (Batra et al. 1981). There have been no reports of this fly on any species of *Cirsium*. Like *R. conicus, U. solstitialis* may have biotypes that specialize on different species of *Carduus*. For example, in western Canada, establishment on plumeless thistle (*Carduus acanthoides*) was successful only when gall flies were collected from *C. acanthoides* in Europe (P. Tipping, Maryland Dept. of Agriculture, personnel communication). In Maryland, attempts to establish a population collected from musk thistle on plumeless thistle failed (P. Tipping, Maryland Dept. of Agriculture, personal communication).

The flea beetle, *Psylliodes chalcomera,* exhibits a strong preference for musk thistle. This conclusion is based on evidence from a two-year, open-field experiment in which beetles were attracted to and oviposited on only musk thistle when offered a choice between members of the *Carduus nutans* species group and various *Cirsium*, including the native North American species, *C. undulatum, C., andrewsii,* and *C. douglasii,* and two European species, *C. eriophorum* and *C. palustre* (Dunn and Campobasso 1993). Furthermore, in a separate cage experiment in which beetles were offered a choice of plants, heavy adult feeding and oviposition was noted on musk thistle and on Italian thistle (*Carduus pycnocephalus*), whereas light feeding and few to no eggs were laid on artichoke (*Cynara scolymus*) and Sweet Sultan (*Centaurea moschata* var. *imperalis*), respectively. Safflower *(Carthamus tinctorius)* was avoided by adult beetles (Dunn and Rizza 1977). In the field, adults feed occasionally on *Onopordum illyricum* L, although there is no evidence that adults oviposit on this plant.(Dunn and Rizza 1976). In tests where *P. chalcomera* was confined with only one plant species (no-choice experiments), a limited amount of adult feeding and oviposition occurred on plants in other genera, including three *Cirsium* species [*C. vulgare, C. monspessulanum,* and *C. arvense*] (Dunn and Rizza 1977). The ability of artificially placed larvae to survive on *Cirsium* or on species of *Carduus* other than *C. nutans* has not been tested. However, Dunn and Campobasso (1993) suggest that non-preferred thistles may be nutritionally suitable for *P. chalcomera* larvae.

Feeding Niches, Seasonal Activity and Competition

Currently, the head weevil, *R. conicus,* and the rosette weevil, *T. horridus*, are the only imported musk thistle natural enemies that have overlapping distributions in the United States. They are considered to be complementary biological control agents because they feed on different plant parts at different times of the year (spatial and temporal niche separation) and, thus, are able to impose different forms of stress and damage. The tephritid fly, *U. solstitialis,* is established in Canada on both plumeless thistle and musk (nodding) thistle. This fly is not known to be established anywhere in the United States despite repeated efforts to colonize it, largely in Maryland (P. Tipping, Maryland Department of Agriculture, personal communication). Because *U. solstitialis* feeds in flower heads, it has the potential for direct competition with *R. conicus*. For example, in Australia, where both species are established, *R. conicus* outcompetes *U. solstitialis* during their seasonal overlap even though the latter is considered the more important species because it completes more than one generation and, thus, covers a broader part of the thistle's flowering period (Woodburn 1996). In Maryland, the presence of *R. conicus* in field cages where *U. solstitialis* had been released may have prevented colonization of the latter species because of competitive exclusion (P. Tipping, Maryland Dept. of Agriculture, personal communication).

Three imported species, *T. horridus, C. corydon,* and *P. chalcomera,* feed exclusively on vegetative and pre-flowering structures. These include rosette leaves, stems, buds and roots. Periods of seasonal activity also overlap. For example, in the middle latitudes of the Mediterranean, adults of *T. horridus* and *P. chalcomera* terminate aestivation in autumn and begin ovipositing in the centers of rosettes. Oviposition in rosettes resumes in the spring. Because larvae begin development in the same plant tissues during similar periods, the likelihood for interspecific competition appears high. In addition, both beetles could compete directly with *C. corydon*, whose larvae develop in the young stems of rosettes following oviposition in early spring. The life histories of the three species diverge after plants bolt, thus reducing direct competition. For example, *T. horridus* larvae remain in the crowns of rosettes where they mature and emerge in late spring., whereas *P. chalcomera* larvae develop in leaf and floral buds after stem elongation, with adults emerging in early summer. In *C. corydon,* larvae mine in the stem before and after bolting. When feeding ceases, mature larvae migrate down into the root zone where they overwinter. Pupation and emergence occur in the spring. Direct competition may be avoided by late-reproducing *P. chalcomera* if oviposition occurs on secondary shoots of plants that have survived an early infestation by the rosette weevil. Likewise, *P. chalcomera* may be able to infest buds of plants whose stems are infested by *C. corydon* larvae. Although *R. conicus* larvae feed predominantly in the maturing flower heads, some larvae may feed on buds and within stems, especially when populations are high and preferred feeding sites are limited. However, competition with other species would probably be minimal.

In areas where multiple biological control agents are present, indirect interspecific competition may be a limiting factor. However, predictions are difficult because of the large number of possible interactions (see review by Wootton 1994), the many biotic and abiotic factors that may influence these interactions, and the limited number of studies. In general, few studies have been made of indirect competition involving temporally displaced herbivores or the consequences of these interactions on biological control programs (but see Crawley and Pattrasudhi 1988, Story et al. 1991, James et al. 1992, Pilson 1996).

OPTIONS FOR BIOLOGICAL CONTROL OF MUSK THISTLE

Various options exist with respect to biological control of musk thistle. These include (1) continued redistribution of established species; (2) continued release of approved species that are not yet established; (3) suspending further releases of some or all species that have been imported to date, and (4) conducting searches for new species. Exploration for new highly host-specific natural enemies may represent a long-term option for musk thistle biological control. However, the benefits of this approach—assuming that undiscovered, potentially useful species exist—would not be realized for many years. At the other extreme, the most immediate and convenient course of action—to collect and redistribute the two established weevils—poses a serious conflict of interest. On the one hand, reductions in musk and plumeless thistle seed quantity and quality caused by *R. conicus* have been well documented (Kok and Surles 1975, Rees 1977, McCarty and Lamp 1982, Kok and Pienkowski 1985; also see review by Andres and Rees 1995 and data contained therein). On the other hand, the ability of this herbivore to use a wide range of thistles creates an added risk for rare and endangered native thistles in North America, whose inherent life history traits and environmental circumstances already threaten their survival (see Louda and Potvin 1995).

The rosette weevil, *T. horridus*, is generally believed to play a complementary suppressive role on musk and plumeless thistle in areas where both weevils exist. In addition, it may aid in the suppression of weeds like plumeless thistle whose seasonal reproductive period is not well synchronized with that of the head weevil, *R. conicus* (Surles and Kok 1977). However, the broad range of thistles on which *T. horridus* feeds and develops suggests that it is quite capable of attacking native *Cirsium* species in North America. *T. horridus* has a New World range that is smaller than that of the head weevil. Consequently, redistribution of this weevil has been actively pursued (e.g., see Hilbert and Brooks 1993, Puttler 1994).

The third option, to continue to import and release species that are not yet established on musk thistle, deserves careful consideration. For example, while *Urophora solstitialis*, *Cheilosia corydon*, and *Psylliodes chalcomera* all have host ranges that are narrower than those of the head and rosette weevils, none is restricted to the genus *Carduus*, and two of the three—*C. corydon* and *P. chalcomera*—appear to be physiologically capable of developing on thistles within the genus *Cirsium*. Thus, some concern remains about potential nontarget impacts. With respect to biological control, establishment of new natural enemies would likely result in interspecific competition with the already-present weevils because of the overlap in seasonal activity and shared plant tissues attacked. Because competition may reduce the level of biological control, new introductions should not be made without giving careful consideration to this issue. Known interactions between *R. conicus* and *U. solstitialis* in Australia (see Woodburn 1996) and Canada (P. Harris, Agriculture Canada, personal communication) justify this concern.

ISSUES AND RECOMMENDATIONS FOR FUTURE ACTION

After more than a quarter of a century as a target of classical biological control, a comprehensive evaluation of the status of musk thistle in North America is needed. At present, the *Rhinocyllus conicus* is established on musk thistle in more than 20 states across the U.S. mainland (Andres and Rees 1995) as well as in the Canadian provinces of Quebec, Ontario, Manitoba, Alberta, Saskatchewan and British Columbia (Julien 1992, Harris and Shorthouse 1996). *Trichosirocalus horridus* is established in six states plus Alberta, British Columbia, Manitoba and Saskatchewan (Harris 1984, Julien 1992, Andres and Rees 1995). The actual ranges of both weevils are likely to be larger than that reported because of subsequent releases and natural dispersal. Future range expansion is expected, even without human assistance. *Urophora solstitialis* is established in at least two Canadian provinces (Ontario and Saskatchewan) and apparently coexists with *R. conicus* (Möller-Joop and Schroeder 1986). The potential of this fly to expand its present North American distribution is unknown.

Experimental data to support the strong circumstantial evidence (Louda 1998, Louda et al. 1997) of *R. conicus'* impact on native *Cirsium* species currently are lacking. Notwithstanding, if a case can be made that prolific seed-producing weeds like musk thistle and plumeless thistle are under suppression from this biological control agent (Rees 1977, Kok 1986, Kok and Surles 1975, Kok and Pienkowski 1985), then it is difficult to argue that native species like Platte thistle *(Cirsium canescens* [Nuttall] Sprengel) and others, which have very limited seed production (Louda and Potvin 1995, Louda et al. 1990), are not at risk from *R. conicus*, and at risk in the future from *T. horridus* and *U. solstitialis*. In part because of this risk, deliberate movement of these species should be suspended indefinitely. Additional justification derives from the absence of data to show how the introduced biological control agents interact with each other on target weeds. For example, no experiment has been done to quantify the impact of *R. conicus* and *T. horridus* singly and together. Therefore, the assumption that the two weevils act in a complementary way to increase the level of biological control has not been tested. Moreover, because the rosette weevil infests thistles before the head weevil, logical but unresolved questions are: Does prior feeding by *T. horridus* alter musk thistle in ways that reduce the effectiveness of *R. conicus*? And, if so, does this decrease the overall level of biological control? For example, could changes in flowering phenology increase the amount of seasonal asynchrony with *R. conicus* beyond that which has been reported (see Kok and Surles 1975). Experimental evaluation over a broad geographic area would reveal regional/climatic differences in how the two weevils interact with each other and with musk thistle. Similar questions need to be answered in Canada in areas where *R. conicus* and *U. solstitialis* overlap. Because both species feed in floral heads, direct competition can occur (see Woodburn 1996). The consequences of this competition on the effectiveness of biological control are unknown.

The use of interspecific plant competition to manage musk thistle should be given more consideration. For example, in the early 1980s, Kok et al. (1986) determined that competition from tall fescue grass played a relatively greater role in the suppression of musk thistle in Virginia than that of the head and rosette weevils. Likewise, results of other studies (Austin et al. 1985, Hamrick and Lee 1987, Popay and Medd 1990, Bourdot 1996) support the idea that musk thistle densities are reduced by competing grasses. Thus, improved land management (e.g., steps to prevent overgrazing, elimination of fallow fields, land restoration, etc.) should play an integral part in future management plans for musk thistle. Other non-chemical procedures, including mowing thistles before flowering, can be very effective tools for thistle management (Fick 1995).

The logistical and economic feasibility of land management and other non-chemical alternatives in all areas infested by musk thistle remains to be demonstrated. The impact of already-established musk thistle natural enemies throughout North America needs further attention. The introduction of new natural enemies for musk thistle, and other species of exotic *Carduus,* continues to be an option for further consideration. However, immediate release of these enemies does not appear to be warranted

Instead, efforts should be directed toward reassessing the current situation at both the state and regional levels. Specific questions and recommendations include:

1. What is the status of the weed problem? Surveys of current distributions and densities of musk thistle should be made. These should go beyond simple "presence-absence" reports. Long-term population data should be reviewed where they exist, and the collection of new data should be made to assess whether average musk

thistle populations are increasing, decreasing, or remaining at current levels. Reasons for population changes should be determined.

2. Are head and/or rosette weevils already established locally? If so, what impact are they having on musk thistle populations? If not, what is the likelihood for future colonization by these species through passive dispersal? A comprehensive survey of the current North American distribution of the head and rosette weevils should be made. Experiments should be conducted to document which combination of weevils would provide the maximum impact on the weed in different geographic/climatic areas. Factors that should be included in the evaluation are possible indirect competitive effects of the rosette weevil on the head weevil and phenological matching between the herbivores and the weed. Direct competition between *R. conicus* and *U. solstitialis* needs to be studied. Finally, the potential effects of other thistles (both weeds and non-weeds) on the population dynamics of all biological control agents should be considered.
3. What management options other than importing natural enemies exist? Are they feasible given the status of the weed?
4. What is the risk of newly-imported insect species to native *Cirsium* in areas that lack musk thistle? This assessment should include additional no-choice tests with specific, high-risk thistles to determine if oviposition occurs and, if so, the suitability of the nonpreferred plants for larval development and survival. Ideally, additional tests would be done in Europe using a modification of the open-field protocol developed and used by USDA scientists there (Dunn and Campobasso 1993; also see review by Clement and Cristofaro 1995).
5. How might the introduction of new species, or biotypes of existing species, affect the overall level of musk thistle biological control? Would competitive interference reduce the overall impact? Or would the species act in a complementary manner? How much seasonal overlap can one predict between currently established and the newer weed biological control agents? And, how well synchronized would new herbivores be with the target weeds? Experiments with different combinations of species are needed, as well as phenological studies.
6. What biological differences occur in populations of insects that are collected from different geographical areas, or from different thistle species? The issue of biotypes may have very significant implications, both for biological control and for the conservation of native species (see Zwölfer and Preiss 1983, Goeden et al. 1985).

CHALLENGES FOR CONFLICT RESOLUTION AND IMPROVED BIOLOGICAL CONTROL

The musk thistle program has been the subject of intense discussion and debate. Much of the criticism of this program has focused on past decisions to import and release weevils whose host plant ranges were broad enough to include native North American species of thistles. The release of additional insect species also has been challenged even though the documented host ranges are considerably narrower than those of the head and rosette weevils. At this juncture, rather than debate what constitutes an acceptable risk for releasing new species, perhaps the key question simply is "Are more natural enemies needed?" The protocol outlined above provides an opportunity to look objectively at the musk thistle program. Adopting a more conservative approach with respect to natural enemy introductions serves the mutual interests of those concerned with biological control of musk thistle and those concerned with nontarget impacts on native thistles.

This general approach could serve as a paradigm for other classical biological control of weeds programs, especially programs for which introduced natural enemies are already established. Besides reducing potential risks to nontarget species, a more deliberate approach would allow for additional biological and ecological studies to be conducted before enemies are released. Careful biological evaluation can lead to useful predictions about natural enemy performance (Ehler 1990), as well as identify factors likely to preclude establishment or limit natural enemy performance (Nechols 1999). For example, phenological investigations are critically important to the success of biological control programs (see pp. 305-309 in Tauber et al. 1986). Also, the possible consequences of interspecific interactions between existing and candidate enemies need to be considered. These include not only competition, but intraguild predation as well (see Story et al. 1991; also review by Rosenheim et al. 1995). Because predation and/or parasitism from resident natural enemies pose a potential barrier to the establishment or effectiveness of introduced enemies, unique approaches to predetermine this risk may be possible (see Nechols et al. 1996).

Finally, the issue of acceptable risk is not easily resolved, and future conflicts are inevitable between practitioners of biological control and those who oppose importation of natural enemies. However, recommendations—some of which have been cited for musk thistle—may foster better cooperation, more positive interactions, and alleviate some conflicts for weeds biological control. Of paramount importance is that constructive dialogue must take place. The adversarial approach that prevails on both sides of the importation issue is counterproductive to *all* interests because resources are limited and the problems that need to be solved are biologically very complex, requiring long-term ecological assessments. Therefore, cooperation is needed from ecologists who are not engaged in the practice of weed biological control. For example, besides data collection, these individuals could help to identify ecosystems that are at greatest risk, and those that contain native species that are closely affiliated with pests targeted for biological control. For their input to be useful, however, representation is needed during both the planning and decision phases. Only through open discussion, debate and the exchange of knowledge between scientists working in managed and unmanaged ecosystems will it be possible to move beyond rhetoric to the point where the issue of importing natural enemies can be examined rigorously, objectively, and comprehensively to protect the mutual interests of biological control and conservation biology.

ACKNOWLEDGEMENTS

I thank the following colleagues for sharing their knowledge of thistle biology and biological control of *Carduus* weeds: Paul Boldt, Gaetano Campobasso, Walt Fick, Peter Harris, Bill Hilbert, Ernst Horber, Bill Scott and Phil Tipping. I also thank Peter Follett and Jian-Jun Duan for editorial assistance and an anonymous reviewer for comments on an earlier draft of this manuscript. I especially thank Svata Louda for critically reviewing the manuscript, for sharing her expertise on thistle ecology, and for many challenging discussions about nontarget issues relative to biological control.

DEDICATION

This chapter is dedicated to the memory of Charles E. Turner, whose botanical expertise and career-long research on weeds served the mutual interests of biological control and conservation biology well.

REFERENCES

Andres, L. A., and N. E. Rees. 1995. Musk thistle. Chap. 67. *In* J. R. Nechols, L. A. Andres, J. W. Beardsley, R. D. Goeden, and C. G. Jackson (eds.), Biological control in the western United States: accomplishments and benefits of Regional Research Project W–84, 1964–1989. No. 3361. Univ. California, Div. Agric. Nat. Res., Oakland. p. 248–251.

Austin, M. P., R. H. Groves, L. M. F. Fresco, and P. E. Kaye. 1985. Relative growth of six thistle species along a nutrient gradient with multispecies competition. J. Ecology 73: 667–684.

Batra, S. W. T. 1978. *Carduus* thistle distribution and biological control in northeastern states. p. 18–22. *In* K. E. Frick (ed.), Biological control of thistle in the genus *Carduus* in the United States: a progress report. U.S. Dept. Agric., Sci. and Educ. Admin.

Batra, S. W. T., J. R. Coulson, P. H. Dunn, and P. E. Boldt. 1981. Insects and fungi associated with *Carduus* thistles (Compositae). U.S. Dept. Agric. Tech. Bull. 1616.

Boldt, P. E., G. Campobasso, and E. Colonnelli. 1980. Palearctic distribution and host plants of *Ceutorhynchus trimaculatus* and *Trichosirocalus horridus* (Coleoptera: Curculionidae). Ann. Entomol. Soc. Am. 73: 694–698.

Bourdot, G. W. 1996. Interference between pasture plants and thistles; B, a review. Plant Prot. Quart. 11: 265–270.

Cartwright, B., and L. T. Kok. 1985. Growth responses of musk and plumeless thistles (*Carduus nutans* and *C. acanthoides*) to damage by *Trichosirocalus horridus* (Coleoptera: Curculionidae). Weed Sci. 33: 57–62.

Clement, S. L., and M. Cristofaro. 1995. Open-field tests in host-specificity determination of insects for biological control of weeds. Biocontrol Sci. Tech. 5: 395–406.

Coombs, E. 1996. Biological Control of Weeds Project Summaries, Oregon Dept. Agric., Commodity Inspection Div., Noxious Weed Control Program, Salem. p. 9.

Crawley, M. J., and R. Pattrasudhi. 1988. Interspecific competition between insect herbivores: asymmetric competition between cinnibar moth and the ragwort seed-head fly. Ecol. Entomol. 13: 243–249.

De Quattro, J. 1997. Two Italian imports tackle musk thistle. Agric. Res. (March) p. 10–11.

Desrochers, A. M., J. F. Bain, and S. I. Warwick. 1988. A biosystematic study of the *Carduus nutans* complex in Canada. Can. J. Bot. 66: 1621–1631.

Dunn, P. H. 1976. Distribution of *Carduus nutans, C. acanthoides, C. pycnocephalus,* and *C. crispus* in the United States. Weed Science 24: 518–524.

Dunn, P. H., and A. Rizza. 1976. Bionomics of *Psylliodes chalcomera*, a candidate for biological control of musk thistle. Ann. Entomol. Soc. Am. 69: 395–398.

Dunn, P. H., and A. Rizza. 1977. Host specificity of *Psylliodes chalcomera*, a candidate for biological control of musk thistle. Environ. Entomol. 6: 449–454.

Dunn, P. H., and G. Campobasso. 1993. Field test of the weevil *Hadroplonthus trimaculatus* and the fleabeetle *Psylliodes chalcomera* against musk thistle (*Carduus nutans*). Weed Sci. 41: 656–663.

Ehler, L. E. 1990. Introduction strategies in biological control of insects. *In* M. Mackauer, L. E. Ehler, and J. Roland (eds.), Critical issues in biological control. Intercept, Andover, UK. p. 111–134.

Fick, W. H. 1995. Ecology and control of musk thistle. Proc. 5th Intl. Rangeland Cong. p. 145–146.

Goeden, R. D. 1978. Initial analyses of *Rhinocyllus conicus* (Froelich) (Coleoptera: Curculionidae) as an introduced natural enemy of milk thistle (*Silybum marianum* [L.] Gaertner) and Italian thistle (*Carduus pycnocephalus* L.) in southern California. U.S. Dept. Agric. Publ. 1978-771-106/02, Washington, D.C. p. 39–50.

Goeden, R. D. 1995a. Italian thistle. *In* J. R. Nechols, L. A. Andres, J. W. Beardsley, R. D. Goeden, and C. G. Jackson (eds.), Biological control in the western United States: accomplishments and benefits of Regional Research Project W-84, 1964–1989. No. 3361, Univ. California, Div. Agric. Nat. Res., Oakland. p. 242–244.

Goeden, R. D. 1995b. Milk thistle. *In* J. R. Nechols, L. A. Andres, J. W. Beardsley, R. D. Goeden, and C. G. Jackson (eds.), Biological control in the western United States: accomplishments and benefits of Regional Research Project W-84, 1964–1989. No. 3361, Univ. California, Div. Agric. Nat. Res., Oakland. p. 245–248.

Goeden, R. D., and D. W. Ricker. 1985. Seasonal asynchrony of Italian thistle, *Carduus pycnocephalus,* and the weevil, *Rhinocyllus conicus* (Coleoptera: Curculionidae), introduced for biological control in southern California. Environ. Entomol. 14: 433–436.

Goeden, R. D., D. W. Ricker, and B. A. Hawkins. 1985. Ethological and genetic differences among three biotypes of *Rhinocyllus conicus* (Coleoptera: Curculionidae) introduced into North America for the biological control of asteraceous thistles. p. 181–189. *In* E. S. Delfosse (ed.), Proc. 6th Intl. Sym. Biological Control of Weeds, Vancouver, Canada, 1984.

Hamrick, J. L., and J. M. Lee. 1987. Effect of soil surface topography and litter cover on the germination, survival, and growth of musk thistle (*Carduus nutans*). Am. J. Bot. 74: 451–457.

Harris, P. 1984. *Carduus nutans* L., nodding thistle and *C. acanthoides* L. plumeless thistle (Compositae) *In* J. S. Kelleher and M. A. Hulme (eds.), Biological control programmes against insects and weeds in Canada 1969–1980. Commonwealth Agric. Bur., Farnham Royal, UK. p. 115–126.

Harris, P., and J. D. Shorthouse. 1996. Effectiveness of gall inducers in weed biological control. Can. Entomol. 128: 1021–1055.

Harris, P., and H. Zwölfer. 1971. *Carduus acanthoides* L., welted thistle, and *C. nutans* L., nodding thistle. *In* Biological control programmes against insects and weeds in Canada 1959–1968. Commonwealth Inst. Biol. Control Tech. Commun. no. 4. p.76–79.

Hilbert, B., and H. L. Brooks. 1993. Biological control of musk thistle in Kansas. Pub. No. L-873, Cooperative Extension Service, Kansas State Univ., Manhattan. 6 p.

Hodgson, J. M., and N. E. Rees. 1976. Dispersal of *Rhinocyllus conicus* for biocontrol of musk thistle. Weed Sci. 24: 59–62.

James, R. R., P. B. McEvoy, and C. S. Cox. 1992. Combining the cinnibar moth (*Tyria jacobaeae*) and the ragwort flea beetle (*Longitarsus jacobaeae*) for control of ragwort (*Senecio jacobaea*): an experimental analysis. J. Appl. Ecol. 29: 589–596.

Julian, M. H. (ed.). 1992. Biological control of weeds: a world catalogue of agents and their target weeds. Commonwealth Agric. Bur. Intl., Canberra, Australia.

Kazmi, S. M. A. 1964. Revision der gaaattung *Carduus* (Compositae). Teil II. Bot. Staats. Munchen, 5: 279–550.

Kok, L. T. 1975. Host specificity studies on *Ceuthorhynchidius horridus* (Panzer) (Coleoptera: Curculionidae) for the biocontrol of musk and plumeless thistle. Weed Res. 15: 21–25.

Kok, L. T. 1986. Impact of *Trichosirocalus horridus* (Coleoptera: Curculionidae) on *Carduus* thistles in pastures. Crop Protection 5: 214–217.

Kok, L. T., and W. T. Mays. 1991. Successful biological control of plumeless thistle, *Carduus acanthoides* L. [Campanulatae: Asteraceae (= Compositae)], by *Trichosirocalus horridus* (Panzer) (Coleoptera: Curculionidae) in Virginia. Biol. Control 1: 197–202.

Kok, L. T., and R. L. Pienkowski. 1985. Biological control of musk thistle by *Rhinocyllus conicus* (Coleoptera: Curculionidae) in Virginia from 1969 and 1980. *In* E. S. Delfosse (ed.), Proc. 6th Internat. Symp. Biol. Contr. Weeds, August 19–25, 1984, Vancouver, Canada. Agric. Can. p. 433–438.

Kok, L. T., and W. W. Surles. 1975. Successful biocontrol of musk thistle by an introduced weevil, *Rhinocyllus conicus*. Environ. Entomol. 4: 1025–1027.

Kok, L. T., and J. T. Trumble. 1979. *Ceuthorhynchidius horridus* (Coleoptera: Curculionidae): life cycle and development on *Carduus* thistles in Virginia. Ann. Entomol. Soc. Am. 72: 563–564.

Kok, L. T., T. J. McAvoy, and W. T. Mays. 1986. Impact of tall fescue grass and *Carduus* thistle weevils on the growth and development of musk thistle (*Carduus nutans*). Weed Sci. 34: 966–971.

Louda, S. M. 1998. Population growth of *Rhinocyllus conicus* (Coleoptera: Curculionidae) on two species of native thistles in prairie. Environ. Entomol. 27: 834–841.

Louda, S. M., and M. A. Potvin. 1995. Effect of inflorescence-feeding insects on the demography and lifetime fitness of a native plant. Ecology 76: 229–245.

Louda, S. M., D. Kendall, J. Connor, and D. Simberloff. 1997. Ecological effects of an insect introduced for

the biological control of weeds. Science 277: 1088–1090.
Louda, S. M., M. A. Potvin, and S. K. Collinge. 1990. Predispersal seed predation, postdispersal seed predation and competition in the recruitment of seedlings of a native thistle in Sandhills prairie. Am. Midl. Nat. 124: 105–113.
McCarty, M. K. 1982. Musk thistle (*Carduus) thoermeri* seed production. Weed Sci. 30: 441–445.
McCarty, M. K., and W. O. Lamp. 1982. Effect of a weevil, *Rhinocyllus conicus*, on musk thistle (*Carduus thoermeri*) seed production. Weed Sci. 30: 136–140.
McCarty, M. K., C. J. Scifres, A. L. Smith and G. L. Horst. 1969. Germination and early seedling development of musk and plumeless thistle. Nebraska Agric. Exp. Sta. Res. Bull. 229.
McGregor, R. L. 1986. *Carduus*. *In* T. M. Barkley (ed.), Flora of the Great Plains. Great Plains Flora Assoc., Univ. Kansas Press, Lawrence. p. 895–897.
Möller–Joop and D. Schroeder. 1986. *Urophora solstitialis* (L.) (Diptera: Tephritidae). A candidate for the biological control of plumeless thistle (*Carduus acanthoides* L.) in Canada. Commonwealth Inst. Biol. Control Report. Delemont, Switzerland.
Moore, W. B., C. J. Doyle, and A. Rahman. 1989. Economics of controlling *Carduus nutans* on grazed pasture in New Zealand. Crop Protection 8: 16–24.
Nechols, J. R. 1999. Challenges in searching for, releasing, and evaluating natural enemies of native versus exotic pests. *In* L. D. Charlet and G. J. Brewer (eds.), Biological control of native or indigenous insect pests: challenges, constraints, and potential. Thomas Say Publications in Entomol., Entomol. Soc. Am., Lanham, Maryland. In press.
Nechols, J. R., L. A. Andres, J. W. Beardsley, R. D. Goeden, and C. G. Jackson (eds.), 1995. Biological control in the western United States: accomplishments and benefits of Regional Research Project W-84, 1964–1989. No. 3361. Univ. California, Div. Agric. and Nat. Res., Oakland.
Nechols, James R., John J. Obrycki, Catherine A. Tauber, and Maurice J. Tauber. 1996. Potential impact of native natural enemies on *Galerucella* spp. (Coleoptera: Chrysomelidae) imported for biological control of purple loosestrife: a field evaluation. Biol. Control 7:60–66.
Nurse, C. G. 1910. Notes regarding the breeding of *Cheilosia grossa*. Entomologist 43: 313–314.
Pilson, D. 1996. Two herbivores and constraints on selection for resistance in *Brassica rapa*. Evolution 50: 1492–1500.
Popay, A. I., and R. W. Medd. 1990. The biology of Australian weeds 21. *Carduus nutans* L. Plant Prot. Qtrly. 5: 3–13.
Puttler, B. 1994. Redistribution and status of musk thistle weevils to enhance the reduction of musk thistle (*Carduus nutans*) populations in Missouri. Dept. Entomol. Spec. Rpt. Ser., Univ. Missouri.
Reece, P. E., and R. G. Wilson. 1983. Effect of Canada thistle (*Cirsium arvense*) and musk thistle (*Carduus nutans*) control on grass herbage. Weed Sci. 31: 488–492.
Rees, N. E. 1977. Impact of *Rhinocyllus conicus* on thistles in southwestern Montana. Environ. Entomol. 6: 839–842.
Rees, N. E. 1992. Biological control of two thistle species. *In* Biological control of forest pests in the Great Plains: status and needs; B, a symposium. Great Plains Agric. Council Pub. 145. p. 109–115.
Rizza, A., G. Campobasso, P. H. Dunn, and M. Stazi. 1988. *Cheilosia corydon* (Diptera: Syrphidae), a candidate for the biological control of musk thistle in North America. Ann. Entomol. Soc. Am. 81: 225–232.
Robinson, A., Jr. 1976. Musk thistle: an eco-problem. Trans. Kans. Acad. Sci. 79: 127–129.
Rosenheim, J. A., H. K. Kaya, L. E. Ehler, J. J. Marois, and B. A. Jaffee. 1995. Intraguild predation among biological-control agents: theory and evidence. Biol. Control 5: 303–335.
Sieburth, P. J., and L. T. Kok. 1982. Ovipositional preference of *Trichosirocalus horridus* (Coleoptera: Curculionidae). Can. Entomol. 114: 1201–1202.
Story, J. M., K. W. Boggs, W. R. Good, P. Harris, and R. M. Nowierski. 1991. *Metzneria paucipunctella* Zeller (Lepidoptera: Gelechiidae), a moth introduced against spotted knapweed: its feeding strategy and impact on two introduced *Urophora* spp. (Diptera: Tephritidae). Can. Entomol. 123: 1001–1007.
Stuckey, R. L., and J. L. Forsyth. 1971. Distribution of naturalized *Carduus nutans* (Compositae) mapped in relation to geology in northeastern Ohio. Ohio J. Sci. 71: 1–15.
Surles, W. W., and L. T. Kok. 1977. Ovipositional preference and synchronization of *Rhinocyllus conicus* with *Carduus nutans* and *C. acanthoides*. Environ. Entomol. 6: 222–224.
Surles, W. W., L. T. Kok, and R. L. Pienkowski. 1974. *Rhinocyllus conicus* establishment for biocontrol of thistles in Virginia. Weed Sci. 22: 1–3.
Tauber, M. J., C. A. Tauber, and S. Masaki. 1986. Seasonal adaptations of insects. Oxford, New York.

Thompson, A., A. E. Saunders, and P. Martin. 1995. The effect of nodding thistle (*Carduus nutans*) on pasture production. *In* M. T. K. Arroyo, P. H. Zedler, and M. D. Fox (eds.), Ecology and biogeography of Mediterranian ecosystems in Chile, California and Australia. Springer, New York. p. 311–344.

Tipping, P. W., and C. A. Holko. 1995 Annual report, Plant Protection Section, Maryland Dept. Agric., Annapolis. p. 12–13.

Trumble, J. T., and L. T. Kok. 1982. Integrated pest management techniques in thistle suppression in pastures of North America. Weed Res. 22: 345–359.

Turner, C. E., and J. C. Herr. 1996. Impact of *Rhinocyllus conicus* on a non-target, rare, native thistle (*Cirsium fontinale*) in California. *In* Proc. 9th Intl. Sym. on Biological Control of Weeds, January 19–26, 1996, Stellenbosch, South Africa. Univ. Cape Town, Cape Town, South Africa. p. 103.

Turner, C. E., R. W. Pemberton, and S. S. Rosenthal. 1987. Host utilization of *Cirsium* thistles (Asteraceae) by the introduced weevil *Rhinocyllus conicus* (Coleoptera: Curculionidae) in California. Environ. Entomol. 111–115.

van Leeuwen, B. H. 1983. The consequences of predation in the population biology of the monocarpic species *Cirsium palustre* and *Cisium vulgare*. Oecologia 58: 178–187.

Ward, R. H., R. L. Pienkowski, and L. T. Kok. 1974. Host specificity of the first-instar of *Ceuthorhynchidius horridus*, a weevil for biological control of thistles. J. Econ. Entomol. 67: 735–737.

Woodburn, T. L. 1996. Interspecific competition between *Rhinocyllus conicus* and *Urophora solstitialis*, two biocontrol agents released in Australia against *Carduus nutans*. *In* V. C. Moran and J. H. Hoffmann (eds.), Proc. 9th Intl. Sym. on Biological Control of Weeds. Univ. Capetown, Stellenbosch, South Africa. p. 409–415.

Wootton, J. T. 1994. The nature and consequences of indirect effects in ecological communities. Annu. Rev. Ecol. Syst. 25: 443–466.

Zwölfer, H. 1967. The host-range, distribution and life-history of *Rhinocyllus conicus* Froel. Commonwealth Inst. Biol. Control Prog. Rpt. 18: 1–21.

Zwölfer, H., and M. Preiss. 1983. Host selection and oviposition behaviour in West-European ecotypes of *Rhinocyllus conicus* Froel. (Col., Curculionidae). Z. angew. Ent. 95: 133–122.

PART 4

PATHOGENS

15

PREDICTING THE HOST RANGE OF ENTOMOPATHOGENIC FUNGI

Ann E. Hajek
Department of Entomology, Cornell University, Ithaca, New York

Linda Butler
Division of Plant and Soil Sciences, West Virginia University, Morgantown, West Virginia

INTRODUCTION

For many years the development and use of natural enemies to control insect pests has primarily focused on entomophagous insects that are predators and parasitoids. As it has become easier to work with microbes and increasing numbers of scientists have cross-trained in studies of microorganisms as well as entomology, the number of programs evaluating use of pathogens to control insect pests has increased. Among the diverse groups of microbes pathogenic to invertebrates, there are >700 species of entomopathogenic fungi. Fungi are unique among invertebrate pathogens because the major site of infection is through the cuticle, while most other pathogens must be ingested and subsequently invade through the gut (Hajek and St. Leger 1994). Because fungi invade through the cuticle, fungal pathogens are especially important for control of insects that suck plant sap and therefore would not ingest other types of entomopathogens. However, entomopathogenic fungi are being investigated for control of virtually all groups of pestiferous arthropods

There is great diversity among the fungi pathogenic to invertebrates, including representatives from all of the major fungal groups but dominated by species from the Hyphomycetes and Entomophthorales. Fungal pathogens of insects range from obligate pathogens that cannot be grown *in vitro* and whose life cycles are often finely tuned to hosts to facultative pathogens that are only opportunistically pathogenic.

Use of entomopathogenic fungi for insect control has been pursued for numerous fungal species due to the dramatic epizootics caused by these pathogens. Fungi could potentially be employed for pest control utilizing several different strategies, but greatest efforts have been directed toward inundative augmentation, i.e., use as mycoinsecticides (Roberts and Hajek 1991). Several fungal species have been developed as mycoinsecticides and are used for pest control in numerous countries. Fungal species are also being investigated for inoculative augmentation, in which a pathogen is released, becomes established, and causes cycles of infection in the host population. In comparison with parasitoid and predators, fungal pathogens have seldom been used for classical biological control (Maddox et al. 1992), and relatively few systems for enhancement of fungal populations via environmental manipulation have been developed (Fuxa 1998).

To use entomopathogenic fungi for biological control responsibly, we need to be able to predict their effects on nontarget organisms. Thus, knowledge of host specificity under field conditions (= ecological host range) is essential. Unfortunately, the factors that determine host specificity probably differ among the different taxonomic groups of entomopathogenic fungi and are generally poorly understood for all groups. We will present a brief overview of factors that determine host specificity of entomopathogenic fungi at the physiological level and the resulting host ranges observed in the laboratory and field. Studies in which host range has been evaluated both in the laboratory and the field will be described in detail. Laboratory host range (= physiological host range) usually includes an expanded spectrum of hosts in comparison with pathogen activity in the field and certainly provide little help in predicting indirect effects of entomopathogenic fungi. Yet decisions about safety of entomopathogens generally are based on laboratory data because ecological host range can only be determined accurately after introduction. This, of course, especially constitutes a problem for classical biological control introductions or for introductions of genetically engineered organisms. We will discuss ideas regarding how to more accurately predict specificity in the field in order to encourage and strengthen the ability to use entomopathogenic fungi for insect control and simultaneously minimize environmental impacts.

HOST SPECIFICITY OF ENTOMOPATHOGENIC FUNGI

The majority of entomopathogenic fungi are thought to infect hosts by direct penetration through the external cuticle. The steps required for successful penetration of hosts by fungi include adhesion of spores to host cuticle, spore germination, fungal growth, often differentiation to produce penetrant structures (= appressoria), and finally, cuticular penetration (Hajek and St. Leger 1994). Penetration of the cuticle is thought to involve a combination of enzymatic degradation and mechanical pressure. After penetration of the cuticle, the fungus must grow within the host. The entire suite of factors making a host permissive or non-permissive has not been identified for any host or entomopathogenic fungus. For many fungi, the cuticle is considered as the principal barrier to invasion. Inability of a fungus to penetrate the cuticle can include either lack of nutrients or presence of inhibitory chemicals on the cuticle, unacceptable cuticular morphology, or possibly the simple inability to penetrate, e.g., as on highly sclerotized coleopteran elytra (St. Leger 1991). However, studies have shown that sometimes, even when a fungal pathogen is injected directly into the insect hemocoel, thus bypassing the need for cuticular penetration, the fungus does not survive (e.g., Hajek et al. 1995). We hypothesize that this could be caused by inability of the fungus to grow either because the insect constitutes a poor source of nutrients, contains compounds inhibitory to fungal growth, or because the fungal cells are recognized as non-self and an effective immune response is activated. Examples are not known in which successful fungal development and reproduction could occur but specific cues necessary to initiate infection are lacking.

Laboratory Studies of Host Specificity

As a general rule, the first evaluations of the specificity of an entomopathogenic fungus are undertaken in the laboratory. However, due to the emphasis in insect pathology on pest control, studies comparing the activity of many fungal isolates against one pest species to test efficacy for pest control are much more common than studies simultaneously challenging numerous host species with one fungus to test specificity. Entomogenous fungi range in host breadth from those that are known from only one host species or genus (e.g., *Furia crustosa*; MacLeod and Tyrrell 1979) to those species that have been isolated from over 200 host species (e.g., *Beauveria bassiana*; Feng et al. 1994). For 19 entomopathogenic fungi, a list has been compiled of primary hosts, nontarget species that can become infected, and nontargets challenged but not infected (Goettel et al. 1990). Although the list of nontarget species that can or cannot become infected is in some instances extensive, caveats are essential. For entomopathogenic fungi, techniques that allow exact fungal identification are only now being developed and thus, many of the older studies are not based on accurate fungal identifications. In addition, some records are based on few specimens and frequently major groups of nontargets have not been challenged.

As a general rule, for fungal species having broad host ranges, individual isolates can differ significantly in specificity (= pathotypes), with the highest virulence often against the host from which the strain was isolated (Goettel et al. 1990, Glare and Milner 1991). As is standard, the summarization of nontargets that do or do not become infected (Goettel et al. 1990) does not include information about the individual fungal strains but only fungal species. Hyphomycetes that are facultatively saprophytic are frequently thought to have broader host ranges than Entomophthorales that tend more to be obligate pathogens. As a generalization, for Hyphomycetes, ability to infect a host species has often been found to be quantitative, with relatively high doses creating an ability to infect species that are less susceptible. However, even within the more obligately pathogenic Entomophthorales, *Zoophthora radicans* has a broad host range including numerous insect orders, although individual pathotypes range in specificity from attacking different insect orders to infecting only one species or insect family (Glare and Milner 1991). Molecular studies have confirmed that this fungal species includes a very heterogeneous group of strains (Milner and Mahon 1985). The importance of identifying variability in specificity among pathotypes is demonstrated in the *Entomophaga grylli* species complex. Although all members are morphologically identical and can only be differentiated using molecular techniques, the five pathotypes vary significantly in their ability to infect species within the Orthoptera (Ramoska et al. 1988). Unfortunately, molecular studies of pathotypes to date have used general genes, e.g., rDNA, for pathotype identification while the actual genes directly responsible for differential virulence in different hosts have not been identified and therefore cannot be used. In this light, use of molecular techniques to predict host specificity will not be optimal until genes associated with host species have been identified.

Laboratory host specificity tests are generally conducted using conditions that optimize fungal infection. Although laboratory testing thus controls pathogen exposures, results are not always clear. Pathogens cause typical infections in susceptible hosts and have no effect in completely refractory hosts. However, for marginal hosts, mortality can occur in some proportion of treated insects yet the fungus either cannot reproduce or reproduces only minimally (Maddox et al. 1992). Obviously it is difficult to interpret whether suboptimal infections under optimized conditions are laboratory artifacts or whether these would actually occur in the field.

The majority of studies evaluating nontarget effects compare pathogenicity among numerous pest insect species or evaluate effects on beneficial insects, e.g., parasitoids, predators, silkworms, and honeybees, and it is rare that other members of the insect fauna are tested. Rare and endangered species of invertebrates are seldom tested, usually due to restricted availability. In addition, the greatest emphasis has been on nontarget insects, with minimal consideration of other invertebrates in the environment. However, in recent years interest has grown regarding the impact of biological control agents on all members of the insect fauna, and not only pests, beneficials, and rare species. To date, few studies investigating faunal effects of entomopathogenic fungi have been undertaken.

In keeping with the trend toward investigation of a greater diversity of faunal members, preliminary studies testing the host range of the entomophthoralean pathogen *Entomophaga maimaiga* demonstrated that this pathogen infected only Lepidoptera (Soper et al. 1988, Vandenberg 1990). These studies were continued by challenging 78 species of nontarget Lepidoptera, predominantly reared from West Virginia forest endemics with optimal doses of *E. maimaiga* conidia (Hajek et al. 1995). *E. maimaiga* killed and sporulated from 35.6% of the species tested although infection levels were generally much lower than 50%. There was little phylogenetic relatedness among the groups that were infected at low levels. Steinkraus and Kramer (1987), testing *E. muscae* against 16 species of Diptera, also found little relationship between phylogeny and ability to infect. They suggested that variation in the surfaces of the insect body could be important in determining specificity instead of phylogenetic relatedness. In agreement with suggestions by Steinkraus and Kramer (1987), a significant association was observed between species with little setation or body sculpturing and lack of infection by *E. maimaiga* (Hajek et al. 1995). For *E. maimaiga*, the only phylogenetically related group that was consistently infected at high levels was the lymantriids, the same lepidopteran family from which this fungus was initially isolated.

Field Studies of Host Specificity

One method for evaluation of the effects of an entomogenous fungus on nontargets is to document naturally occurring infection levels in the field. Such studies are conducted with the assumption that if infections normally occur and the pathogen subsequently becomes more active (either by augmentation or environmental manipulation) the effects of the pathogen on nontargets would increase. However, quantitative studies encompassing many nontarget species have seldom been conducted. One study evaluated 19 species of overwintering carabids and staphylinids, collected from agricultural fields and reared to detect the extent of naturally occurring fungal infection (Steenberg et al. 1995). The three entomogenous hyphomycete species isolated were more abundant than two species of Entomophthorales but, in general, infection levels were low. Although an epizootic of *Beauveria bassiana* was documented among two staphylinid species during laboratory rearing, this occurred among adults that had not been surface sterilized prior to laboratory rearing for eight weeks. In this case it seems possible that *B. bassiana* conidia on surfaces of field-collected beetles could have germinated and caused infection only after insects were inside warm and humid rearing containers, demonstrating one of the pitfalls in trying to evaluate infection of field-collected insects.

When fungi are considered for application as mycoinsecticides, nontargets have been collected after field applications to detect infections. In an unpublished study by Goettel et al. (personal communication), *B. bassiana* was applied to both rangeland and alfalfa for grasshopper control. Leaves from the rangeland exposed to inoculum and subsequently fed to the target species, *Melanoplus sanguinipes*, documented that the fungus was active. In the rangeland none of the 2500 arthropods (solfugids, spiders and at least 18 families of insects from six orders) collected in pitfall traps and placed in humid chambers showed signs of infection. However, in alfalfa approximately 20% of leafcutting bees (Megachilidae) collected 10 days after application showed overt colonization by *B. bassiana* when placed under high humidity (Goettel et al., unpubl. data). The most common nontarget species in each habitat were collected, surface sterilized and processed to detect growth of *B. bassiana* on selective media. The objective of this part of the study was to provide a very liberal estimate of the effect of the fungal application by detecting the presence of the fungus within the digestive tract as well as within infected insects. For those insects in alfalfa fields, the prevalence of *B. bassiana* in coccinellids and harvestmen (Phalangidae) increased two days after application. These increases declined during the remainder of the study, suggesting that this effect was transient. In both habitats and for the other groups of nontargets evaluated, *B. bassiana* was generally only present at very low levels and absent from scarab larvae. Because this fungus was generally detected only at very low levels, it is hypothesized that although *B. bassiana* spores were ubiquitous and would grow on selective media, insects were not succumbing to infections in the field possibly because fungal doses were too low (Goettel et al., unpubl. data).

To evaluate the effect of applications of the hyphomycete, *Beauveria brongniartii*, against the scarab pest, *Melolontha melolontha*, in Switzerland, the insect and spider fauna in the forest/pasture ecotone were sampled after fungal application (Baltensweiler and Cerutti 1986). A total of 10,165 insects and spiders were collected and reared.

Overall infection was only 1.1% with no infections in coccinellids and neuropterans although up to 9% of the spiders were infected. The authors concluded that these applications did not endanger nontarget invertebrates because overall infection was low and the infections were transient in the field.

In one of the few studies of indirect ecological effects of entomopathogenic fungi applied as mycoinsecticides, soil nematodes were examined after applications of a beetle-derived isolate of *B. bassiana* (Boverol) to fallow areas (Hanel 1994). Phytophagous and bacteriophagous nematodes, in particular, increased four months after application. The cause is suggested as being an indirect effect due to insect mortality resulting in increased plant food for phytophagous nematodes and changes in soil organisms yielding increased bacterial growth for bacteriophagous nematodes. Few studies of indirect community level effects have been conducted and linkage between cause and effect demands a thorough understanding of individual systems. Indirect effects, therefore, can be predicted for few systems.

PHYSIOLOGICAL VERSUS ECOLOGICAL HOST RANGE

It is commonly stated that many species that can be infected in the laboratory (physiological host range) are not found infected by the same pathogens in the field (ecological host range) (e.g., Goettel et al. 1990, Glare and Milner 1991). Unfortunately, it is rare that hosts and pathogens are tested in both the laboratory and field; thus, the majority of comparisons of ecological and physiological host ranges are anecdotal. For example, seven of twelve species of mosquitoes were susceptible to *Coelomomyces psorophorae* in the laboratory but none of these potential hosts are generally found infected in the field (Zebold et al. 1979).

Curiously, naturally occurring epizootics caused by entomophthoralean fungi have frequently been reported as extremely host specific; only one host species is infected although other insects that are closely related are present in the area (see Wilding 1981). For example, when flies of the *Delia platura/florilega* complex and *Delia brassicae* occurred in mixed populations, *D. platura/florilega* were only infected by *Entomophthora muscae* while *D. brassicae* were only infected by *Strongwellsea castrans* (Nair and McEwen 1973). However, both of these flies are known to support infection of the alternate pathogens at other times (Wilding 1981). Studies comparing fungal strains at the molecular level support the existence of species-specific strains (or pathotypes) within a fungal species. Comparisons of different strains of the lepidopteran-specific *Entomophaga aulicae* have demonstrated a high degree of similarity among strains isolated from the spruce budworm, *Choristoneura fumiferana*, in contrast to low levels of similarity to strains recovered from other sympatric Lepidoptera (Walsh 1996). Based on an analysis of isoenzymes, *E. aulicae* isolated during an epizootic in spruce budworm in Ontario was composed of 15 different phenotypes at the beginning of the epizootic but, by the end of the epizootic, one strain predominated (Tyrrell and Harvey, unpublished data; cited in Walsh 1996). This apparent change in fungal population structure throughout an epizootic could constitute an adaptation to the principal host species being attacked and could help to explain lack of agreement between physiological and ecological host range. Host specialization during epizootics would also be supported by the "epidemic" model of population structure proposed by Maynard Smith et al. (1993) for asexually reproducing organisms in which one or a few highly successful individuals reproduce clonally and increase to high frequencies.

Examples of comparisons between physiological and ecological host ranges for several fungal species are given below, with greatest detail on Entomophthorales:

Entomophthorales

Entomophaga maimaiga

Cadavers usually are attached to tree trunks for some time when late instar *Lymantria dispar* larvae die from *E. maimaiga* infections. Based on results from laboratory specificity tests, Hajek et al. (1995) assumed that they would occasionally find larvae of other species of Lepidoptera killed by *E. maimaiga* hanging on tree trunks but this is not the case. Therefore, a study of the gypsy moth fungal pathogen *E. maimaiga* was conducted with the express objective of comparing physiological and ecological host range because field observations did not match predictions from laboratory bioassays (Hajek et al. 1996), i.e., infected nontargets were not found in the field even during epizootics.

L. dispar larvae were collected from seven sites in Virginia and the simultaneous occurrence of high levels of infection by *E. maimaiga* at all sites was documented (Hajek et al. 1996). Over 1500 nontarget lepidopteran larvae of 52 species from 7 families were also collected from these same sites and were reared to detect infection by *E. maimaiga*. Aside from *L. dispar* larvae, only two individuals were found to be infected by *E. maimaiga*: one *Malacosoma disstria* (one of 318 collected = 0.3% infection) and one *Catocala ilia* (one of 96 collected = 1.0% infection). Curiously, neither of these insect species belong to the family that was most heavily infected during nontarget laboratory bioassays, the Lymantriidae, and no lymantriids collected were infected. Unfortunately, nontarget lymantriids were not abundant during the year of this study, demonstrating the merit of long-term evaluations. One of these two infected species, *M. disstria*, had been tested in the laboratory also but the percent infection documented in the field was much lower than the infection level in the laboratory. Of the 19 other species challenged both in the laboratory and collected in the field, 5 had become infected during laboratory bioassays, yet none of these were infected in the field.

The question was asked as to how this fungus could be causing epizootics in *L. dispar* populations and not infecting larvae of sympatric lepidopteran species, especially those found susceptible in the laboratory. A closer examination demonstrated that *L. dispar* larvae display very specialized behaviors; neonates balloon and then wander before settling on an acceptable host plant while later instars wander down trees in the early morning to rest in dark locations during daylight hours (Campbell et al. 1975). Therefore, *L. dispar* larvae wander or rest in the leaf litter, thereby putting themselves at heightened risk of infection by fungal spores in the soil (A.E.H., unpublished data). Studies are in progress (A.E.H. unpubl. data) to evaluate to what extent nontargets share these risky behaviors but, based on low levels of nontarget infections in the field, it is hypothesized that few susceptible species share these behaviors with *L. dispar*.

Zoophthora radicans

Evidence of host specificity of this species having a broad host range agrees with a model of specialization by fungal isolates. Eight isolates of *Z. radicans* from Lepidoptera, Diptera, and Homoptera can each attack the pea aphid, *Acyrthosiphon pisum*, in the laboratory (Papierok et al. 1984), yet this host is rarely found infected by this fungus in the field (Glare and Milner 1989).

Among lepidopteran-infective strains of this same species, larvae of *Heliothis armigera* and *Heliothis punctigera* are susceptible to numerous Australian isolates in the laboratory. However, these *Heliothis* species were surprisingly not infected when an epizootic occurred among larvae of the lucerne leafroller (*Merophyas divulsana*) occupying the same fields in New Zealand (Glare and Milner 1989).

Additional Aphid-pathogenic Entomophthorales

Laboratory studies demonstrated that *Acyrthosiphon kondoi* and *A. pisum* were equally susceptible to *Pandora neoaphidis* (= *Entomophthora aphidis*), *Conidiobolus obscurus* (= *Entomophthora obscura*), and *Zoophthora phalloides* (Cameron and Milner 1981). During field collections in dryland alfalfa, total infection in *A. pisum* ranged from 28–94% while in *A. kondoi* only 2–17% infection occurred. This difference primarily was due to infection of *A. pisum* by *P. neoaphidis* (1487%) (infection in *A. kondoi* populations by *P. neoaphidis* was low (0–13%)). The authors suggested that the tendency of *A. pisum* to readily drop from plants, leading to increased exposure to spores on the soil surface, could help explain this higher infection in *A. pisum* (Cameron and Milner 1981).

In the field, the entomophthoralean *Zoophthora phalloides* infected 72% of the aphid *Myzus ascalonicus* but only 6% of *Myzus ornatus*. On the same plants. *P. neoaphidis* killed 71% of *M. ornatus* and only 1% of *M. ascalonicus* (Remaudière et al. 1981). Therefore, while each fungal species can infect both of the aphid species, the aphid species are differentially infected by these fungal species in the field. These two aphid species often occur on different plants but have somewhat different microhabitats even when occurring on the same plant (O.E. Heie, personal communication). Whether differences in aphid behavior could also play a part in this differential infection is not known. However, *Z. phalloides* infects only via capilliconidia that require small, mobile hosts while *P. neoaphidis* infects via primary conidia and therefore could be associated with more stationary hosts.

Remaudière (1971) suggested that differences in host specificity between aphid species and fungi occur not only due to intrinsic differential susceptibility of hosts to pathogens but also due to microenvironments inhabited by different aphid species. When *Macrosiphum avenae* and *Schizaphis graminum* were reared on the same plants. *P. neoaphidis* (= *E. aphidis*) rapidly killed the former and secondarily infected the latter. Because *M. avenae* colonizes the apical leaves of wheat while *S. graminum* forms colonies around the basal portion of the leaf, Remaudière (1971) suggested that *M. avenae* was infected first because its potential exposure to fungal spores was greater. Ecological host range can also differ from physiological host range because entomophthoralean fungi and potentially susceptible aphid species can differ in phenology so that both fungi and insects are not active simultaneously (Remaudière et al. 1981).

Hyphomycetes

Beauveria bassiana

Since *B. bassiana* can cause epizootics in silkworm, *Bombyx mori*, colonies and because this fungus is applied for pest control in many areas of the People's Republic of China, the effect of field applications on silkworm colonies was tested (Anonymous 1981, cited in Goettel et al. 1990). Silkworms were reared in a pine plantation where *B. bassiana* was applied to 110 ha six times to control pine caterpillars (*Dendrolimus* spp.). While good control of *Dendrolimus* spp. was achieved, only 4% infection was found in the silkworm rearings.

A strain of *B. bassiana* isolated from the aphid *S. graminum* was tested against first instar coccinellids in the laboratory and found to be highly infective toward the coccinellid *Hippodamia convergens* (James and Lighthart 1994). When this same strain was applied against pea aphid populations in alfalfa, aphids were not affected but densities of *Hippodamia convergens* placed on caged plants decreased by 75–93% during a spring trial but were not affected during a summer trial (James et al. 1995). However, these authors did not confirm that the population decrease was actually caused by fungal infections. In another study of coccinellids, *Coleomegilla maculata* was neither susceptible to a lepidopteran active strain of *B. bassiana* in the laboratory, infected when placed on plants to which *B. bassiana* was applied in the field, nor infected by indigenous populations of *B. bassiana* in the field (Pingel and Lewis 1996).

Nontarget effects of *B. bassiana* applied as a mycoinsecticide against *Schistocerca gregaria* were evaluated in Niger by caging and sampling tenebrionids in the treated area (Peveling and Weyrich 1992). This group of nontargets was emphasized because of their abundance and central function in the ecosystem, including competition with *S. gregaria* for food. *B. bassiana* infections were not detected in the 5 species of tenebrionids included in this study although for at least one species, topical applications in the laboratory resulted in infection. However, lack of *B. bassiana* infection in *S. gregaria* in the field during this study leads to questions regarding whether this study can provide an accurate test of the effect of *B. bassiana* on nontargets. Was the lack of infection that was documented due to inability to infect nontargets or due to poor survival of the fungal inoculum?

To compare physiological and ecological host range, we can also turn to studies of other pests. Hop aphids were highly susceptible to an aphid-derived strain of *B. bassiana* in the laboratory but applications in the field did not result in epizootics (Dorschner et al. 1991). The authors suggest that weather conditions during field trials were detrimental to survival and persistence of fungal inoculum. In fact, hostile environmental conditions are frequently cited as the cause of failures in the field of entomopathogenic fungi that are known to be virulent in the laboratory. This also demonstrates that for ecological host range to be tested, fungal survival must be evaluated and these studies should be considered complete only if it is known that fungi survived and were infective.

Metarhizium anisopliae

This fungus often occurs as a pathogen of soil dwelling insects and is often associated with immature insect stages. The host range of *M. anisopliae* is very broad based on species found infected in the field (see Goettel et al. 1990). However, laboratory assays demonstrate that potential hosts are often only susceptible to fungal strains isolated from insects of that same species (Ferron et al. 1972; Fargues 1976) and, once again, high doses can cause mortality while more realistic doses are non-infectious (Glare and Milner 1991).

Curiously, although mosquitoes are rarely found infected with *M. anisopliae* in the field, in the laboratory doses of 1 mg conidia/16 cm^2 of 26 out of 52 strains caused >50% infection in *Culex pipiens* larvae (Daoust and Roberts 1982). Eggs, pupae, and adults of *C. pipiens* were also readily infected in the laboratory but these stages are rarely infected in the field (Glare and Milner 1991). It is possible that if egg mortality due to *M. anisopliae* occurs in the field it is probably undetected and that adults would probably not become infected because they do not remain in the soil environment. However, the cause for lack of field records of pupal infection is not clear unless this is once again dose-related and the doses to which pupae are exposed are usually not adequate for infection.

WHY ARE THERE DIFFERENCES BETWEEN PHYSIOLOGICAL AND ECOLOGICAL HOST RANGE?

Of course physiological host range is based on optimized conditions and doses, and it is easy to imagine that such situations might not often occur in the field. Fungi used for laboratory testing can also differ from fungi in the field as in spores produced vitro-produced can differ from spores produced on cadavers (St. Leger et al. 1991). Laboratory manipulation of fungi can also decrease virulence through extensive subculturing, storage, etc. (e.g., Hajek et al. 1990)

A suite of host-related factors also differ between laboratory and field. Some insects are known to be able to actively thermoregulate to cure themselves of fungal infections, e.g., grasshoppers (Inglish et al. 1996, 1997) and flies (Watson et al. 1993). Molting and preening can remove fungal spores before infection occurs. Burrowing in soil has been shown to influence the sites where fungi penetrate the cuticle, e.g., while spores penetrated most readily through the mouth and anus of cockchafers, these were not the most frequent penetration sites supposedly due to mechanical spore removal during burrowing (Butt and Goettel in press). Diet can also influence susceptibility, e.g., laboratory-reared insects fed artificial diet can be more susceptible to fungal infection than field-collected insects fed natural diets (Butt and Goettel in press). This would agree with the general theory by Steinhaus (1958) that stress enhances susceptibility to infection.

Spatial and temporal divergence between pathogens and hosts can also help to explain the divergence between infection levels in the laboratory and field. However, our imprecise knowledge of exactly when and where pathogen propagules are successfully transmitted jeopardizes our ability to estimate which permissive nontargets might actually be at risk of becoming infected. In addition, while we frequently understand seasonality of insect activity, the actual locations occupied by many insects 24 hours a day are frequently not known, especially for very mobile species. For example, ultraviolet radiation and desiccation are commonly associated with mortality of fungal spores (Hajek 1997). Therefore, fungal spores survive longer on the protected and more humid undersides of leaves than on upper sides of leaves. Differential utilization of such microenvironments by potential insect hosts would clearly put insects dwelling on undersides of leaves at greater risk of infection compared to those remaining predominantly on leaf surfaces. Knowledge of the actual numbers of viable pathogen propagules encountered by insects succumbing to infection in the field would aid in our understanding of requirements for field infections. This knowledge is especially important for Hyphomycetes for which dose can strongly influence susceptibility (see above). In summary, a myriad of factors, influencing both fungus and insect, can help explain why laboratory-susceptible species do not become infected in the field.

CONCLUSIONS

Physiological and ecological host ranges of entomopathogenic fungi have infrequently been critically compared yet, based on our present knowledge, these host ranges are seldom in agreement. Because arthropods that can be infected in the laboratory are often not found infected in the field, physiological host range does not provide a realistic picture of the potential impact of an entomopathogen on nontarget organisms. While we have no perfect solution at this time, we suggest that the following information should be gathered in order to better predict ecological host range.

Identify Those Species Susceptible to the Pathogen under Laboratory Conditions

Laboratory testing is clearly not enough to predict which species will succumb to infections in the field. However, data from such studies provide general guidelines regarding potentially susceptible species. These studies are often conducted only with laboratory colonies of a limited number of species that are often inbred and for which response to a pathogen could be much more homogenous than wild populations. We suggest attempting to include insects reared from field-collected populations of a greater diversity of species to more closely simulate the spectrum of nontargets in the field.

In particular, laboratory studies should be conducted with attention to the origin of the fungal pathotype used due to differential specificity by fungal isolates. To predict physiological host range, both the mechanisms determining host specificity of entomopathogenic fungi as well as the resulting host range urgently require more study. If we understand the mechanisms determining host specificity, we can potentially predict which members of the community could be permissive for infection. For example, in the *E. maimaiga*/*L. dispar* system, results to date suggest that knowledge of insect morphology (e.g., larval setation and sculpturing) and behavior are essential for prediction of susceptibility in the field. Such information is important because it is not possible to conduct laboratory bioassays with all of the species occurring in the field.

Is the Pathogen to be Applied Inoculatively or Inundatively?

The invertebrate species that should be evaluated differ based on application strategy. The two very different strategies for application, inoculative release with expectations that the fungus will subsequently cycle in the host population or habitat manipulation versus application of large fungal doses for immediate impact, could have differing impacts on nontarget species (Maddox et al. 1992). For the former types of control strategy, exposure to inoculum would potentially be at lower levels and would occur at the specific locations and times when transmission normally occurs. Improving our understanding of the locations and times when transmission is greatest as well as which nontarget species are present at these locations and times would facilitate evaluation of nontarget effects. Only those potentially susceptible hosts occurring at locations and times when and where they would be exposed to viable inoculum could become infected. To develop an ability to begin to predict which nontargets would be exposed, it is clear that we need to understand the epizootiology of the disease, and especially the dynamics of pathogen transmission.

Alternatively, if fungi are applied as mycoinsecticides, all invertebrates in the application area would encounter high doses and therefore evaluation of all arthropods in the community would be more appropriate. However, even with inundative applications, epizootiology is important to understand because the pathogen could persist and spread in the field.

It is clear that at this time there is no precise method for predicting the effects of entomopathogenic fungi on the entire ecosystem. However, evidence agrees that while laboratory bioassays provide an indication of species susceptibility, they are poor predictors of host range under natural conditions. Further studies of the epizootiology of entomopathogenic fungi, including both when and where transmission occurs, as well as what environmental conditions are associated with successful transmission of infection by spores, are necessary to improve our ability to predict which species would potentially become infected in the field.

ACKNOWLEDGMENTS

We thank M. Goettel and O. Heie for sharing unpublished data with us and M. Goettel and T. Glare for providing helpful reviews of this manuscript. This study was supported, in part, by NRICGP # 96-35312-3883 (AEH) and USDA Cooperative Agreements # 42-96-0009 (AEH.) and # 42-793 (LB).

REFERENCES

Baltensweiler, W., and F. Cerutti. 1986. Bericht über die Nebenwirkungen einer Bekämpfung des Maikäfers (*Melolontha melolontha* L.) mit dem Pilz *Beauveria brongniartii* (Sacc.) Petch auf die Arthropodenfauna des Waldrandes. Mitt. Schw. Ent. Ges. 59: 267–274.

Butt, T. M., and M. S. Goettel. Bioassays of entomogenous fungi. *In* A. Navon and K. R. S. Ascher (eds.), Bioassays of entomopathogenic microbes and nematodes. CABI Publ., UK. In press.

Cameron. p. J., and R. J. Milner. 1981. Incidence of *Entomophthora* spp. in sympatric populations of *Acyrthosiphon kondoi* and *A. pisum*. N. Z. J. Zool. 8: 441–446.

Campbell, R. W., D. L. Hubbard, and R. J. Sloan. 1975. Patterns of gypsy moth occurrence within a sparse and numerically stable population. Environ. Entomol. 4: 535–542.

Daoust, R. A., and D. W. Roberts. 1982. Virulence of natural and insect passaged strains of *Metarhizium anisopliae* to mosquito larvae. J. Invertebr. Pathol. 40: 107–117.

Dorschner, K. W., M.-G. Feng, and C. R. Baird. 1991. Virulence of an aphid-derived isolate of *Beauveria bassiana* (Fungi: Hyphomycetes) to the hop aphid, *Phorodon humuli* (Homoptera: Aphididae). Environ. Entomol. 20: 690–693.

Fargues, J. 1976. Spécificité des champignons pathogènes imparfaits [Hyphomycètes] pour les larves de coléoptères [Scarabaeidae et Chrysomelidae]. Entomophaga 21: 313–323.

Feng, M. G., T. J. Poprawski, and G. G. Khachatourians. 1994. Production, formulation and application of the entomopathogenic fungus *Beauveria bassiana* for insect control: current status. Biocon. Sci. Technol. 4: 3–34.

Ferron, P., B. Hurpin, and P. H. Robert. 1972. Sur la spécificité de *Metarrhizium anisopliae* (Metsch.) Sorokin. Entomophaga 17: 165–178.

Fuxa, J. R. 1998. Environmental manipulation for microbial control of insects. *In* P. Barbosa (ed.), Conservation biological control. Academic Press, San Diego, CA. p. 255–268.

Glare, T. R., and R. J. Milner. 1991. Ecology of entomopathogenic fungi. p. 547–612. *In* D. K. Arora, L. Ajello and K. G. Mukerji (eds.), Handbook of Applied Mycology, vol. 2. Dekker, New York.

Goettel, M. S., T. J. Poprawski, J. D. Vandenberg, Z. Li, and D. W. Roberts. 1990. Safety to nontarget invertebrates of fungal biocontrol agents. *In* M. Laird, L. A. Lacey, and E. W. Davidson (eds.), Safety of microbial insecticides. CRC Press, Boca Raton, Florida. p. 209–240.

Hajek, A. E. 1997. Ecology of terrestrial fungal entomopathogens. Adv. Microb. Ecol. 15: 193–249.

Hajek, A. E., R. A Humber, and M. H. Griggs. 1990. Decline in virulence of *Entomophaga maimaiga* (Zygomycetes: Entomophthorales) with repeated *in vitro* subculture. J. Invertebr. Pathol. 56: 91–97.

Hajek, A. E., and R. J. St. Leger. 1994. Interactions between fungal pathogens and insect hosts. Annu. Rev. Entomol. 39: 293–322.

Hajek, A. E., L. Butler, and M. M. Wheeler. 1995. Laboratory bioassays testing the host range of the gypsy moth fungal pathogen *Entomophaga maimaiga*. Biol. Control 5: 530–544.

Hajek, A. E., L. Butler, S. R. A. Walsh, J. C. Silver, F. P. Hain, F. L. Hastings, T. M. Odell, and D. R. Smitley. 1996. Host range of the gypsy moth (Lepidoptera: Lymantriidae) pathogen *Entomophaga maimaiga* (Zygomycetes: Entomophthorales) in the field versus laboratory. Environ. Entomol. 25: 709–721.

Hanel, L. 1994. Effects of Boverol (*Beauveria bassiana*) application on soil nematodes (Nematoda) in a fallow. Ochr. Rost. 30: 121–132 (In Czech.).

Inglis, G. D., D. L. Johnson, and M. S. Goettel. 1996. Effects of temperature and thermoregulation on mycosis by *Beauveria bassiana* in grasshoppers. Biol. Control 7: 131–139.

Inglis, G. D., D. L. Johnson, and M. S. Goettel. 1997. Effects of temperature and sunlight on mycosis (*Beauveria bassiana*) (Hyphomycetes: Sympodulosporae) of grasshoppers under field conditions. Environ. Entomol. 26: 400–409.

James, R. R., and B. Lighthart. 1994. Susceptibility of the convergent lady beetle (Coleoptera: Coccinellidae) to four entomogenous fungi. Environ. Entomol. 23: 190–192.

James, R. R., B. T. Shaffer, B. Croft, and B. Lighthart. 1995. Field evaluation of *Beauveria bassiana*: Its persistence and effects on the pea aphid and a non–target coccinellid in alfalfa. Biocontr. Sci. Tech. 5: 425–437.

MacLeod, D. M., and D. Tyrrell. 1979. *Entomophthora crustosa* n. sp. as a pathogen of the forest tent caterpillar, *Malacosoma disstria* (Lepidoptera: Lasiocampidae). Can. Ent. 111: 1137–1144.

Maddox, J. V., M. L. McManus, M. R. Jeffords, and R. E. Webb. 1992. Exotic insect pathogens as classical biological control agents with an emphasis on regulatory considerations. *In* W. C. Kauffman and J. E. Nechols (eds.), Selection criteria and ecological consequences of importing natural enemies. Entomol. Soc. Amer., Lanham, Maryland. p. 27–39.

Maynard Smith, J., N. H. Smith, and M. O'Rourke. 1993. How clonal are bacteria? Proc. Natl. Acad. Sci. 90: 4384–4388.

Milner, R. J., and R. J. Mahon. 1985. Strain variation in *Zoophthora radicans*, a pathogen on a variety of insect hosts in Australia. J. Austral. Ent. Soc. 224: 195–198.

Nair, K. S. S., and F. L. McEwen. 1973. *Strongwellsea castrans* (Phycomycetes: Entomophthoraceae), a fungal parasite of the adult cabbage maggot, *Hylemya brassicae* (Diptera: Anthomyiidae). J. Invertebr. Pathol. 22: 442–449.

Papierok, B., B. V. L. Torres and M. Arnault. 1984. Contribution a l'étude de la specificité parasitaire du champignon entomopathogène *Zoophthora radicans* (Zygomycètes, Entomophthorales). Entomophaga 29: 109–119.

Peveling, R., and J. Weyrich. 1992. Effects of neem oil, *Beauveria bassiana* and dieldrin on non–target tenebrionid beetles in the desert zone of the Republic of Niger. *In* C. J. Lomer and C. Prior (eds.), Biological control of locusts and grasshoppers. CAB Intl., UK. p. 321–336.

Pingel, R. L., and L. C. Lewis. 1996. The fungus *Beauveria bassiana* (Balsamo) Vuillemin in a corn ecosystem: its effects on the insect predator *Coleomegilla maculata* De Geer. Biol. Control 6: 137–141.

Ramoska, W. A., A. E. Hajek, M. E. Ramos and R. S. Soper. 1988. Infection of grasshoppers (Orthoptera: Acrididae) by members of the *Entomophaga grylli* species complex (Zygomycetes: Entomophthorales). J. Invertebr. Pathol. 52: 309–313.

Remaudière, G. 1971. Vers l'utilisation pratique des Entomophthorales parasites de pucerons. Parasitica 27: 116–126.

Remaudière, G., J.-P. Latgé, and M. F. Michel. 1981. Ecologie comparée des Entomophthoracées pathogènes de pucerons en France littorale et continentale. Entomophaga 26: 157–178.

Roberts, D. W., and A. E. Hajek. 1992. Entomopathogenic fungi as bioinsecticides. *In* G. F. Leatham (ed.), Frontiers in industrial mycology. Chapman and Hill, New York. p. 144–159.

St. Leger, R. J. 1991. Integument as a barrier to microbial infections. *In* K. Binnington and A. Retnakaran (eds.), Physiology of the insect epidermis. CSIRO, Australia. p. 184–306.

St. Leger, R. J., M. Goettel, D. W. Roberts, and R. C. Staples. 1991. Prepenetration events during infection of host cuticle by *Metarhizium anisopliae*. J. Invertebr. Pathol. 58: 168–179.

Steinkraus, D. C., and J. P. Kramer. 1987. Susceptibility of sixteen species of Diptera to the fungal pathogen *Entomophthora muscae* (Zygomycetes: Entomophthoraceae). Mycopathologia 100: 55–63.

Soper, R. S., M. Shimazu, R. A. Humber, M. E. Ramos, and A. E. Hajek. 1988. Isolation and characterization of *Entomophaga maimaiga* sp. nov., a fungal pathogen of gypsy moth, *Lymantria dispar*, from Japan. J. Invertebr. Pathol. 51: 229–241.

Steenberg, T., V. Langer, and P. Esbjerg. 1995. Entomopathogenic fungi in predatory beetles (Col.: Carabidae and Staphylinidae) from agricultural fields. Entomophaga 40: 77–85.

Vandenberg, J. D. 1990. Safety of four entomopathogens for caged adult honey bees (Hymenoptera: Apidae). J. Econ. Entomol. 83: 755–759.

Walsh, S. R. A. 1996. Development of molecular markers for the detection and differentiation of *Entomophaga* strains pathogenic for insects. Ph.D. dissertation, Univ. Toronto, Canada.

Watson, D. W., B. A. Mullens, and J. J. Petersen. 1993. Behavioral fever response of *Musca domestica* (Diptera: Muscidae) to infection by *Entomophthora muscae* (Zygomycetes: Entomophthorales) J. Invertebr. Pathol. 61: 10–16.

Wilding, N. 1981. Pest control by Entomophthorales. *In* H. D. Burges (ed.), Microbial control of pests and plant diseases 1970–1980. Academic Press, London. p. 539–554.

Zebold, S. L., H. C. Whisler, J. A. Shemanchuk, and L. B. Travland. 1979. Host specificity and penetration in the mosquito pathogen *Coelomomyces psorophorae*. Can. J. Bot. 57: 2766–2770.

16

MONITORING THE EFFECTS OF *BACILLUS THURINGIENSIS KURSTAKI* ON NONTARGET LEPIDOPTERA IN WOODLANDS AND FORESTS OF WESTERN OREGON

Jeffrey C. Miller
Department of Entomology, Oregon State University, Corvallis, Oregon

INTRODUCTION

The management of lepidopterous pests can be multifaceted. Many techniques are available for suppression or eradication such as (1) habitat manipulation, (2) plant breeding, (3) genetic alteration of the pest, (4) pheromone-based trapping, (5) natural enemy enhancement, and (6) chemical pesticides. Historically, lepidopterous pests of forests and woodlands, such as, gypsy moth, eastern spruce budworm, western spruce budworm, and Douglas fir tussock moth, have been managed using chemical pesticides and/or natural enemies. However, since the mid-1980s a bacterial biological control agent, *Bacillus thuringiensis* Berliner var. *kurstaki* (Btk; applied as live spores) has been widely used for suppression or eradication of lepidopterous forest pests (Beegle and Yamamoto 1992, Martin 1994, Milner 1994).

Although microbial biological control agents provide benefits as a means for biorational alternatives in insect pest control they also present their own set of ecological risks (Laird 1973, Podgewaite 1986, Flanagan 1989, Tiedje et al. 1989). Within the last decade, efforts have been launched to assess the impact of microbial biological control agents against nontarget and beneficial organisms (Croft 1990, Flexner et al. 1986), with numerous studies investigating the impact of Btk on species of nontarget Lepidoptera (Miller 1990a, 1990b, 1994, 1996; Butler 1992a; Peacock et al. 1993; Sample et al. 1993, 1996; Duthie 1994; Waley 1994; Carter et al. 1995; Johnson et al. 1995; Wagner and Miller 1995; Carter 1996; Herms 1996; Lih et al. 1996; Stephen 1996; Wagner et al. 1996). All of these studies have demonstrated negative impacts on species of nontarget Lepidoptera.

Ecological, economic, and legal issues are part of the justification for assessing nontarget Lepidoptera. In particular, many species of Lepidoptera serve as biological control agents against noxious weeds. In the role of a biological control agent Lepidoptera species are highly beneficial and should be conserved. In fact, the broad scale use of Btk could result in a conflict of interest between weed pest managers and insect pest managers (Huang and Tamashiro 1966, James et al. 1993). Furthermore, many species of Lepidoptera have been placed on the Federal List of Threatened and Endangered species. Thus, listed species have legal standing requiring protection. In the context of legal ramifications, nontarget impacts involving threatened and endangered species may require special customized approaches to pest management programs. In the context of community and ecosystem ecology, nontarget species serve a function related to being a provider, eliminator, and/or a facilitator, all of which may have an influence on the composition and dynamics of plant communities and ecosystems (Miller 1993).

Once the decision is made to implement a Btk-based pest suppression program, a generalized approach to assessing nontarget impacts should consider at least three interdependent perspectives. One perspective involves the empirical documentation of field effects to document ecological impacts on nontarget organisms. Another approach would be to model environmental conditions in an attempt to predict ecological effects on single species, communities, and ecosystems. A third perspective might focus on the decisions relating to if, when, and how to mitigate the activity of Btk to curtail predictable and unwanted ecological impacts. Mitigation procedures would consider the need to enhance recovery of nontarget species populations through such measures as identifying no treatment spots within the overall treatment zone, buffer corridors (often considered along riparian zones), and refaunation attempts. This report addresses the first perspective: the empirical documentation of field effects on nontarget species, specifically Lepidoptera. The studies focus on Lepidoptera because this taxon represents a group of high-risk nontarget species in the context of Btk as the pest control agent.

METHODS AND MATERIALS

Four field studies were conducted in different woodlands and forests of western Oregon to monitor the ecological effects of Btk on larvae of nontarget Lepidoptera feeding on certain host plants. The studies were conducted as part of large-scale suppression or eradication projects conducted by state or federal agencies regarding either the spruce budworm or the gypsy moth. Each project provided a unique set of circum-

Table 1. Characteristics of four study sites in western Oregon used to assess the impact of Btk on nontarget Lepidoptera, 1986–1995.

Site	Years	Area (ha)	Annual dose (BIU)	Dominant plant species	No. Lep. species
I	1986–88	2300	54	*Quercus garryana*	35
II	1989–90	4000	18	*Ceanothus velutinus*	32
III	1992–93	275	72	*Arctostaphylos olumbiana*	13
				Ceanothus cuneatus	23
				Ceanothus integerrimus	28
IV	1992–93	30,000	24	*Arctostaphylos patula*	17
				Ceanothus velutinus	25
				Holodiscus discolor	18

stances involving year, place, dose, timing of treatment(s), plant communities, nontarget species, and measurable indicators of impacts. Field monitoring of nontarget impacts was conducted for two or three years depending upon the location and year of the particular project (Table 1).

Description of the Study Sites

Site I. The study site was near Elmira, Lane Co., 9 km west of Eugene at the southern end of the Willamette Valley, 300–400 m elevation, where an infestation of the gypsy moth was located. The treated and untreated plots were located in a woodland dominated by Oregon white oak within an area characterized by a rural farmland and native forest/woodland landscape. Additional information on this site and the overall study can be found in Miller (1990a).

Site II. An outbreak of the western spruce budworm resulted in the location of this study 50 km south of Estacada, Clackamas Co. The site was located on the western slope of the Cascade Mountains at 1000–2000 m. The plant community was dominated by Douglas-fir with a mix of alder, willow, and maple. The understory consisted of numerous shrubs including snowbrush, blueberry, rhododendron, and ocean spray. Additional information on this site and the overall study can be found in Miller (1990b).

Site III. The study site occurred 10 km southeast of Cave Junction, Josephine Co. The project goal was gypsy moth eradication. This area was at an elevation of 1000 m in the Siskiyou Mountains in the southwestern region of the state. The mixed coniferous-hardwood forest of this region was characterized by the presence of Douglas-fir, Ponderosa pine, Oregon white oak, black oak, bigleaf maple, and madrone. The understory was composed of numerous trees and shrubs including manzanita, currant, blueberry, deerbrush, common buckbrush, poison oak, and hazelnut.

Site IV. The Warm Springs Indian Reservation is located on the east slope of the central Cascade Mountains in northern Oregon. The north boundary occurs 20–30 km south of Mt. Hood, the south boundary occurs along the Metolius River, and the west boundary is near the crest of the Cascade Mountains. The eastern boundary of the treated area was delimited by the lower edge of the true fir/Douglas-fir zone at 1000 m. These boundaries delimit the forested area of the Reservation where the western spruce budworm was a concern and the area within which the application of Btk occurred. The dominant conifers were Douglas fir and grand fir. White fir, incense cedar, and Engelmann spruce were present as well. The more abundant broadleaf trees were vine maple, mountain alder, mountain ash, chinquapin, bittercherry, Douglas maple, and serviceberry. The most abundant shrubs were snowberry, greenleaf manzanita, ocean spray, Pacific ninebark, willows, huckleberries, currants, thimbleberry, and bitterbrush.

Treatments

All of the projects involved aerial application of Btk either once or three times in the spring, either in May or June. The year, date of treatment, and dose depended upon the particular project.

Site I: Three treatments of Btk were applied to 2000 ha at 18 BIU per 0.4 ha: May 15–16, May 26, and June 5, 1986. The annual dose totaled 54 BIU per 0.4 ha.

Site II: One treatment of Btk was applied to 7600 ha at 18 BIU per 0.4 ha on June 26, 1989. Thus, the annual dose was 18 BIU per 0.4 ha

Site III: Three treatments of Btk at 24 BIU per 0.4 ha were applied to 216 ha: April 21, May 5, and May 19, 1992. The annual dose of Btk was 72 BIU per 0.4 ha.

Site IV: One treatment of Btk at 24 BIU per 0.4 ha was applied to 28,000 ha on June 19-21, 1993, providing an annual dose of 24 BIU per 0.4 ha.

Sampling

The impact of Btk was assessed by sampling the caterpillars associated with particular plant species. Caterpillars were used as the life stage for data compilation and analysis because they are the life stage directly impacted by the Btk treatment. Plot locations for untreated and treated comparisons, and the decision on which plants to sample for immature for caterpillars, considered (1) physical aspects of the habitat (i.e., slope, canopy, and elevation), (2) size of plants (i.e., height and foliage density), (3) the number of plants within patches of 100 x 100 m., and (4) previous experience with Lepidoptera associated with certain plant species in other regions of Oregon (see Miller 1995). Either two or three pairs of plots were matched to represent a treated and untreated sample area. The area of each plot was determined by the dispersion of plants suitable for sampling and varied from 400 x 400 m to 800 x 800 m.

The caterpillars associated with the study plants were sampled at a given position by shaking foliage above a 75 x 75 cm sheet for a timed interval of ca. 5 sec. No samples were obtained from previously sampled positions within any given year. The position of each sample ranged between 0.5 and 2.0 m above ground. After 5 sec. the sheet was moved to another position under foliage of the same plant species and the timed sampling technique was repeated. Sampling continued until a total of 100–110 sec was achieved at each plot in each of four plots per treatment. Thus, a total of 400–440 sec of sample effort was conducted within each treatment at each site. At each site the data for each treatment were totaled for statistical analyses and are presented on an adjusted basis of values per 100 sec sampling effort for standardization among plots and among sites. Caterpillars were collected alive, counted, photographed, weighed, and individually reared to adults. Adults were required for accurate identification of the species. The samples were assessed to determine species richness, abundance, and live mass of all immature Lepidoptera.

At least one pretreatment sample was obtained at sites I, II, and IV. The emergency need for an application of Btk at site III did not allow time for a pretreatment sample. The first post-treatment samples were taken two weeks after treatment and then at four week intervals throughout the summer of the treatment year and from May-September throughout the next one or two years depending on the study.

Site I: Three pairs of plots were sampled over three consecutive years. One plant species, Oregon white oak, was sampled.

Site II: Three pairs of plots were sampled for a two-year period. One plant species, snowbrush, was sampled.

Site III: Two pairs of plots were sampled for a two-year period. Three species of plants were sampled: deerbrush, buckbrush, and hairy manzanita.

Site IV: Three pairs of plots were sampled for a three-year period. Three species of plants were sampled the year of treatment: ocean spray, green leaf manzanita, and snowbrush. The latter two plant species were also sampled for the next two consecutive years following treatment.

Analysis

Data were analyzed using a chi-square or G-test (Sokal and Rohlf 1981) based on absolute counts, adjusted to 100 sec of sampling effort, regarding the number of species, abundance, and total weight of caterpillars. These tests provide one degree of freedom based on a chi-square distribution in a comparison of observed and expected values. For this study the expected values were the average of observed values from the treated and untreated sites. The hypothesis for these tests was H_0: A (treated) = B (untreated) = 50%. This hypothesis states that samples from treated and untreated sites will exhibit equal values for either species richness, number of caterpillars, or live caterpillar mass. In select situations where the chi-square value was near 3.64 (P=0.05) a G-test was performed. A G-test operates on the same hypothesis as the Chi-square test but is more sensitive to small differences between treatments when overall numbers are small (i.e., few species, few individuals, low mass).

RESULTS AND DISCUSSION

A summary of the most common species, their host plants, peak abundance, and differences (percent) between treated and untreated plots are listed in Table 2. More specific details regarding sites I and II were presented in Miller (1990a) and Miller (1990b), respectively. Additional details concerning sites III and IV will be published in separate manuscripts. Overall the nontarget Lepidoptera consisted of numerous species whose caterpillars were feeding on hairy manzanita, 13 species; greenleaf manzanita, 17 species; or ocean spray, 18 species snow brush, 32 species; deerbrush, 23

Table 2. List of nontarget Lepidoptera species exhibiting peak abundance values of at least three larvae/100 sec sampling effort in pretreatment (to-be-treated) plots and the percent decline in the first post-treatment sample. The sites were treated with *Bacillus thuringiensis kurstaki* for suppression of either gypsy moth or western spruce budworm, Oregon, 1986–1995.

Lepidoptera	Site	Host plant[a]	Peak abundance[b]	% decline[c]
GELECHIIDAE				
Telphusa sedulitella	I	Qgar	14	93
GEOMETRIDAE				
Drepanulatrix monicaria	III	Ccun	16	100
Drepanulatrix unicalcararia	III	Cint	17	100
Drepanulatrix spp.	II	Cvel	10	90
Eudrepanulatrix rectifascia	III	Cint	25	96
Eupithecia misturata	III	Cint	17	59
Hesperumia sulphuraria	IV	Arct	3	100
	III	Cint	6	100
Lambdina fiscellaria	I	Qgar	5	100
Nemoria darwiniata	I	Qgar	9	100
Synaxis cervinaria	IV	Arct	5	100
LYCAENIDAE				
Satyrium saepium	III	Ccun, Cint	9	88
LYONETIIDAE				
Bucculatrix sp."A"	I	Qgar	10	80
NOCTUIDAE				
Acronicta marmarota	I	Qgar	10	100
Aseptis ethnica	IV	Apat	3	100
Cosmia calami	I	Qgar	7	100
Lithophane contenta	I	Qgar	3	100
Lithophane georgii	IV	Hdis	3	85
Oncocnemis dunbari	IV	Hdis	15	100
Orthosia pulchella	III	Apat	16	94
TORTRICIDAE				
Archips rosana	I	Qgar	7	86
Choristoneura rosaceana	I	Qgar	12	84
Ditula angustiorana	I	Qgar	10	80
Epinotia emarginana	I	Qgar	27	82
Tortricid spp.	IV	Apat	21	87

[a] Apat = *Arctostaphylos patula*, Cint = *Ceanothus integerrimus*, Ccun = *Ceanothus cuneatus*; Cvel = *Ceanothus velutinus*, Hdis = *Holodiscus discolor*, Qgar = *Quercus garryana*

[b] Number of larvae collected per 100 sec of sampling effort with a 75x75 cm beating sheet.

[c] % loss = [1 – (no. treated / no. untreated)] x 100

Table 3. Percent difference in species richness, caterpillar abundance, and live caterpillar mass between plots treated with *Bacillus thuringiensis kurstaki* and untreated plots. Sites I, II, III, and IV (see text) combined, 1986–1995, Oregon.

Variable	Year posttreatment 0	1	2
Species richness	55–67	20–50	0–43
Caterpillar abundance	72–86	55–70	0–44
Live caterpillar mass	84–97	64–84	0–39

species; buckbrush, 23 species; or Oregon white oak, 35 species.

Particular species of Lepidoptera were more abundant on certain host plants and therefore served as important indicators of change in the Lepidoptera fauna (Table 2). For instance, while 35 species of Lepidoptera were noted to be associated with Oregon white oak, only eleven of the species were present in relatively high numbers. In fact, the most abundant species on Oregon white oak was *Epinotia emarginana* which occurred in pretreatment peak numbers of 27 larvae/100 sec sample effort in the to-be-treated plots at site I. Similarly, 28 species were collected from *C. integerrimus* but only five species occurred in relatively high numbers with the most abundant pretreatment species being *Eudrepanulatrix rectifascia* which exhibited a peak abundance of 25 larvae/100 sec sample effort in to-be-treated plots at site III. Overall, six families were represented by 23 species that were most notable by their occurrence at relatively high numbers, that is, at least three larvae per 100 sec of sample effort in untreated plots. The relevance of these species as indicators of nontarget impacts is evidenced by the range of 80–100% difference (percent loss) in their abundance seen by comparing untreated and treated plots following the application of Btk.

The total number of species, caterpillar abundance, and caterpillar live mass in treated versus untreated plots differed by 50–65, 79–93, and 84–99%, respectively (Table 3). The reduction in the values of each variable in the treated plots was most apparent within three to four weeks after the final Btk application. In the one or two years following treatment a trend in the reduction of differences for values of each variable was evident among the caterpillars on all plant species at each site. Furthermore, the values of each variable in the treated sites demonstrated a trend toward pretreatment values in the first and second year after treatment.

Variability in impacts is expected with respect to place, time, and species concerned. The susceptibility of a caterpillar to Btk is variable and influenced by such factors as diet, temperature, and instar. For instance, certain plants contain allelochemicals that confer a degree of reduced susceptibility to Btk. Thus, the impact of Btk on nontarget species in part may be influenced by host plant associations. Some reports of this phenomenon can be found in Appel and Schultz (1994), Maksymiuk (1970), Meade and Hare (1993), and Moldenke et al. (1994). Similarly, conditions relating to temperature in the microhabitat of a caterpillar may influence the effect of

Btk on its host. Thus, nontarget impacts, which can be instar dependent (Li et al. 1995, van Frankenhuyzen et al. 1997) will be influenced by geographic, daily, and seasonal climatic patterns. Nonetheless, the impacts noted in the projects reported upon here demonstrated relatively consistent reduction in species richness, abundance, and live mass within the year of treatment.

One of the main purposes of these studies was to document the impact of Btk on nontarget Lepidoptera because moths appear to be an important component in the diet of birds and bats. The data regarding species richness, abundance, and live mass of caterpillars in Btk pest management programs are critical to addressing possible impacts regarding the availability of food for insectivorous vertebrates. Because Lepidoptera are an important source of food for certain birds and bats (Csuti et al. 1997, Hammond and Miller 1998) a reduction in species richness, abundance, and live mass could have a dramatic impact on food availability which in turn will affect foraging, reproduction, and fitness (see Lih et al. 1996).

A number of messages emerge from a consideration of Btk-based pest management programs and a concern for maintaining biodiversity that are germane to the general issue of ecosystem sustainability and the conservation of species. Some of these messages are as follows: (1) Documentation of the impact on nontarget species needs to be conducted in real field conditions. Furthermore, it would greatly improve the assessment of nontarget impacts if observations included species representing multiple trophic levels and continued for an extended period of time (3–5 years); (2) Research on nontarget species is dependent upon the availability of taxonomic expertise, thus, any nontarget assessment program must include a strong component involving personnel skilled in identification; (3) The conservation and protection of rare and endangered species may be threatened if Btk is used, but it is equally important to consider how threatened or rare species may be at risk if Btk is not used and the target pest is not suppressed or suppressed by the employment of other pest control measures; (4) Programs involving surveys of the insect fauna of an area, or larger scale bioinventory projects, are needed as baseline studies; (5) The scientific community and the public needs to support multifaceted research aimed at improving information about natural history of pertinent organisms, as well as relationships between population trends, ecosystem function, and impacts of various management practices.

ACKNOWLEDGMENTS

Field sites were located with the assistance of Art Weber and Beth Wilhite (USFS), Portland, Oregon, and Dan Hilburn (Oreg. Dept. Agric). Funds for this research were provided in part by U.S. Environmental Protection Agency, Willamette Institute for Biological Control, Oregon State University Agricultural Experiment Station, and the U.S. Forest Service. Field assistance was provided by J.E. Miller, M. LaMana, and B. Scaccia. Taxonomic assistance for Lepidoptera was provided by P. Hammond, J. Powell, and D. Ferguson. Verification of plant identifications was provided by K. Chambers and S. Sundberg. This is Oregon State University Agricultural Experiment Station Technical Paper No. 11483.

REFERENCES

Appel, H. M., and J. C. Schultz. 1994. Oak tannins reduce effectiveness of thuricide (*Bacillus thuringiensis*) in the gypsy moth. J. Econ. Entomol. 87: 1736–1742.

Beegle, C. C., and T. Yamamoto. 1992. Invitational paper (C.P. Alexander fund): History of *Bacillus thuringiensis* Berliner research and development. Can. Entomol. 124: 587–616.

Butler, L. 1992a. Non-target impact of dimilin and *Bacillus thuringiensis*. Gypsy Moth News 28: 4–5.

Carter, J. L. 1996. The effects of one or two applications of *Bacillus thuringiensis* on nontarget Lepidoptera. *In* R. F. Billing and T. E. Nebeker (eds.), Proceedings: North American Forest Insect Work Conference, Forest Entomology; Vision 20:21. San Antonio, Texas, Texas For. Serv., Texas A&M Univ. Publ. 160. p. 142

Carter, J. L., D. F. Schweitzer, and R. C. Reardon. 1995. The effects of two applications of *Bacillus thuringiensis* on nontarget Lepidoptera. Final Rept., U.S. Dept. Agric., Forest Svc., NE Forest Expt. Sta., Hamden, Connecticut.

Croft, B.A. 1990. Arthropod biological control agents and pesticides. Wiley and Sons.

Csuti, B., A. J. Kimerling, T. A. O'Neil, M. M. Shaughnessy, E. P. Gaines, and M. M. P. Huso. 1997. Atlas of Oregon wildlife. Oreg. State Univ. Press, Corvallis, Oregon.

Duthie, M. 1994. Effects of *Bacillus thuringiensis* var. *kurstaki* on nontarget Lepidoptera. Final research report: Ministry of Environment, Lands, and Parks, Pesticide Mgt. Branch, Nanaimo, British Colombia, Canada.

Flanagan, P. W. 1989. The need for basic research on genetically engineered microorganisms. Bull. Ecol. Soc. Am. 70: 14–19.

Flexner, J. L., B. Lighthart, and B. A. Croft. 1986. The effects of microbial insecticides on nontarget, beneficial arthropods. Agric., Ecosyst., Envir. 16: 203–254.

Hammond, P. C., and J. C. Miller 1998. Comparison of the biodiversity of Lepidoptera within three forested ecosystems. Ann. Entomol. Soc. Am. 91: 323–328.

Herms, C. P. 1996. The endangered Karner blue butterfly in Michigan: habitat suitability, potential impacts of gypsy moth suppression, and laboratory rearing. MS Thesis. Mich. State Univ., East Lansing, Michigan.

Huang, S.-S., and M. Tamashiro. 1966. The susceptibility of *Cactoblastis cactorum* (Berg) to *Bacillus thuringiensis* var. *thuringiensis*. Proc. Hawaiian Entomol. Soc. 19: 213–221.

James, R., J. C. Miller, and B. Lighthart. 1993. *Bacillus thuringiensis* var. *kurstaki* affects a beneficial insect, the cinnabar moth. J. Econ. Entomol. 86: 334–339.

Johnson, K. S., J. M. Scriber, J. K. Nitao, and D. R. Smitley. 1995. Toxicity of *Bacillus thuringiensis* var. *kurstaki* to three nontarget Lepidoptera in field studies. Environ. Entomol. 24: 288–297.

Laird, M. 1973. Environmental impact of insect control by microorganisms. Ann. N.Y. Acad. Sci. 217: 218–226.

Li, S. Y., S. M. Fitzpatrick and M. B. Isman. 1995. Effect of temperature on toxicity of *Bacillus thuringiensis* to the obliquebanded leafroller. Can. Entomol. 127: 271–273.

Lih, M., K. Smith, and F. Stephen. 1996. Impact of Bt on canopy Lepidoptera and their neotropical migrant bird predators. *In* R. F. Billing and T. E. Nebeker (eds.), Proceedings: North American Forest Insect Work Conference, Forest Entomology; Vision 20:21. San Antonio, Texas, Texas Forest Svc., Texas A&M Univ. Publ. 160. p. 143.

Maksymiuk, B. 1970. Occurrence and nature of antibacterial substances in plants affecting *Bacillus thuringiensis* and other entomogenous bacteria. J. Invert. Pathol. 15: 356–371.

Martin, P. A. 1994. An iconoclastic view of *Bacillus thuringiensis* ecology. Am. Entomol. 40: 85–90.

Miller, J. C. 1990a. Field assessment of the effects of a microbial pest control agent on nontarget Lepidoptera. Am. Entomol. 36: 135–139.

Miller, J. C. 1990b. Effects of a microbial insecticide, *Bacillus thuringiensis kurstaki*, on nontarget Lepidoptera in a spruce budworm-infested forest. J. Res. Lepidop. 29: 267–276.

Miller, J. C. 1993. Insect natural history, multispecies interactions and biodiversity in ecosystems. Biodiversity and Conservation 2: 233–241.

Miller, J. C. 1994. Protocol for assessing effects of *Bacillus thuringiensis* var. *kurstaki* on nontarget arthropods. *In* Proceedings of the 1994 Annu. Gypsy Moth Rev. Portland, Oregon. p. 202–208.

Miller, J. C. 1995. Caterpillars of Pacific Northwest forests and woodlands. FHM-NC-04-95. Morgantown, WV: U.S. Dept. Agric., Forest Svc., Natl. Ctr. Forest Health Mgt.

Miller, J. C. 1996. Assessing effects *of Bacillus thuringiensis kurstaki* on notarget Lepidoptera in western

Oregon. *In* R. F. Billing and T. E. Nebeker (eds.), Proceedings: North American Forest Insect Work Conference, Forest Entonology; Vision 20:21. San Antonio, Texas, Texas For. Serv., Texas A&M Univ. Publ. 160. p. 144–145

Milner, R. J. 1994. History of *Bacillus thuringiensis*. Agric., Ecosys., Envir. 49: 9–13.

Moldenke, A. F., R. E. Berry, J. C. Miller, J. G. Wernz, and X.-H. Li. 1994. Toxicity of *Bacillus thuringiensis* subsp. *kurstaki* to gypsy moth, *Lymantira dispar*, fed with alder or douglas-fir. J. Invert. Pathol. 64: 143–145.

Peacock, J. W., D. L. Wagner and D. E Schweitzer. 1993. Impacts of Bt on nontarget Lepidoptera. Newsl. Michigan Entomol. Soc. 38: 1–5.

Podgewaite, J. D. 1986. Effects of insect pathogens on the environment. Fortschritte der Zool. 32: 279–287.

Sample, B. C., L. Butler, C. Zivkovich and R. C. Whitmore. 1993. Evaluation of *Bacillus thuringiensis* and defoliation effects on native Lepidoptera. U.S. Dept. Agric. Forest Svc., NE Area, Forest Health Prot., Appalachian Integrated Pest Mgt. NA-TP-10-93.

Sample, B.C., L. Butler, C. Zivkovich, R. C. Whitmore and R. C. Reardon. 1996. Effects of *Bacillus thuringiensis* and defoliation by the gypsy moth on native arthropods. Can. Entomol. 128: 573–592.

Sokal, R. R., and F. J. Rohlf. 1981. Biometry. W. H. Freeman and Co., San Francisco.

Stephen, F. 1996. Nontarget effects of suppression projects. In: R. F. Billing and T. E. Nebeker (eds.), Proceedings: North American Forest Insect Work Conference, Forest Entomology; Vision 20:21. San Antonio, Texas, Texas Forest Svc., Texas A&M Univ. Publ. 160. p. 142.

Tiedje, J. M., R. K. Colwell, Y. L. Grossman, R. E. Hodson, R. E. Lenski, R. N. Mack, and P. J. Regal. 1989. The planned introduction of genetically engineered organisms: ecological considerations and recommendations. Ecology 70: 298–315.

van Frakenhuyzen, K., L. Gringorten, J. Dedes, and D. Gauthier. 1994. Susceptibility of different instars of the spruce budworm (Lepidoptera: Tortricidae) to *Bacillus thuringiensis* var. *kurstaki* estimated with a droplet-feeding method. J. Econ. Entomol. 90: 560–565.

Wagner, D., and J. C. Miller. 1995. Must butterflies die for the gypsy moth's sins? Amer. Butterflies 3: 19–23.

Wagner, D. L., J. W. Peacock, J. L. Carter and S. E. Talley. 1996. Field assessment of *Bacillus thuringiensis* on nontarget Lepidoptera. Environ. Entomol. 25: 1444–1454.

Waley, W. H. 1994. The monitoring of sensitive Utah Lepidoptera during the gypsy moth eradication preogram, 1990 to 1993. Final Tech. Rep., U.S. Dept. Agric, Forest Svc., State and Private Forestry, Forest Pest Management, Ogden, Utah.

17

ENVIRONMENTAL IMPACTS OF ENTOMOPATHOGENIC NEMATODES USED FOR BIOLOGICAL CONTROL IN SOIL

Mary E. Barbercheck and Leah C. Millar
Department of Entomology, North Carolina State University, Raleigh, North Carolina

INTRODUCTION

Biological control is a powerful tactic that has been used to obtain many significant, well documented economic and public health benefits against pests in agroecosystems and against disease vectors (Caltagirone 1981, DeBach 1974, Pimentel et al. 1984). The use of biological control has generally been considered environmentally safe and risk-free (Batra 1982, DeBach 1974, Hokkanen and Pimentel 1989, Hoy 1992), especially in comparison to the widespread use of synthetic insecticides (Pedigo and Higley 1992, Pimentel et al. 1992). However, data to support the assumption of absence of environmental impacts are not regularly collected (Howarth 1991), and it cannot be assumed that the lack of published evidence of negative environmental impacts is evidence of their absence. Indications of the growing concern about this complex issue include the recent debate in the literature over the use of introduced biological control agents to manage rangeland grasshoppers (Carruthers and Onsager 1993, Lockwood 1993a, b).

The term "nontarget effects" encompasses a wide variety of impacts. For example, nontargets could include vertebrates, non-arthropod invertebrates, pest or non-pest arthropods, and even plants. Exotic natural enemies may also adversely affect native natural enemies (Bennett 1993; Obrycki et al. 1998; Simberloff 1992; Simberloff and Stiling 1996, 1998; Waage and Greathead 1988). Biological pollution is yet another effect that could result from the introduction of a species. We are using entomopathogenic nematodes as a model to study the environmental impact of introduced biological control agents in the soil. The commercial availability of entomopathogenic nematodes as biological control agents for soil-inhabiting pests is continually expanding (Gaugler and Kaya 1990, Kaya and Gaugler 1993), and this growth in use represents an increasing potential for nontarget effects. The potential for environmental impacts have been discussed earlier (Ehlers and Hokkanen 1996, Ehlers and Peters 1995, Georgis et al. 1991) However, results from experiments designed explicitly to test for nontarget effects in the field are uncommon and nontarget effects are an unknown risk that must be evaluated. In this chapter, we will review the current literature on the nontarget effects of entomopathogenic nematodes (EPN), and summarize research on the effects of applications of exotic on endemic nematodes. We consider a nematode species exotic if it did not originate from the site or region to which it is being applied.

ENTOMOPATHOGENIC NEMATODE BIOLOGY

The term "entomopathogenic" is used to describe nematodes in the rhabditid nematode families Steinernematidae and Heterorhabditidae, which are unique because of their mutualistic association with bacteria of the genera *Xenorhabdus* and *Photorhabdus* (Akhurst and Boemare 1990, Boemare et al. 1993). These bacteria are carried monoxenically within the intestine of the infective juvenile (IJ). All stages of these nematodes except the soilborne IJ occur within a host insect. The IJ enters the host insect through natural openings; however, heterorhabditid IJ may also be able to enter directly through the integument (Bedding and Molyneux 1982). Entry by nematodes into the insect hemocoel is associated with suppression or destruction of components of the host insect's immune response (Götz et al. 1981, Simões and Rosa 1996). In the hemocoel the nematodes release their bacterial symbiont, which proliferates and kills the host. Conditions resulting from the fatal septicemia by the symbionts favor nematode reproduction by providing nutrients and antibiotics that inhibit the growth of other microbial flora in the cadaver. Nematodes mature and complete multiple generations in the cadaver. As resources are depleted and crowding occurs, IJ are produced, which leave the cadaver and are available in the soil to infect new hosts. Currently, it is believed that the bacterial symbiont is unable to survive in the soil outside of its nematode host (Boemare et al. 1996).

As a group, EPN are virulent against a broad range of insects, but do not normally infect vertebrate fauna. Because of these attributes, as well as their ease of mass production and exemption from U.S. EPA registration, a number of commercial enterprises are producing entomopathogenic nematodes as biological "insecticides" for soil dwelling insect pests. The commercial products contain the IJ stage. These products are being promoted in a wide variety of pest control markets including ornamental and nursery, home owner/urban, turf, and perennial and annual agriculture.

Entomopathogenic nematodes are widely distributed throughout the world (Akhurst and Bedding 1986, Akhurst and Brooks 1984, Ehlers et al. 1991, Gaugler et al. 1992, Glazer et al. 1991, Griffin et al. 1991, Hara et al. 1991, Hominick and Briscoe 1990, Stuart and Gaugler 1994, Vänninen et al. 1989). There are currently about 14 and 8 described species of *Steinernema* and *Heterorhabditis*, respectively; however, the literature contains reports of several undescribed species and new populations are frequently reported. As of this writing, the following species of EPN are available commercially in the USA: *S. feltiae*, *S. carpocapsae*, *S. scapterisci*, *S. riobrave*, *S. glaseri*, *H. bacteriophora*, and *H. megidis*.

Use of EPN has been assumed to be environmentally safe (Gaugler 1988). Even though these nematodes have been isolated from an extensive range of habitats, little is known about their biology in natural or agricultural ecosystems. The focus of most studies has been on factors affecting efficacy of a small number of isolates of commercial interest rather than on environmental impacts. Commercially produced nematodes are applied inundatively, like a pesticide, but their effectiveness in the field has been variable (Georgis and Gaugler 1991). Compared with efficacy trials of introduced EPN, research on the effects of local or native isolates on local insect populations are relatively recent and infrequent (Akhurst et al. 1992, Beavers et al. 1983, Brust 1991, Cabanillas and Raulston 1994, Campbell et al. 1995, Glazer et al. 1991, Mracek 1982, Raulston et al. 1992, Sexton and Williams 1981, Strong et al. 1995, 1996a, b). More information on native nematodes is critical to understanding their biology and nontarget impacts, and to improving the predictability of EPN as introduced or conserved biological control agents.

BIOLOGY RELATED TO NONTARGET IMPACTS

The environmental impact of biological control agents has been classified according to a number of factors: duration (short or long term), predictability, outcome (positive or negative), and magnitude (none to minor to considerable) (Harris 1990). Howarth (1991) related the relative degree of risk to factors such as permanency in the environment, host range, habitat range, genetic plasticity, and vulnerability of the target region.

Permanency in the Environment

Successful introductions for classical biological control are usually irreversible, and this attribute has generally been considered desirable for pest insect control (DeBach 1974, Fuxa 1987). Generally, inundative releases of organisms unable to reproduce or permananently establish should pose less environmental risk compared with classical agents. Permanency or establishment of EPN at a particular site is related to persistence (survival of IJ in the soil), infectiousness, reproduction within an infected insect, and dispersal. EPN for biological control are typically applied inundatively at a rate of 2.5 billion/hectare. Despite knowledge of physical factors that affect survival under standardized laboratory conditions, it is still unresolved by what processes and at what rate EPN are "lost" in the soil after application. Recoveries directly after application can be less than 50% (Van der Werf et al. 1995). Many commercially produced nematodes are not detectable more than 4 weeks after inundative application to soil and epizootics appear to rarely occur (Georgis 1992). Smits (1996) estimated that in most cases less than 1% of the applied population is still present after two to six weeks, but through recycling in insects these low levels may persist for years. If large numbers of IJ are required to successfully parasitize an insect, then the probability of establishment may diminish over time after application. However, because one infected insect cadaver can produce thousands of IJ, very high local populations can temporarily occur. There is little information available in the literature on seasonal population dynamics or possible linkages with host population dynamics (Strong et al. 1995, Strong et al. 1996a, b), adding to the difficulty in assessing potential risks of introducing non-indigenous strains or species.

Long term studies (longer than one growing season) of persistence or establishment are relatively uncommon. Some examples of observed persistence for more than one growing season follow. Fleming (1968) reported that *S. glaseri* (a native nematode) maintained itself in the field for 24 years after application. *Steinernema scapterisci*, imported from South America to control mole crickets, has persisted over 6 years after application in northern Florida (Parkman et al. 1993), and is currently considered the only example of classical biological control using EPN. Two isolates of *H. bacteriophora* persisted for over 300 d after application to turf (Klein 1990). The North Carolina strain of *H. bacteriophora* applied to a cranberry bog in Washington to control black vine weevil was still detectable after 1 year (Shanks and Agudelo-Silva 1990). Jansson et al. (1991, 1993) showed that *H. bacteriophora* (HP88 strain) and *Heterorhabditis* sp. persisted for over 230 days in southern Florida after application to sweetpotato fields. Monitoring was terminated before nematodes were no longer detected. Grub mortality from introduced *H. bacteriophora* in New Zealand increased from 9 to 56% in two years after release (Jackson and Wouts 1987).

Infectivity of IJ has been described as "phased" (Bednarek and Nowicki 1986, Fan and Hominick 1991). Only a portion of IJ appears to be infectious at any given time, and these proportions are dynamic over time. Because IJ exist as aggregated groups, phased infectivity may reduce intraspecific competition and deleterious density dependent and genetic effects (Bohan and Hominick 1995, Bednarek and Nowicki 1986, Fan and Hominick 1991, Selvan et al. 1993, Stuart and Gaugler 1994).

Success in detection of EPN may fluctuate over time and give the impression of failure to establish. *S. riobrave* from Texas, and *S. carpocapsae* (Mexican strain) applied to control southern corn rootworm in peanuts were undetectable in the middle of the growing season, but appeared to be increasing late in the season when monitoring was stopped (Barbercheck, unpubl. data). This may have been due to factors such as fluctuating soil moisture and temperatures, sampling error, proportion of non-infectious nematodes in the population (phased infectivity), or nematode and host insect population dynamics. These are factors that influence all applications of EPN and may give the impression of failure to establish when low levels of EPN are present.

All soil nematodes can actively disperse or be moved passively on or in an organism, or with soil or water. Passive dispersal with ship's ballast has been suggested as a long-range disperal mechanism for *Heterorhabditis* sp., which have frequently been isolated in coastal areas (Akhurst and Bedding 1986, Gaugler et al. 1992, Hara et al. 1991, Griffin et al. 1991, 1994). Timper et al. (1988) demonstrated that *S. carpocapsae*

(All strain) can be transported by flying beet armyworm adults that were infected during emergence. *Steinernema scapterisci*, introduced from South America, was passively dispersed by its mole cricket host at least 150 m, to occupy an area of 4.2 ha, 2 years after its release (Parkman et al. 1993a, b). Entomopathogenic nematodes may be transported on non-host soil biota, e.g., oribatid and mesostigmatid mites (Epsky et al. 1988).

Active dispersal of EPN in soil is well documented (Georgis and Poinar 1983a, b, c; Poinar and Hom 1986). Active dispersal may be affected by the search behavior of the IJ. From laboratory assays, *S. carpocapsae* and *S. scapterisci* are generally considered to be relatively immobile nematodes which tend to stay near the soil surface, employing a sit-and-wait strategy for host finding (Campbell and Gaugler 1993; Lewis et al. 1992, 1993). *Heterorhabditis bacteriophora*, *H. megidis*, *S. anomali*, and *S. glaseri* tend to be more mobile in the soil and are considered to employ an active host-finding strategy. *Steinernema riobrave* and *S. feltiae* appear to be intermediate in their motility (Campbell and Gaugler 1993; Grewal et al. 1995; Lewis et al. 1992, 1993). However, also in laboratory assays, both steinernematid and heterorhabditid nematodes find and infect insects over a range of horizontal and vertical distances from the point of application in soil (Alatorre-Rosas and Kaya 1990; Barbercheck and Kaya 1991a, b; Koppenhöfer et al. 1995b; Schroeder et al. 1993; Schroeder and Beavers 1987; Westerman 1995).

It is not known what proportion of a population of *S. carpocapsae* or other relatively sedentary nematode does not actively disperse under natural conditions. In an unreplicated field trial, Poinar and Hom (1986) calculated that *S. carpocapsae*, considered to be a relatively non-motile species, dispersed horizontally at a rate of 4.35 cm/day over a seven week period following application to a clay loam soil. Nine weeks after placement of sweetpotato roots containing sweetpotato weevils infected with steinernematid and heterorhabditid nematodes in soil, wax moth larvae placed 15 to 60 cm away became infected (Mannion and Jansson 1992b). *Steinernema riobrave* is fairly efficient at vertical and horizontal dispersal and moves approximately 4 cm/day (Cabanillas and Raulston 1994). The ways in which these assays have been conducted, usually by the placement of an aqueous solution of IJ on soil, may have biased our interpretation of EPN dispersal behavior. It appears that dispersal is greater from an infected cadaver than from inundative applications of an aqueous EPN suspension (Shapiro and Glazer 1996).

Host Range

Many researchers have suggested that polyphagous species are more efficient than monophagous or oligophagous species as biological control agents (Fuxa 1987, Gaugler 1988), but generalists also have greater potential for negative environmental impacts. Soil is the natural habitat of EPN. Early discoveries of EPN were made by finding infected insects in the field, but more recently, nematodes are usually isolated from soil using a "bait" insect (Bedding and Akhurst 1975) without determination of their natural host associations (Peters 1996). Therefore, there are species of EPN for which the natural hosts are unknown. Laboratory studies indicate that many steinernematid and heterorhabditid nematodes are generalist insect pathogens with very broad host ranges (Poinar 1979). The probability of infection of a particular insect under natural conditions by a nematode is influenced by a myriad of biotic and abiotic factors (reviewed in Barbercheck 1992, Gaugler and Kaya 1990, Jansson 1993, Kaya and Gaugler 1993, Peters 1996). Insects vary considerably in their susceptibility to infection by a specific nematode species or isolate (Barbercheck 1993, Barbercheck et al. 1995, Bedding et al. 1983, Epsky and Capinera 1993, Jackson and Brooks 1989, Mannion and Jansson 1992a, Molyneux et al. 1983). EPN are subject to destruction by desiccation and UV radiation that limits their persistence on exposed surfaces such as foliage without special formulation. Therefore, not surprisingly, the actual (ecological) host range has proven to be much narrower than the experimental (laboratory) host range and risk assessment for nontarget arthropods should not be based on laboratory assays alone.

Peters (1996) reviewed the known natural hosts for EPN but cautioned that most research emphasis has been placed on a few nematode and insect species of economic interest. The natural host range of *S. carpocapsae*, which is of great commercial interest, appears to be broad, including four insect orders and ten families. Natural infections have been commonly observed in Coleoptera and Lepidoptera. *Steinernema feltiae*, also commercially produced, has been detected mainly in soil dwelling Coleoptera (Scarabaeidae, Curculionidae), Lepidoptera (Noctuidae) and Diptera. In more specialized nematodes, e.g., *S. scapterisci* (host: Gryllotalpidae) and *S. glaseri* (host: Scarabaeidae), the risk of nontarget effects is probably lower. It is interesting to observe that only a specialized nematode, *S. scapterisci*, has been successfully used as a classical biological control agent (Parkman et al. 1993a, b).

Effects of EPN and their symbionts on birds, mammals, reptiles and amphibians, parasitoids, predatory arthropods, honey bees, nontarget soil insects and earthworms have been studied (Akhurst 1990, Bathon 1996, Gaugler and Boush 1979, Georgis et al. 1991, Jansson 1993, Kaya et al. 1982, Obendorf et al. 1983, Poinar 1979, 1989; Poinar et al. 1982, Poinar and Thomas 1985a, b, 1988). There is no evidence that EPN are able to kill and reproduce in vertebrates under natural conditions. There is some evidence of infection of young amphibians under laboratory conditions with very high doses of IJ, but in general, EPN are considered safe to non-arthropods (Kermarrec et al. 1991).

The risk of adverse effects on nontarget arthropods is probably the greatest in insects with at least one soil-dwelling stage that are in the same order or family as the natural host. Larvae, but not adults, of several predatory beetles were susceptible to infection by *H. bacteriophora* and *S. carpocapsae* in laboratory assays (Georgis et al. 1991). In the same study, field applications of EPN did not affect the abundance of nontarget arthropods captured in pitfall traps (Georgis et al. 1991). No significant impact was found on population densities of arthropod groups of higher taxonomic ranks and most species after applications of *S. feltiae*, H. megidis, and *Heterorhabditis* sp. to beech forest, pine forest, orchard and wheat field (Bathon 1996, Buck and Bathon 1993, Koch and Bathon, 1993). The application of nematodes in these experiments never resulted in the extinction of local populations of any pest or of any nontarget species (Bathon 1996). In laboratory assays, egg and larval stages of parasitoids inside insect hosts were susceptible to infection by EPN (Kaya 1978, Kaya and Hotchkin 1981). Adult parasitoids are unlikely to be infected because they do not generally occur in soil.

Habitat Range

Entomopathogenic nematodes have been found on all continents except the Antarctic (Poinar 1990). As more surveys are conducted, more information about the distribution of genera and species is becoming available. A major problem in understanding the biogeography of these nematodes is that it is not known whether the current distribution of EPN is due to natural dispersal, intentional application, or the accidental distribution with soil moved by people. A comprehensive interpretation of the biogeography of EPN is lacking (Hominick et al. 1996). More widespread use of EPN will make the determination of the native home of any particular species more difficult over time.

EPN occur in many types of natural and managed habitats in a wide variety of soil types. Within a site they have a patchy or aggregated distribution (Cabanillas and Raulston 1994, Spiridonov and Voronov 1995, Stuart and Gaugler 1994). Stuart and Gaugler (1994) found that naturally-occurring heterorhabditid and steinernematid nematodes are aggregated on a scale smaller than 4 m along a transect, and Cabanillas and Raulston (1994) found that *S. riobrave* is aggregated on a scale of less than 1 m^2. Spiridonov and Voronov (1995) studied the small scale distribution of *S. feltiae* and found that the characteristic dimension of an aggregation of infective juveniles is 15 to 20 cm. They suggested that the aggregations represent the results of a recent insect infection, while a low-level background of nematodes results from "old" IJ. Patchy distributions may relate to the spatial or temporal availability of hosts, the large number of infective juveniles produced by a single host, dispersal behavior, establishment and persistence of nematodes under different local conditions, local extinctions and reintroductions, natural enemies, or competition from other nematodes or insect pathogens (Barbercheck and Kaya 1990; 1991a, b; Ishibashi and Kondo 1990; Mracek 1980; Rosenheim et al. 1995; Strong et al. 1995, 1996a, b; Timper et al. 1988, 1991; Timper and Kaya 1992; Stuart and Gaugler 1994).

Genetic Diversity

Considerable variation exists in virulence of specific nematode species and geographic isolates (hence the strain or isolate names in published reports) to specific insects or stages of insects (Barbercheck et al. 1995, Jackson and Brooks 1989, Bedding et al. 1983, Gaugler and Molloy 1981, Geden et al. 1985). Variation in virulence of the symbionts has also been demonstrated (Griffin et al. 1989, Akhurst and Boemare 1990, Han et al. 1991). Selection experiments for traits considered useful for increasing effectiveness of *S. carpocapsae* as a biological control agent have been conducted (Gaugler et al. 1989a, b; 1991). Glazer et al. (1991) found considerable variation in four traits associated with survival in the soil among populations derived from 20 inbred lines of *H. bacteriophora* (HP88 strain). *Steinernema carpocapsae* is able to adapt to an insect based on the insect's host plant (Barbercheck and Wang, unpubl.). If reduction of nontarget effects is a priority, this variability in many features observed among isolates could be used to produce more host-specific EPN for use as biocontrol agents. Downes and Griffin (1996) suggested that commercial producers could be encouraged to develop strains of nematodes that are incapable of long persistence in the environment to reduce nontarget effects.

Vulnerability of the Target Region

Relatively little is known about the field ecology of natural or introduced EPN, or what characteristics would render an area vulnerable to negative nontarget effects. Soil is the home of diverse assemblages of organisms interacting in complex trophic webs (Hawksworth 1991, Moore et al. 1988). Even when EPN are applied inundatively, their densities are relatively low compared to those of other soil organisms (Wasilewska 1979). Research on soil invertebrates has revealed that omnivory is common, and many organisms include nematodes as a food resource (Gilmoreand Raffensperger 1970; Mueller et al.1990; Murphy and Doncaster 1957; Small 1987; Stirling 1991; Walter 1987a, b; 1988a, b; Walter et al. 1986, 1987). Interactions among EPN and other soil biota affect nematode survival and efficacy in simplified laboratory arenas (Epsky et al. 1988, Gilmore and Potter 1993).

Applications of EPN have the potential to alter the populations of other types of nematodes (Grewal et al. 1997). Addition of *S. carpocapsae* or *S. glaseri* to soil resulted in an initial decline in native plant parasitic nematode populations and a temporary increase in predatory and free-living rhabditids (Ishibashi and Kondo 1986, 1987). Application of *S. glaseri* adversely affected the root-knot nematode, *Meloidogyne javanica*, in pots (Bird and Bird 1986). The extent and ecological significance of these effects are currently unknown. There is a commercial EPN product being promoted as a "nematicide" (Biovector® 355, Devour®, ThermoTrilogy, Columbia, MD). Free-living nematodes are an important component of decomposition and nutrient cycles in the soil. Therefore, the effects of possible shifts in nematode communities resulting from the application of EPN on nutrient cycles in soil should be investigated.

Displacement of native by exotic entomopathogenic nematodes through competition is an unknown risk. Competition may be an important factor determining establishment or persistence, as well as levels of insect control. More than one nematode species has been isolated from a single site, although a single species from one site is common (Akhurst et al. 1992, Akhurst and Bedding 1986, Akhurst and Brooks 1984, Campbell et al. 1995, Ehlers et al. 1991, Hara et al. 1991, Stuart and Gaugler 1994). Insect cadavers infected by more than one species simultaneously are rare (Ehlers et al. 1991). In laboratory studies, heterorhabditid and steinernematid nematodes generally cannot coexist in the same host because they cannot develop on each other's symbiont (Alatorre-Rosas and Kaya 1990, Alatorre-Rosas and Kaya 1991). Koppenhöfer et al. (1995a) found that *S. carpocapsae* and *S. glaseri* could coinfect and produce progeny from a single *G. mellonella*, but *S. glaseri* was less negatively affected than *S. carpocapsae* by the mixed infection. This superior competitiveness of *S. glaseri* inside the host was attributed to its faster development in the insect and the ability to use the symbiont of *S. carpocapsae* as well as its own. When EPN compete with another insect pathogen for a shared host, the host usually will not survive and therefore competition can result in a high level of host exploitation. However, nematode progeny production may not be successful in dually infected hosts (Barbercheck and Kaya 1990, 1991b).

In the field, factors affecting the ecology of nematodes and their host insects may influence competition between nematodes and allow coexistence. Differences in search behaviors may reduce competition among nematodes that in the laboratory possess broad host ranges (Campbell and Gaugler 1993, Grewal et al. 1994, 1995, Kaya and Gaugler 1993, Lewis et al. 1992, 1993). The potential for infection of nontarget arthropods that occur in different niches may depend on EPN host-finding behaviors, e.g., *S. carpocapsae* may infect more insects at the soil surface and *H. bacteriophora* may infect more insects deeper in the soil profile. In pot studies, an application of these two nematodes suppressed the black cutworm, *Agrotis ipsilon*, on the soil surface and black vine weevil, *Otiorhynchus sulcatus*, near roots (Koppenhöfer and Kaya 1996). However, introduced isolates of *Steinernema carpocapsae* applied in a field trial to control southern corn rootworm reduced the number of infections of bait insects by a naturally occurring population of *H. bacteriophora* (Lewiston strain) (Barbercheck, unpubl. data).

Because nematodes have their own suite of antagonists that include predatory microarthropods and nematodes and nematophagous fungi (Kerry 1995), uncontrolled growth of nematode populations is unlikely. The capacity of a predator to exert a regulatory effect on a population of nematodes is determined partly by their ability to increase their population level and/or predation rate as prey density increases. Even when EPN are applied inundatively, their population densities are relatively low compared to the densities of other soil organisms (Wasilewska 1979). Many nematophagous organisms have rapid development times and high reproductive capacities, and many species, e.g., predatory mesostigmatid mites, exhibit at least some degree of specificity towards nematodes and many are capable of reproducing rapidly by parthenogenesis (Walter and Kaplan 1990). Mites had faster development times, lower mortality and higher egg-laying rates when feeding on nematodes in comparison to when feeding on arthropod prey (Walter et al. 1987; Walter 1988a, b). Even specialist nematophagous invertebrates will attack a variety of nematode prey (Small 1987; Walter et al. 1987).

EFFECTS OF INTRODUCED EPN ON ENDEMIC EPN

Displacement of native EPN may or may not be a risk of using introduced nematodes, but it should be considered as a possible consequence of releasing exotic species or isolates. Native nematodes are an important resource because of their diversity in many characteristics important to the control of insects. We can exploit this resource as isolates that are more effective than current ones, or as a genetic resource for traits desirable for biological control.

We are currently conducting field experiments to determine the effects of an introduced EPN on native EPN. Two species, *H. bacteriophora* and *S. carpocapsae,* are endemic to the research site. *Steinernema riobrave*, endemic to Texas, is being used as the introduced nematode. Different combinations of native and introduced nematodes were applied to field plots in no-till and conventionally-tilled corn, and the soil was sampled periodically throughout the growing season to determine the relative "success" of each nematode in terms of their ability to cause infections of test insects. We are currently in the third year of the experiment.

We included tillage as a variable in this research as an example of a production practice that could shift the advantage to a particular nematode species and affect the persistence of native and introduced EPN. Reduced tillage and its associated surface

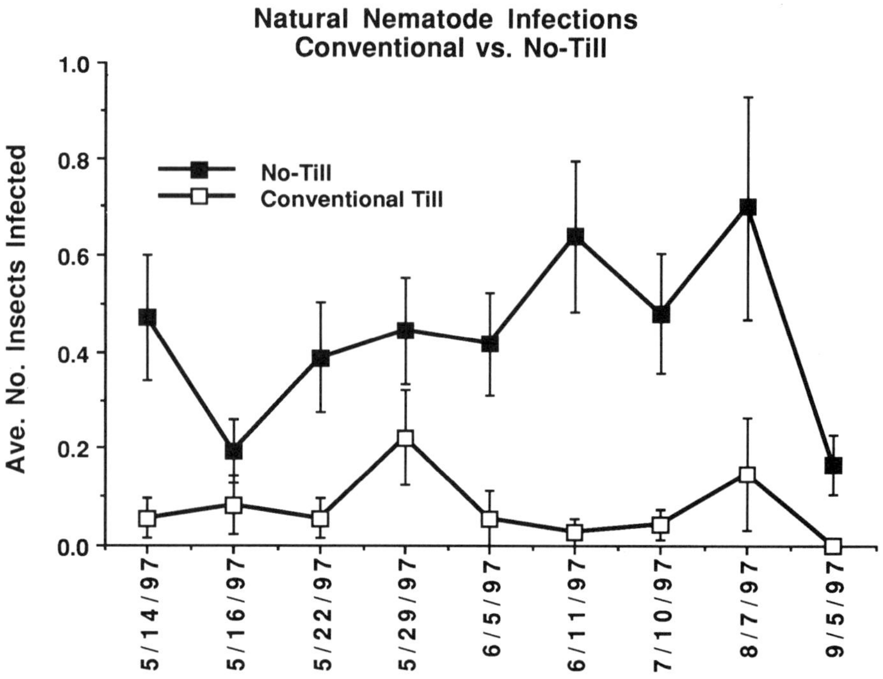

Figure 1. Average number of *Galleria mellonella* infected per soil core (three insects each exposed to 10 cm x 2 cm soil core for 4 days; 144 soil cores/date/tillage type) by an endemic population of *Steinernema carpocapsae* in no-till and conventionally-tilled corn.

residues result in lower temperatures and higher soil moisture levels than in conventional tillage, which may favor the development of disease in insect populations (Barbercheck 1992, Burges and Hussey 1971, Sloderbeck and Yeargan 1983). Therefore, we hypothesized that no-till may create conditions conducive to nematode survival and recycling (Brust 1991). If we understand how specific production practices affect endemic and introduced natural enemies in the soil, we may be able to suggest production practices that enhance natural soil "suppressiveness" against soilborne stages of pests, or minimize practices that disrupt natural control of crop pests and therefore reduce the need for introduced agents.

Tillage affected the occurrence of native nematodes (Fig. 1). As plots were tilled 5 months and disked 1 month before application of EPN, the effects are probably due to the physical, chemical and biotic conditions created by tillage in addition to direct destruction. In a baiting assay with *Galleria mellonella*, total numbers of infected insects were greater in no-till than in conventionally tilled soil throughout the growing season. In another study, a greater percentage of bait insects was infected with native *S. carpocapsae* (Fletcher strain) in no-till than in clean-till corn, tomatoes, and cabbage (Hummel, Barbercheck and Walgenbach, unpubl. data).

Tillage did not affect all of the nematodes species in the same way. Non-native *S. riobrave* and endemic *H. bacteriophora* were relatively insensitive to or were favored by the conditions created by conventional tillage in comparison to endemic *S. carpocapsae*. Numbers of insects killed by *S. riobrave* and *H. bacteriophora* were greater, and by *S. carpocapsae* were lower, in conventionally-tilled than in no-till soil. From these results we concluded that there is variability in sensitivity to conditions created by tillage, in order from least to most sensitive, was *S. riobrave* (introduced) < *H. bacteriophora* (endemic) < *S. carpocapsae* (endemic). The insensitivity of *S. riobrave* to conditions created by tillage may favor its persistence at higher levels than native *S. carpocapsae* in tilled soils.

The application of an introduced EPN had a negative impact on two endemic EPN. Application of introduced *S. riobrave* resulted in the highest levels of insect mortality but was also associated with a reduction of proportion of infection by both of the native nematodes (Fig. 2). The number of insects killed by endemic *S. carpocapsae* in control plots was greater than the number of insects killed by endemic *S. carpocapsae* in *S. riobrave*-treated plots or by augmented *S. carpocapsae* in plots where both *S. riobrave* and *S. carpocapsae* had been applied. The number of insects killed by endemic *H. bacteriophora* in the *H. bacteriophora*-treated plots was greater than in plots where *S. riobrave* and *H. bacteriophora* were both applied.

We determined the frequency of samples in which we detected no, single, or multiple EPN species. The detection of multiple species from a single core was rare relative to the number of cores from which only one nematode species was detected (Fig. 3). Detection of *S. riobrave* and *S. carpocapsae* was more frequent at 0–10 cm than at 10–20 cm. There was a significant negative association between *S. carpocapsae* and *S. riobrave* at both depths. Out of 1896 soil cores from *S. riobrave*-treated plots, *S. carpocapsae* and *S. riobrave* were detected together in only 29 cores, and *H. bacteriophora* and *S. riobrave* were detected together in only 7 cores. Out of 3716 soil cores, the two native nematodes were detected together in only 15 cores. There were no cores in which all three nematodes were detected together, and no insect cadavers infected by more than one nematode species. These results support previous observations on the lack of co-occurrence of multiple species in a single sample or cadaver (Campbell et al. 1995, 1998; Ehlers et al. 1991; Hara et al. 1991; Stuart and Gaugler 1994). If baiting assays accurately reflect the occurrence of EPN, there may be some behavioral or other mechanism that reduces spatial overlap and could allow an introduced species to establish without leading to the complete displacement of the native species. An inferior competitor may persist in a patchy environment without possessing compensating advantages in vagility or intrinsic fecundity provided that the superior competitor has a sufficiently great propensity to aggregate among patches, and resources exist in patches that are discrete and ephemeral (Ives and May 1985). Patchiness of EPN distribution is probably favorable for co-existence of multiple species at a site, reducing the risk of displacement of native by non-native nematodes (at least for these three species), but this requires long-term monitoring.

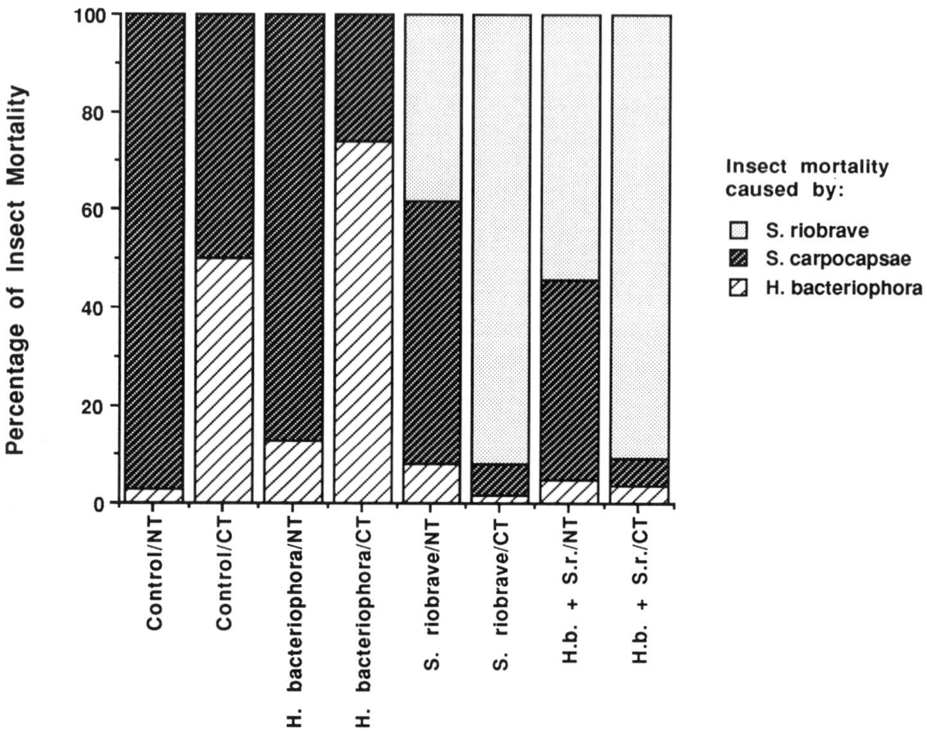

Figure 2. Average proportion of *Galleria mellonella* infected per core (three insects each exposed to 10 cm x 2 cm soil core for 4 days; 144 soil cores/date/tillage type/treatment) by introduced *S. riobrave*, and native *S. carpocapsae* and *H. bacteriophora* over 9 sample dates in 1997 in no-till (NT) and conventionally tilled (CT) corn. Treatments: Control = no nematodes applied, natural infection only; *H. bacteriophora* = native *H. bacteriophora* applied at a rate of two buried cadavers/m^2; *S. riobrave* = introduced *S. riobrave* applied in water drench at a rate of 250,000/m^2; H.b. + S.r. = combined introduced and native nematode application at same rates as other treatments.

Outcomes of the interactions between EPN may be influenced by several other factors, e.g., preferred microclimates, preferred host range of each EPN and the occurrence/distribution of those hosts and occurrence/preferences of predators. Because we have never found a naturally infected insect, we do not know the natural hosts of the endemic EPN species at our research site. The reported host range of *S. carpocapsae* is broad, with natural infections commonly observed in Coleoptera and Lepidoptera at the soil surface (Peters 1996). *Heterorhabditis bacteriophora* has been reported in naturally infected insects in Lepidoptera and Coleoptera (Scarabaeidae, Chrysomelidae,

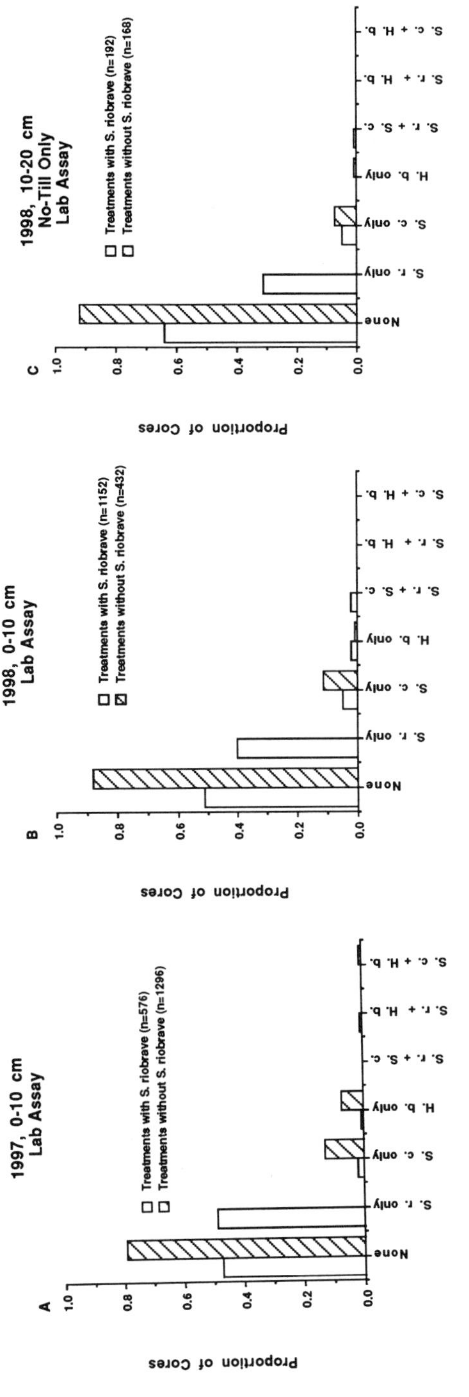

Figure 3. Proportion of cores in which no, a single, or multiple species of entomopathogenic nematodes were detected in 0–10 cm cores in 1997 (A) and 1998 (B), and in 10–20 cm cores in 1998 (C). Soil cores were collected from a research site with endemic populations of *Steinernema carpocapsae* and *Heterorhabditis bacteriophora*. Non-native *S. riobrave* was applied in some areas.

Curculionidae). *Steinernema riobrave* was discovered in noctuid larvae, but also has been effective as a biocontrol agent against the citrus root weevil, so it may also possess a broad host range.

In the soil, omnivory is common and it is likely that trophic webs based on detritus and those based on primary production overlap and interact, which could result in impacts that are indirect or diffuse (Walter 1987a, b, 1988a). Entomopathogenic nematodes could provide a link between these food webs when they are preyed upon by organisms that also function in decomposition and nutrient cycling food webs. Changes in populations of predators caused by the inundative application of EPN could therefore indirectly affect other soil organisms, e.g., free-living and plant-parasitic nematodes, microarthropods. To determine if there are more subtle effects associated with the application of introduced EPN, we are monitoring populations of soil microarthropods and other soil nematodes where EPN are applied.

CONCLUSIONS

There are risks and benefits associated with all methods of insect management. Biodiversity issues (Wilson 1985, Hokkanen and Lynch 1995,Hominick et al. 1996) have become an important factor to consider when evaluating the potential risks and benefits of pest management strategies. Documentation and evaluation of naturally occuring biological control is critical because the knowledge gained can be relevant to applied biological control and pest management. In most areas we expect to find native nematodes that are adapted to survive in those habitats. One topic that has not received much research attention is how to conserve or enhance populations of native nematodes. The widespread occurrence of nematodes indicates that they could be significant factors in the natural control of some insect populations. The natural "value" of nematodes, the service they provide in the regulation of insect pests, has not been estimated but may be considerable. Mracek (1986) found that the native nematode *S. kraussei* annually kills 24–27% of *Cephalia abietis*, the false spruce webworm, in the Czeck Republic. *Steinernema carpocapsae* caused 8–15% mortality of *C. lariciphila*, the larch sawfly, in the UK (Georgis and Hague 1981). Mortality rates of 23–68% of bibionid larvae were attributed to natural infection by *S. feltiae* and *S. affinis* in Denmark (rev. in Peters, 1996). Mortality of *Spodoptera frugiperda* and *Heliothis zea* in the Rio Grande Valley is approximately 10% (Raulston et al. 1992). Short-term natural epizootics have also been reported. In Australia, 71% of collected white grubs in sugar cane fields were infected by two *Heterorhabditis* sp. (Akhurst et al. 1992). Natural populations of *Heterorhabditis* and *Steinernema* were associated with a short-term reduction in numbers of *Graphognathus* sp. in California (Klein 1990).

Soil is an extremely complex ecosystem and it is difficult to predict from simplified laboratory systems what will occur in the field. Because of the complexity of trophic webs in the soil, we should expect interactions between native and introduced nematodes as well as with other soil biota (Hawksworth 1991, Moore et al. 1988). Our ability to develop successful biological control programs and minimize environmental risk will be enhanced by field studies that address the complexity of trophic interactions in agricultural systems (Cohen et al. 1993). Insect pests and their natural enemies do not exist in isolation, but are a part of a system of interacting components

(Stimac and O'Neil 1985, Stimac 1990). Because of the rapid expansion of commercial interest in EPN, research on nontarget/environmental impacts has not kept pace with other types of research. Field trials almost always have a focus on efficacy and, as with other types of biocontrol, follow-up studies to help explain establishment (or lack of) are not common. As the cost of production of EPN and availability of soil pesticides decreases, the use of EPN may become more widespread, and the need for ecological studies will grow with the necessity of evaluating the risk of nontarget effects. There are many factors in the field that can affect nematodes, and by nature some effects will be site specific. However, careful field observations and manipulations will help generate new hypotheses, and help us know the impact of native nematodes on insect pests and under what conditions native nematodes will be efficacious.

There is a need to evaluate and balance the relative risks and advantages of using any pest management tactic, including biological control (Ehlers and Hokkanen 1996). Because some EPN are generalists, nontarget effects should be at least considered as a possibility and not dismissed out-of-hand. It appears that the most likely potential risks associated with the use of EPN are biological pollution or local displacement of native nematode species. These appear to be relatively minor in comparison to risks associated with the use of chemical pesticides, especially when the safety to vertebrate fauna is considered. Because of the relatively high cost, EPN are currently promoted for use in highly managed, as opposed to natural, systems. Susceptible rare and endangered species are probably not likely to occur in the highly disturbed habitats where EPN are applied. As soil organisms, their dispersal potential is not as great as aerial organisms, e.g., parasitoids. Severe negative impacts do not seem likely because of rapid decline in detectable populations after application. Indeed, apparent failure to persist and establish has been considered a limitation to the use of EPN for biological control. If technological developments in EPN production or formulation allow a decrease in cost, an increase in efficacy or changes in use pattern, risk of nontarget effects may change. Research that includes long-term monitoring is needed to assess the more subtle effects from the long persistence of low populations of introduced nematodes on soil organisms and processes.

REFERENCES

Akhurst, R. J. 1990. Safety to nontarget invertebrates of nematodes of economically important pests. *In* M. Laird, L. A. Lacey, and E. W. Davidson (eds.), Safety of microbial insecticides. CRC Press, Boca Raton, Florida. p. 233–240.

Akhurst, R. J., and R. A. Bedding. 1986. Natural occurrence of insect pathogenic nematodes (Steinernematida and Heterorhabditidae) in soil in Australia. J. Aust. Entomol. Soc. 25: 241

Akhurst, R. J., R. A. Bedding, R. M. Bull, and D. R. J. Smith. 1992. An epizootic of *Heterorhabditis* sp. (Heterorhabditidae: Nematoda) in sugar cane Scarabaeidae (Coleoptera). Fundam. Appl. Nematol. 15: 71–73.

Akhurst, R. J., and N. E. Boemare. 1990. Biology and taxonomy of *Xenorhabdus*. *In* R. Gaugler and H. K. Kaya (eds.), Entomopathogenic nematodes in biological control. CRC Press, Boca Raton, Florida. p. 75–90.

Akhurst, R. J., and W. M. Brooks. 1984. The distribution of entomophilic nematodes (Heterorhabditidae and Steinernematidae) in North Carolina. J. Invertebr. Pathol. 44: 140–145.

Alatorre-Rosas, R., and H. K. Kaya. 1990. Interspecific competition between entomopathogenic nematodes in the genera *Heterorhabditis* and *Steinernema* for an insect host in sand. J. Invertebr. Pathol. 55: 179–188.

Alatorre–Rosas, R., and H. K. Kaya. 1991. Interaction between two entomopathogenic nematode species in the same host. J. Invertebr. Pathol. 57: 1–6.
Baker, K. F., and R. J. Cook. 1974. Biological control of plant pathogens. W. H. Freeman, San Francisco. 433p.
Barbercheck, M. E., and H. K. Kaya. 1990. Interactions between *Beauveria bassiana* and the entomogenous nematodes, *Steinernema feltiae* and *Heterorhabditis heliothidis*. J. Invertebr. Pathol. 55: 225–234.
Barbercheck, M. E., and H. K. Kaya. 1991a. Effect of host condition and soil texture on host finding by the entomogenous nematodes *Heterorhabditis bacteriophora* (Rhabditida:Heterorhabditidae) and *Steinernema carpocapsae* (Rhabditida: Steinernematidae). Environ. Entomol. 20:582–589.
Barbercheck, M. E., and H. K. Kaya. 1991b. Competitive interactions between entomopathogenic nematodes and *Beauveria bassiana* (Deuteromycotina: Hyphomycetes) in soilborne larvae of *Spodoptera exigua* (Lepidoptera: Noctuidae) Environ. Entomol. 20: 707–712.
Barbercheck, M. E. 1992. Effect of soil physical factors on biological control agents of soil insect pests. Florida Entomol. 75:539–548.
Barbercheck, M. E. 1993. Tritrophic level effects on entomopathogenic nematodes. Environ. Entomol. 22: 1166–1171.
Barbercheck, M. E., J. Wang, and I. S. Hirsh. 1995. Host plant effects on entomopathogenic nematodes. J. Invertebr. Pathol. 66: 169–177.
Bathon, H. 1996. Impact of entomopathogenic nematodes on non-target hosts. Biocontrol Sci. Tech. 6: 421–434.
Batra, S. W. T. 1982. Biological control in agroecosystems. Science 215: 134–139.
Beavers, J. B., C. W. McCoy, and D. T. Kaplan. 1983. Natural enemies of subterranean *Diaprepes abbreviatus* (Coleoptera: Curculionidae) larvae in Florida. Environ. Entomol. 12: 840–843.
Bedding, R. A., and R. J. Akhurst. 1975. A simple technique for the detection of insect parasitic rhabditid nematodes in soil. Nematologica 21: 109–110.
Bedding, R. A., and A. S. Molyneux. 1982. Penetration of insect cuticle by infective juveniles of *Heterorhabditis* sp. (Heterorhabditidae: Nematoda). Nematologica 28: 354–359.
Bedding, R. A., A. S. Molyneux, and R. J. Akhurst. 1983. *Heterorhabditis* sp., *Neoaplectana* sp., and *Steinernema kraussei*: interspecific and intraspecific differences in infectivity for insects. Exp. Parasitol, 55:249–257.
Bednarek, A., and T. Nowicki. 1986. New estimation method for the density of entomogenous nematodes (Rhabditidae: Steinernematidae) in the soil. Rev. Nematol. 14: 638–39.
Bennett, F. D. 1993. Do introduced parasitoids displace native ones? Florida Entomol. 76: 54–63.
Bird, A. F., and J. Bird. 1986. Observations on the use of insect parasitic nematodes as a means of biological control of root-knot nematodes. Int. J. Parasitol. 16: 511–516.
Boemare, N. E., R. J. Akhurst, and R. G. Mourant. 1993. DNA relatedness between *Xenorhabdus* sp. (Enterobacteriaceae), symbiotic bacteria of entomopathogenic nematodes, and a proposal to transfer *Xenorhabdus luminescens* to a new genus, *Photorhabdus* gen. nov. Int. J. Syst. Bacteriol. 43: 249–255.
Boemare, N., C. Laumond, and H. Mauleon. 1996. The entomopathogenic nematode–bacterium complex: biology, life cycle and vertebrate safety. Biocontrol Sci. Tech. 6: 333–345.
Bohan, D. A., and W. M. Hominick. 1995. Intra-population infectious structure and temporal variation in *Steinernema feltiae*. *In* C. T. Griffen, R. L. Gwynn and J. P. Masson (eds.), Ecology and transmission strategies of entomopathogenic nematodes. European Commission, Luxembourg. p. 83–94.
Brust, G. E. 1991. Augmentation of an endemic entomogenous nematode by agroecosystem manipulation for the control of a soil pest. Agric. Ecosyst. Environ. 36:175–184.
Buck, M., and H. Bathon. 1993. Auswirkungen des Einsatzes entomopathogener Nematoden (*Heterorhabditis* sp.) im Freiland auf die Nichtzielfauna, 2. Teil: Diptera. Anzeiger für Schadlingskunde, Pflanzenschutz, Umweltschutz 66: 84–88
Burges, H. D., and N. W. Hussey. 1971. Microbial control of insects and mites. Academic Press, New York.
Cabanillas, H. E., and J. R. Raulston. 1994. Evaluation of the spatial pattern of *Steinernema riobravis* in corn plots. J. Nematol. 26: 25–31.
Caltagirone, L. E. 1981. Landmark examples in classical biological control. Annu. Rev. Entomol. 26: 213–232.
Campbell, J. F., and R. Gaugler. 1993. Nictation behavior and its ecological implications in the host search strategies of entomopathogenic nematodes (Heterorhabditidae and Steinernematidae). Behaviour 126: 155–169.
Campbell, J. F., E. Lewis, F. Yoder, and R. Gaugler. 1995. Entomopathogenic nematodes (Heterorhabditidae

and Steinernematidae) seasonal population dynamics and impact on insect populations in turfgrass. Biol. Control 5: 598–606.

Carruthers, R. I., and J. A. Onsager. 1993. Perspective on use of exotic natural enemies for biological control of pest grasshoppers (Orthoptera: Acrididae). Environ. Entomol. 22: 885–903.

Cohen, J. E., R. A. Beaver, S. H. Cousins, D. L. DeAngelis, L. Goldwasser, K. L. Heong, R. D. Holt, A. J. Kohn, J. H. Lawton, N. Martinez, R. O'Malley, L. M. Page, B. C. Patten, S. L. Pimm, G. A. Polis, M. Rejmanek, T. W. Schoener, K. Schoenly, W. G. Sprules, J. M. Teal, R. E. Ulanowicz, P. H. Warren, H. M. Wilbur, and P. Yodzis. 1993. Improving food webs. Ecology 74: 252–258.

DeBach. P. 1974. Biological control by natural enemies. London: Cambridge Univ. Press. 323 p.

Downes, M. J., and C. T. Griffin. 1996. Dispersal behaviour and transmission strategies of the entomopathogenic nematodes *Heterorhabditis* and *Steinernema*. Biocontrol Sci. Technol. 6: 347–356.

Ehlers, R.-U., K. V. Deseo, and E. Strackebrandt. 1991. Identification of *Steinernema* sp. (Nematoda) and their symbiotic bacteria *Xenorhabdus* sp. from Italian and German soils. Nematologica 37:360–366.

Ehlers, R. –U., and A. Peters. 1995. Entomopathogenic nematodes in biological control: feasibility, perspectives and possible risks. *In* H. M. T. Hokkanen and J. M. Lynch (eds.), Biological control: benefits and risks. Cambridge Univ. Press. p. 119–136.

Ehlers, R. –U., and H. M. T. Hokkanen. 1996. Insect biocontrol with non-endemic entomopathogenic nematodes (*Steinernema* and *Heterorhabditis* sp.): OECD and COST workshop on scientific and regulatory policy issues. Biocontrol Sci. Tech. 6: 295–302.

Epsky, N. D., and J. L. Capinera. 1993. Quantification of invasion of two strains of *Steinernama carpocapsae* (Weiser) into three lepidopteran larvae. J. Nematol. 25:173–180.

Epsky, N. D., D. E. Walter, and J. L. Capinera. 1988. Potential role of nematophagous microarthropods as biotic mortality factors of entomopathogenic nematodes (Rhabditida, Steinernematidae and Heterorhabditidae). J. Econ. Entomol. 81: 821–825.

Fan, X., and W. M. Hominick. 1991. Efficiency of *Galleria* (wax moth baiting technique for recovering infective stage of entomopathogenic rhabditids (Steinernematidae and Heterorhabditidae) from sand and soil. Revue de Nematol. 14:381–387.

Fleming, W. E. 1968. Biological control of the Japanese beetle. USDA Tech. Bull. No. 1383. 78 p.

Fuxa, J. R. 1987. Ecological considerations for the use of entomopathogens in IPM. Annu. Rev. Entomol. 32: 225–251.

Gaugler, R. 1988. Ecological considerations in the biological control of soil–inhabiting insects with entomopathogenic nematodes. Agric. Ecosyst. Environ. 24: 361–60.

Gaugler, R., and G. M. Boush. 1979. Nonsusceptibility to rats to the entomopathogenic nematode, *Neoaplectana carpocapsae*. Environ. Entomol. 8: 658–660.

Gaugler, R, and H. K. Kaya. 1990. Entomopathogenic nematodes in biological control. CRC, Boca Raton. 365 p.

Gaugler, R., J. F. Campbell ,and T. R. McGuire. 1989a. Selection for host-finding in *Steinernema feltiae*. J. Invertebr. Pathol. 54: 363–372.

Gaugler, R., J. F. Campbell, and P. Gupta. 1991. Characterization and basis of enhanced host-finding in a genetically improved strain of *Steinernema carpocapsae*. J. Invertebr. Pathol. 57: 234–241.

Gaugler, R., J. F. Campbell, S. Selvan, and E. E. Lewis. 1992. Large-scale inoculative releases of the entomopathogenic nematode *Steinernema glaseri*: Assessment 50 years later. Biol. Control 2: 181–187.

Gaugler, R., T. McGuire, and J. Campbell. 1989b. Genetic variability among strains of the entomopathogenic nematode *Steinernema feltiae*. J. Nematol. 21: 247–253.

Gaugler, R., and D. Molloy. 1981. Instar susceptibility of *Simulium vittatum* (Diptera: Simuliidae) to the entomogenous nematode *Neoaplectana carpocapsae*. J. Nematol. 13: 1–5.

Geden, C. J., R. C. Axtell, and W. M. Brooks. 1985. Susceptibility of the lesser mealworm *Aliphitobius diaperinus* (Coleoptera: Tenebrionidae) to the entomogenous nematodes *Steinernema feltiae*, *S. glaseri* (Steinernematidae) and *Heterorhabditis heliothidis* (Heterorhabditidae). J. Entomol. Sci. 20: 331–339.

Georgis, R. 1992. Present and future prospects for entomopathogenic nematode products. Biocontrol Sci. Tech. 2: 83–99.

Georgis, R., and R. Gaugler. 1991. Predictability in biological control using entomopathogenic nematodes. J. Econ. Entomol. 84: 713–720.

Georgis, R., and N. G. M. Hague. 1981. Field evaluation of *Steinernema feltiae* against the web–spinning larch sawfly *Cephalia lariciphila*. J. Nematol. 20: 317–320.

Georgis, R., and G. O. Poinar, Jr. 1983a. Effect of soil texture on the distribution and infectivity of *Neoaplectana carpocapse* (Nematoda: Steinernematidae). J. Nematol. 15: 308–311.

Georgis, R., and G. O. Poinar, Jr. 1983b. Effect of soil texture on the distribution and infectivity of *Neoaplectana glaseri* (Nematoda: Steinernematidae). J. Nematol. 15: 329–332.

Georgis, R., and G. O. Poinar, Jr. 1983c. Effect of soil texture on the distribution and infectivity of *Heterorhabditis heliothidis* (Nematoda: Steinernematidae). J. Nematol. 15: 652–654.

Georgis, R., H. K. Kaya, and R. Gaugler. 1991. Effect of steinernematid and heterorhabditid nematodes (Rhabditida: Steinernematidae and Heterorhabditidae) on nontarget arthropods. Environ. Entomol. 20: 815–822.

Gilmore, S. K., and D. A. Potter. 1993. Potential role of Collembola as biotic mortality agents for entomopathogenic nematodes. Pedobiologia 37: 30–38.

Gilmore, S. K., and E. M. Raffensperger. 1970. Foods ingested by *Tomocerus* sp. [Collembola, Entomobryidae], in relation to habitat. Pedobiologia 10: 135–140.

Glazer, I., R. Gaugler, and D. Segal. 1991. Genetics of the nematode *Heterorhabditis bacteriophora* strain HP88: The diversity of beneficial traits. J. Nematol. 23:324–333.

Götz, P., A. Boman, and H. G. Boman. 1981. Interactions between insect immunity and an insect pathogenic nematode with symbiotic bacteria. Proc. R. Soc. London, Ser. B. 212:333– 350.

Grewal, P. S., E. E. Lewis, R. Gaugler, and J. F. Campbell. 1994. Host finding behaviour as a predictor for foraging strategy for entomopathogenic nematodes. Parasitology 108: 207–215.

Grewal, P. S., R. Gaugler, and R. Georgis. 1995. Predictors of foraging strategy in entomopathogenic nematodes. In C. T. Griffen, R. L. Gwynn, and J. P. Masson (eds.), Ecology and transmission strategies of entomopathogenic nematodes. European Commission, Luxembourg. p. 95–104.

Grewal. p. S., W. R. Martin, R. W. Miller, and E. E. Lewis. 1997. Suppression of plant–parasitic nematode populations in turfgrass by application of entomopathogenic nematodes. Biocontrol Sci. Technol. 7: 393–399.

Griffin, C. T., W. R. Simons, and P. H. Smits. 1989. Activity and infectivity of four isolates of *Heterorhabditis* sp. J. Invertebr. Pathol. 53: 107–112.

Griffin, C. T., J. F. Moore, and M. J. Downes. 1991. Occurrence of insect-parasitic nematodes (Steinernematidae, Heterorhabditidae) in the Republic of Ireland. Nematologica 37: 92–100.

Griffin, C. T., S. A. Joyce, I. Dix, A. M. Burnell, and M. J. Downes. 1994. Characterisation of the entomopathogenic nematode *Heterorhabditis* (Nematoda: Heterorhabditidae) from Ireland and Britain by molecular and cross-breeding techniques, and the occurrence of the genus in these islands. Fund. Appl. Nematol. 17: 245–254.

Han, R., W. M. Wouts, and L. Li. 1991. Development and virulence of *Heterorhabditis* sp. strains associated with different *Xenorhabdus luminescens* isolates. J. Invertebr. Pathol. 58:27–32.

Hara, A. H., R. Gaugler, H. K. Kaya, and L. M. LeBeck. 1991. Natural populations of entomopathogenic nematodes (Rhabditida: Heterorhabditidae, Steinernematidae) from the Hawaiian Islands. Environ. Entomol. 20: 211–216.

Harris. p. 1990. Environmental impact of introduced biological control agents. *In* M. Mackauer, L. E. Ehler and J. Roland (eds.), Critical issues in biological control. Intercept, Andover. p. 289–300.

Hawksworth, D. L. 1991. The biodiversity of microorganisms and invertebrates: Its role in sustainable agriculture. CAB International, Oxon, UK. 302 p.

Hokkanen, H. M. T., and D. Pimentel. 1989. New associations in biological control: theory and practice. Can. Entomol. 121 829–840.

Hokkanen, H. M. T., and J. M. Lynch. 1995. Biological control: Benefits and risks. Cambridge Univ. Press. 304 p.

Hominick, W. M., and B. R. Briscoe. 1990. Occurrence of entomopathogenic nematodes (Rhabditida: Steinernematidae and Heterorhabditidae) in British soils. Parasitology 100: 295–302.

Hominick, W. M., and A. P. Reid. 1990. Perspectives on entomopathogenic nematology. *In* R. Gaugler and H. K. Kaya (eds.), Entomopathogenic nematodes in biological control. CRC Press, Boca Raton, Florida. p. 327–345.

Hominick, W. M., A. P. Reid, D. A. Bohan, and B. R. Briscoe. 1996. Entomopathogenic nematodes: biodiversity, geographical distribution and the Convention on Biological Diversity. Biocontrol Sci. Technol. 6: 317–331.

Howarth, F. G. 1991. Environmental impacts of classical biological control. Annu. Rev. Entomol. 36: 485–509.

Hoy, M. A. 1992. Biological control of arthropods: genetic engineering and environmental risks. Biol. Control 2: 166– 170.

Ishibashi, N., and E. Kondo. 1986. *Steinernema feltiae* (DD–136) and *S. glaseri*: persistence in soil and

bark compost and their influence on native nematodes. J. Nematol. 18: 310–316.
Ishibashi, N., and E. Kondo. 1987. Dynamics of the entomogenous nematode *Steinernema feltiae* applied to soil with and without nematicide treatment. J. Nematol. 19: 404–412.
Ishibashi, N., and E. Kondo. 1990. Behavior of infective juveniles. *In* R. Gaugler and H. K. Kaya (eds.), Entomopathogenic nematodes in biological control. CRC, Boca Raton, Florida. p. 139–150.
Ives, R. A., and R. M. May. 1985. Competition within and between species in a patchy environment: Relations between microscopic and macroscopic models. J. Theor. Biol. 115:65–92.
Jackson, J. J., and M. A. Brooks. 1989. Susceptibility and immune response of western corn rootworm larvae (Coleoptera: Chrysomelidae) to the entomogenous nematode, *Steinernema feltiae* (Rhabditida: Steinernematidae). J. Econ. Entomol. 82: 1073–1077.
Jackson, T. A., and W. M. Wouts. 1987. Delayed action of an entomopathogenic nematode *Heterorhabditis* sp. (V16) for grass grub control. Proc. 40th New Zealand Weed and Pest Control Conf: 3–35.
Jansson, R. K. 1993. Introduction of exotic entomopathogenic nematodes (Rhabditida: Heterorhabditida and Steinernematidae) for biological control of insects: potential and problems. Florida Entomol. 76: 82–96.
Jansson, R. K., S. H. Lecrone, and R. Gaugler. 1991. Comparison of single and multiple releases of *Heterorhabditis bacteriophora* Poinar (Nematoda: Heterorhabditidae) for control of *Cylas formicarius* (Fabricius) (Coleoptera Apionidae) Biol. Control 1:320–328.
Jansson, R. K., S. H. Lecrone, and R. Gaugler. 1993. Field efficacy and persistence of entomopathogenic nematodes (Rhabditida: Steinernematidae, Heterorhabditidae) as biological control agents of the sweetpotato weevil (Coleoptera: Apionidae) in southern Florida. J. Econ. Entomol. 86: 1055–1063.
Kaya, H. K. 1978. Interaction between *Neoaplectana carpocapsae* (Nematoda: Steinernematidae) and *Apanteles militaris* (Hymenoptera: Braconidae), a parasitoid of the armyworm, *Pseudaletia unipuncta*. J. Invertebr. Pathol. 31: 358–364.
Kaya, H. K., and R. Gaugler. 1993. Entomopathogenic nematodes. Ann Rev. Entomol. 38:181–206.
Kaya, H. K., and P. G. Hotchkin. 1981. The nematode *Neoaplectana carpocapsae* Weiser and its effect on selected ichneumonid and braconid parasites. Environ. Entomol. 10: 474–478.
Kaya, H. K. J. Marston, J. Lindegren, and Y. Peng. 1982. Low susceptibility of the honey bee, *Apis mellifera* L. (Hymenoptera: Apidae) to the entomophagous nematode, *Neoaplectana carpocapsae* Weiser. Environ. Entomol. 11: 920–923.
Kermarrec, A., H. Mauleon, C. Sirjusingh, and L. Baud. 1991. Etudes experimentale de la sensibilite de vertebres heterothermes tropicaux (crapauds, grenouilles, lezards) a diverses souches de nematodes entomoparasites des genres *Heterorhabditis* et *Steinernema*. Rencontres de Lutte Biologique: 58: 193–204.
Kerry, B. 1995. The potential impact of natural enemies on the survival and efficacy of entomopathogenic nematodes. *In* C. T. Griffin, R. L. Gwynn, and J. P. Masson, (eds.), COST 819 Biotechnology: ecology and transmission strategies of entomopathogenic nematodes, European Commission, Luxembourg. p. 7–13.
Klein, M. G. 1990. Efficacy against soil-inhabiting insect pests. *In* R. Gaugler and H. K. Kaya (eds.), Entomopathogenic nematodes in biological control. CRC Press, Boca Raton, Florida. p. 195–214.
Koch, U., and H. Bathon. 1993. Auswirkungen des Einsatzes entomopathogener Nematoden im Freiland auf die Nichtzielfauna, 1. Teil: Coleoptera. Anzeiger für Schadlingskunde, Pflanzenschutz, Umweltschutz 66: 65–68.
Koppenhöfer, A. M., H. K. Kaya, S. Shanmugam, and G. L. Wood. 1995a. Interspecific competition between steinernematid nematodes within an insect host. J. Invertebr. Pathol. 66: 99–103.
Koppenhöfer, A. M., H. K. Kaya, and S. Taormino. 1995b. Infectivity of entomopathogenic nematodes (Rhabditida: Steinernematidae) at different soil depths and moistures. J. Invertebr. Pathol. 65:193–199.
Koppenhöfer, A. M., and H. K. Kaya. 1996. Coexistence of two steinernematid nematode species (Rhabditida: Steinernematidae) in the presence of two host species. Appl. Soil Ecol. 4: 221–230.
Lewis, E. E., R. Gaugler, and R. Harrison. 1992. Entomopathogenic nematode host finding: response to contact cues by by cruise and ambush foragers. Parasitology 105: 309–315.
Lewis, E. E., R. Gaugler, and R. Harrison. 1993. Response of cruiser and ambusher entomopathogenic nematodes (Steinernematidae) to host volatile cues. Can. J. Zool. 71: 765–769.
Lockwood, J. A. 1993a. Environmental issues involved in biological control of rangeland grasshoppers (Orthoptera: Acrididae) with exotic agents. Environ. Entomol. 22: 505–518.
Lockwood, J. A. 1993b. Benefits and costs of controlling rangeland grasshoppers (Orthoptera: Acrididae) with exotic organisms: Search for a null hypothesis and regulatory compromise. Environ. Entomol. 22:

904–914.
Mannion, C. M., and R. K. Jansson. 1992a. Comparison of ten entomopathogenic nematodes for control of sweetpotato weevil (Coleoptera: Apionidae). J. Econ. Entomol. 85: 1642–1650.
Mannion, C. M., and R. K. Jansson. 1992b. Movement and postinfection emergence of entomopathogenic nematodes from sweetpotato weevil, *Cylas formicarius* (F.) (Coleoptera: Apionidae). Biol. Control 2:297–305.
Molyneux, A. S., R. A. Bedding, and R. J. Akhurst. 1983. Susceptibility of larvae of the sheep blowfly, *Lucilia cuprina*, to various *Heterorhabditis* sp., *Neoaplectana* sp., and an undescribed steinernematid (Nematoda). J. Invertebr. Pathol. 42:1–7.
Moore, J. C., D. E. Walter, and H. W. Hunt. 1988. Arthropod regulation of micro- and mesobiota in below-ground detrital food webs. Annu. Rev. Entomol. 33: 419–439.
Mracek, Z. 1980. The use of "*Galleria* traps" for obtaining nematode parasites of insects in Czechoslovakia (Lepidoptera: Nematoda, Steinernematidae). Acta ent. bohemoslavica 77: 378–382.
Mracek, Z. 1982. Horizontal distribution in soil and seasona dynamics of the nematode *Steinernema kraussei*, a parasite of *Cephalcia abietis*. J. Appl. Entomol. 94:110–112.
Mracek, Z. 1986. Nematodes and other factors controlling the sawfly, *Cephalcia abietis* (Pamphilidae: Hymenoptera), in Czechoslovakia. Forest Ecol. Mgt. 15: 75–79.
Mueller, B. R., M. H. Beare, and D. A. Crossley. 1990. Soil mites in detrital food webs of conventional and no–tillage agroecosystems. Pedobiologia 34: 389–401.
Murphy, P. W., and C. C. Doncaster. 1957. A culture method for soil meiofauna and its application to the study of nematode predators. Nematologica 2: 202–214.
Obendorf, D. L., B. Peel, R. J. Akhurst, and L. A. Miller. 1983. Non-susceptibility of mammals to the entomopathogenic bacterium *Xenorhabdus nematophilus*. Environ. Entomol. 12:368–370.
Obrycki, J., K. L. Giles, and A. M. Ormond. 1998. Interactions between and introduced and indigenous coccinellid species at different prey densities. Oecologia 117: 279–285.
Parkman, J. P., J. H. Frank, K. B. Nguyen, and G. C. Smart. 1993a. Dispersal of *Steinernema scapterisci* (Rhabditida: Steinernematidae) after inoculative applications for mole cricket (Orthoptera: Gryllotalpidae) control in pastures. Biol. Control 3: 226–232.
Parkman, J. P., W. G. Hudson, J. H. Frank, K. B. Nguyen, and G. C. Smart. 1993b. Establishment and persistence of *Steinernema scapterisci* (Rhabditida: Steinernematidae) in field populations of *Scapteriscus* sp. mole crickets (Orthoptera: Gryllotalpidae). J. Entomol. Sci. 28:182–190.
Pedigo, L. P., and L. G. Higley. 1992. The economic injury level concept and environmental quality: a new prospective. Am. Entomol. 38:12–21.
Peters, A. 1996. The natural host range of *Steinernema* and *Heterorhabditis* sp., and their impact on insect populations. Biocontrol Sci., and Tech. 6:389–402.
Pimentel, D., C. Glenister, S. Fast, and D. Gallahan. 1984. Environmental risks of biological pest controls. Oikos 42: 283–290.
Pimentel, D., H. Acquay, M. Biltonen. p. Rice, M. Silva, J. Nelson, V. Lipner, S. Giordano, A. Horowitz, and M. D'Amore. 1992. Environmental and economic costs of pesticide use. Bioscience 42: 750–760.
Poinar, G. O., Jr. 1979. Nematodes for biological control of insects. CRC Press, Boca Raton, Florida. 277 p.
Poinar, G. O. Jr. 1989. Non–insect hosts for the entomogenous rhabditoid nematodes *Neoaplectana* (Steinernematidae) and *Heterorhabditis* (Heterorhabditidae). Rev. Nematol. 12:423– 428.
Poinar, G. O., Jr. 1990. Biology and taxonomy of Steinernematidae and Heterorhabditidae. *In* R. Gaugler and H. K. Kaya (eds.), Entomopathogenic nematodes in biological control. CRC Press, Boca Raton, Florida. p. 23–62.
Poinar, G. O. Jr., and A. Hom. 1986. Survival and horizontal movement of infective stage *Neoaplectana carpocapsae* in the field. J. Nematol. 18: 34–36.
Poinar, G. O., Jr., and G. M. Thomas. 1985a. Laboratory infection of spiders and harvestmen (Arachnida: Araneae and Opiliones) with *Neoaplectana* and *Heterorhabditis* nematodes (Rhabditoidea). J. Arachnol. 13: 297–300.
Poinar, G. O., Jr., and G. M. Thomas. 1985b. Laboratory infection of *Garypus californicus* (Pseudoscorpionida, Garypidae) with *Neoaplectana* and *Heterorhabditis* nematodes (Rhabditoidea). J. Arachnol. 13: 400–403.
Poinar, G. O., Jr., and G. M. Thomas. 1988. Infection of frog tadpoles (Amphibia) by insect parasitic nematodes (Rhabditida). Experientia 44: 528–531.
Poinar, G. O.,Jr., G. M. Thomas, S. B. Presser, and J. L. Hardy. 1982. Inoculation of entomogenous nematodes, *Neoaplectana* and *Heterorhabditis*, and their associated bacteria, *Xenorhabdus* sp., into chicks

and mice. Environ. Entomol. 12:368–370.

Proctor, D. L. C. 1984. Towards a biogeography of free-living soil nematodes. I. changing species richness, diversity and densities with changing latitudes. J. Biogeography 11: 103–117.

Raulston, J. R., S. D. Pair, J. Loera, and H. E. Cabanillas. 1992. Prepupal and pupal parasitism of *Helicoverpa zea* and *Spodoptera frugiperda* (Lepidoptera: Noctuidae) by *Steinernema* sp. in cornfields in the lower Rio Grande valley. J. Econ Entomol. 85: 1666–1670.

Rosenheim, J. A., H. K. Kaya, L. E. Ehler, J. J. Marois, and B. A. Jaffee. 1995. Intraguild predation among biological control agents: theory and evidence. Biol. Control 5: 303–335.

Schroeder, W. J., and Beavers. 1987. Movement of entomogenous nematodes of the families Steinernematidae and Heterorhabditidae in soil. J. Nematol. 19: 257–259.

Schroeder, P. C., M. G. Villani, C. S. Ferguson, J. P. Nyrop, and E. J. Shields. 1993. Behavioral interactions between Japanese beetle (Coleoptera: Scarabaeidae) grubs and an entomopathogenic nematode (Nematoda: Heterorhabditidae) within turf microcosms. Environ. Entomol. 22: 595–600.

Selvan, S, J. F. Campbell, and R. Gaugler. 1993. Density-dependent effects on entomopathogenic nematodes (Heterorhabditidae and Steinernematidae) within an insect host. J. Invertebr. Pathol. 62:278–284.

Sexton, S. B., and P. Williams. 1981. A natural occurrence of parasitism of *Graphognathus leucoloma* (Boheman) by the nematode *Heterorhabditis* sp. J. Austral. Entomol. Soc. 20:253–255.

Shanks, C. H., Jr., and F. Agudelo-Silva. 1990. Field pathogenicity and persistence of heterorhabditid and steinernematid nematodes (Nematoda) infecting black vine weevil larvae (Coleoptera: Curculionidae) in cranberry bogs. J. Econ. Entomol. 83: 107– 110.

Shapiro, D. I., and I. Glazer. 1996. Comparison of entomopathogenic nematode dispersal from infected hosts versus aqueous suspension. Environ. Entomol. 25: 1455–1461.

Simberloff, D. 1992. Conservation of pristine habitats and unintended effects of biological control. *In* W. C. Kauffman and J. E. Nechols (eds.), Selection criteria and ecological consequances of importing natural enemies. Proc. Thomas Say Publ. Entomol., Entomological Soc. Amer., Lanham, Maryland. p. 101–117.

Simberloff, D., and P. Stiling. 1996. How risky is biological control? Ecology 77: 1965–1974.

Simberloff, D., and P. Stiling. 1998. How risky is biological control?; reply. Ecology 79:1834–1836.

Simões, N., and J. S. Rosa. 1996. Pathogenicity and host specificity of entomopathogenic nematodes. Biocontrol Sci. Tech. 6:403–411.

Sloderbeck, P. E., and K. V. Yeargan. 1983. Green cloverworm populations in conventional and double-cropped no-till soybeans. J. Econ. Entomol. 76: 785–791.

Small, R. W. 1987. A review of the prey of predatory soil nematodes. Pedobiologia 30: 179–206.

Smits, P. H. 1996. Post-application persistence of entomopathogenic nematodes. Biocontrol Sci. Tech. 6: 379–387.

Spiridonov, S. E., and D. A. Voronov. 1995. Small scale distribution of *Steinernema feltiae* juveniles in cultivated soil. *In* C. T. Griffin, R. L. Gwynn and J. P. Masson (eds.), Ecology and transmission strategies of entomopathogenic nematodes, European Commission, Luxembourg. p. 36–41.

Stimac, J. L. 1990. Crop management systems and their effects on biological control of insects. *In* R. R. Baker and P. E. Dunn (eds.), New directions in biological control: Alternatives for suppressing agricultural pests and diseases. A. R. Liss, Inc. New York. p. 449–465.

Stimac, J. L., and R. J. O'Neil. 1985. Integrating influences of natural enemies into models of crop/pest systems. *In* M. A. Hoy and D. C. Herzog. (eds.), Biological control in agricultural IPM systems . Academic Press, New York. p. 323–344.

Stirling, G. R. 1991. Biological control of plant parasitic nematodes. CAB International, Wallingford, UK. 282 p.

Strong, D. R., J. L. Maron. p. G. Connors, A. Whipple, S. Harrison and R. L. Jeffries. 1995. High mortality, fluctuation in numbers, and heavy subterranean insect herbivory in bush lupine, *Lupinus arboreus*. Oecologia 104: 85–92.

Strong, D. R., H. K. Kaya, A. V. Whipple, A. L Child, S. Kraig, M. Bondonno, K. Dyer, and J, L. Maron. 1996a. Entomopathogenic nematodes: natural enemies of root–feeding caterpillars on bush lupine. Oecologia 108:167–173.

Strong, D. R., J. L Maron, and P. G. Connors. 1996b. Top down from the underground: The underappreciated influence of subterranean food webs on above–ground ecology. *In* G. A. Polis, K. O. Winemiller (eds.), Food webs: integration of patterns and dynamics. Chapman and Hall, New York. p. 170–175.

Stuart, R. J., and R. Gaugler. 1994. Patchiness in populations of entomopathogenic nematodes. J. Invertebr. Pathol. 64:39–45.

Timper. p., and H. K. Kaya. 1992. Impact of a nematode– parasitic fungus on the effectiveness of entomogenous nematodes. J. Nematol. 24: 1–8.
Timper, P., H. K. Kaya, and R. Gaugler. 1988. Dispersal of the entomogenous nematode *Steinernema feltiae* (Rhabditida: Steinernematidae) by infected adult insects. Environ. Entomol. 17: 546–50.
Timper, P., H. K. Kaya, and B. A. Jaffee. 1991. Survival of entomogenous nematodes in soil infested with the nematode– parasitic fungus *Hirsutella rhossiliensis* (Deuteromycotina: Hyphomycetes). Biol. Control 1: 42–50.
United Nations. 1992. Convention on Biological Diversity. Environmental Law and Institutions Programme Activity Center, United Nations Environment Program, New York.
Van Der Werf, W. . p. H. Smits, M. Downes, and C. Griffin. 1995. Concepts and prospects for modelling the efficacy and ecology of entomopathogenic nematodes. *In* C. T. Griffin, R. L. Gwynn and J. P. Masson (eds.), Ecology and transmission strategies of entomopathogenic nematodes. European Commission, Luxembourg. p. 42–51.
Vänninen, I, G. –B. Husberg and H. M. T. Hokkanen. 1989. Occurrence of entomopathogenic fungi and entomoparasitic nematodes in cultivated soils in Finland. Acta Entomol. Fenn. 55: 65–71.
Waage, J., and D. Greathead. 1988. Biological control: challenges and opportunities. Phil. Trans. Roy. So Lond. B 318: 111–128.
Walter, D. E. 1987a. Trophic behavior of 'mycophagous' microarthropods. Ecology 68: 226–229.
Walter, D. E. 1987b. Life history, trophic behaviour and description of_*Gamasellodes vermivorax* n. sp. (Mesostigmatoa: Ascidae), a predator of nematodes and arthropods in semiarid grassland soils. Can. J. Zool. 65: 1689–1695.
Walter, D. E. 1988a. Predation and mycophagy by endeostigmatid mites (Acariformes: Prostigmata). Exp. Appl. Acarol. 4:159–166.
Walter, D. E. 1988b. Nematophagy by soil arthropods from the shortgrass steppe, Chihauahuan Desert and Rocky Mountains of the central United States. Agric. Ecosyst. Environ. 24: 307–316.
Walter, D. E., and E. K. Ikonen. 1989. Species, guilds, and functional groups: taxonomy and behavior in nematophagous arthropods. J. Nematol. 21:315–327.
Walter, D. E., and D. T. Kaplan. 1990. Feeding observations on two astigmatic mites, *Schwiebea rocketti* (Acaridae) and *Histiostoma bakeri* (Histiostomatidae) associated with citrus feeder roots. Pedobiologia 34: 281–286.
Walter, D. E., R. A. Hudgens, and D. W. Freckman. 1986. Consumption of nematodes by fungivorous mites, *Tyrophagus* sp. (Acarina: Astigmata: Acaridae). Oecologia 70: 357–361.
Walter, D. E., H. W. Hunt, and E. T. Elliot. 1987. The influence of prey type on the development and reproduction of some predatory soil mites. Pedobiologia 30: 419–424.
Wasilewska, L. 1979. The structure and function of soil nematode communities in natural ecosystems and agrocenoses. Polish Ecological Studies 5: 97–145.
Westerman, P. R. 1995. Comparative vertical migration of twenty one isolates of the insect parasitic nematode *Heterorhabditis* sp. in sand at 20°C. Fund. Appl. Nematol. 18: 149–158.
Wilson, E. O. 1985. The biological diversity crisis: a challenge to science. Issues Sci. Tech. 2: 20–29.

INDEX